C0-AVR-841

Richard John Huggett

Climate, Earth Processes and Earth History

With 71 Figures

Springer-Verlag

Berlin Heidelberg New York
London Paris Tokyo
Hong Kong Barcelona
Budapest

Dr. RICHARD JOHN HUGGETT
University of Manchester
School of Geography
Manchester, M13 9PL
England

ISBN 3-540-53419-9 Springer-Verlag Berlin Heidelberg New York
ISBN 0-387-53419-9 Springer-Verlag New York Berlin Heidelberg

Library of Congress Cataloging-in-Publication Data. Huggett, Richard J. Climate, earth processes, and earth history / Richard John Huggett. p. cm. Includes bibliographical references. 1. Climatology. 2. Paleoclimatology. 3. Biosphere. I. Title. QC981.H84 1991   551.6 – dc 20   91-23129

This work is subject to copyright. All rights are reserved, whether the whole or part of the material is concerned, specifically the rights of translation, reprinting, reuse of illustrations, recitation, broadcasting, reproduction on microfilms or in other ways, and storage in data banks. Duplication of this publication or parts thereof is only permitted under the provisions of the German Copyright Law of September 9, 1965, in its current version, and a copyright fee must always be paid. Violations fall under the prosecution act of the German Copyright Law.

© Springer-Verlag Berlin Heidelberg 1991
Printed in Germany

The use of general descriptive names, registered names, trademarks, etc. in this publication does not imply, even in the absence of a specific statement, that such names are exempt from the relevant protective laws and regulations and therefore free for general use.

Typesetting: International Typesetters Inc., Makati, Philippines
32/3145-543210 – Printed on acid-free paper

551.6
H89/c

For Shelley and Zoë

Foreword

Today, climate-related processes and problems are referred to as Global Change by nearly everyone including scientists, politicians, and economists; citizens worldwide are anxious about the often observed disorientation of our environment under the influence of man. Better information on the Earth's natural systems and their possible alterations is necessary. The topic itself is so wide that sound scientific descriptions of it as a whole are rare. For the non-specialist information from relevant fields is not easy to obtain; and often, the prognostic models presented are contradictory and even for specialists difficult to evaluate. Therefore, this book on Climate, Earth Processes and Earth History by Richard Huggett fills an important gap. It discusses the great, climate-related areas of the Earth's environment. The atmosphere, the hydrosphere, the sediments as products of weathering and geomorphic processes, the relief as landforms and soils, and the biosphere are thoroughly treated as the prominent subsystems which are greatly affected by climate. These subsystems not only control the visual and internal aspects of our landscapes, but they are themselves especially influenced by climatic changes which can be due to either changes in the natural system or anthropogenic changes. Thus, our landscapes will be subject to significant alterations, if climatic variations exceed certain thresholds.

The plan for the present book by Richard Huggett was originally discussed in regard to the Springer Series on Physical Environment. Soon, it became clear that it should be directed towards a broader audience and not only to specialists. Therefore, it is not published as part of the series, in the hope that a wider audience will be reached. The book will help to further our understanding of our physical environment and the climate-related changes which may occur in the years to come.

Dietrich Barsch

Preface

Our knowledge and understanding of the workings of the world climate system has taken a quantum leap in the last couple of decades. The reasons for this lie partly in the building of sophisticated climate models for computers, partly in the vast amounts of data sensed by satellites, and partly in the establishment of a reliable calendar of geological events. This book shows how the great strides being made in climatology and palaeoclimatology are leading to fresh insights into the interaction between the atmosphere and other parts of the biosphere (animals, plants, sediments, soils, and landforms), and to a new view of the relations between the atmosphere and other Earth systems − a view in which climatic control is seen as both more profound and yet more subtle than was previously thought.

After introducing the concept of the world climate system, the book explores the chief components of the biosphere − air, ice and water, sediments, landforms and soils, animals and plants, and biomes and zonobiomes − in relation to climatic factors. For each component, a thumbnail sketch of the historical development of ideas on the role played by climate is provided. Some readers may be surprised to discover the antiquity of many, supposedly new-fangled, hypotheses. A concluding chapter moves towards a synthesis of the preceding material. It opens by developing a general model of the biosphere and then, in the context of the model, discusses two key ideas arising from the material in the earlier chapters, namely, the question of scale and the origin of cyclicity within the biosphere.

The subject matter of the book being so broad, it would have been easy to pen a tome of Brobdingnagian proportions. Keeping to a modest wordage has required a ruthless and somewhat invidious selecting of topics and examples, as well as swingeing excisions of entire sections from earlier drafts. The result is a book which focusses on the interactions of Earth surface systems with climate at regional and global levels of interest. Even under that more closely circumscribed rubric, it was not possible to include all fields of enquiry. Topics included were chosen for their currency and for their being familiar to the author. They are, however, diverse enough to cater for a wide audience. A theme which runs through the topics is the immeasurable value of an interdisciplinary approach to the study of the world climate system (biosphere). The book will have served a useful purpose if it alerts specialists who delve into fairly narrow aspects of climatic connexions — be they biologists, ecologists, environmental scientists, geographers, geomorphologists, or geologists — to developments in fields other than their own and to more general issues which cross the artificial divides of geoscientific disciplines.

I should like to thank several people for helping with the production of the book: for continued faith in my unfashionable style of research, Ian Douglas; for drawing the figures, Graham Bowden (who had a little assistance from Nick Scarle); for smoothing the process of production, the editorial staff at Springer's Heidelberg Office; for acting as guinea-pigs for much of the material presented in the book, all students who were brave enough to opt for my third-year course on the Biogeography and Geomorphology of the Continents; and, as ever, for constant patience and support, my wife Shelley.

Poynton, June 1991   Richard Huggett

Contents

1 Introduction

As the sky is divided into two zones on the right hand, and two on the left, with a fifth in between, hotter than any of the rest, so the world which the sky encloses was marked off in the same way, thanks to the providence of the god: he imposed the same number of zones on earth as there are in the heavens. The central zone is so hot as to be uninhabitable, while two others are covered in deep snow: but between these extremes he set two zones to which he gave a temperate climate, compounded of heat and cold.

Ovid (1955 edn, p. 30)

1.1 From Thermometers to Satellites

The zonal pattern of climate was appreciated by the ancient Greeks. The climatic zones designated torrid, temperate, and frigid (reputedly by Parmenides of Elea but more likely by Eudoxus) are still in use. These zones were marked out chiefly according to the amount of illumination during a revolution of the Earth around the Sun. Clearly, such solar illumination declines from the equator towards the poles. Indeed, the term "climate" comes from the Greek κλιμα (from κλινειν: slope, lean) which refers to a zone occupying a particular elevation on the supposed slope of the Earth or sky from equator to poles. The description of climates was first put on a quantitative footing during the seventeenth century. Ferdinand II, Grand Duke of Tuscany, ordered Mariani, his master glass-blower, to make thermometers which would respond to a wide range of temperature changes (Middleton and Spilhaus 1953). In late 1654, these instruments were sent to several Italian cities including Milan and Bologna, and the first known weather network was established. Evangelista Torricelli, working under the auspices of Ferdinand, took the first steps to the understanding and measurement of air pressure while attempting to produce a vacuum in a tube filled with mercury. The Torricellian tube was dubbed a barometer by Robert Boyle sometime during the 1660s. Later, Gottlieb von Leibnitz invented the aneroid barometer. From these crude instruments and patchy meteorological networks were to evolve more refined instruments and instrument stations (Landsberg 1964; Manley 1974).

Global patterns of basic atmospheric variables were first established by the navigators and naturalist-explorers who sailed the oceans. The location of the chief wind belts (easterly trade winds and the mid-latitude westerlies), the zonal pattern of temperature, and the greater seasonal contrast of temperature over continents than over oceans were all charted by the early years of the nineteenth century. The global pattern of temperature was especially well known and ably summarized in Alexander von Humboldt's (1820–21) fine essay entitled *On Isothermal Lines, and the Distribution of Heat over the Globe*. A more detailed resolution of world climates required the setting up of weather stations around the globe. In the United States, scattered observations had been made since colonial days, but not until 1817 were the first meteorological observations made by government agencies. Forty years later, Lorin Blodget presented a series of rainfall maps in his *Climatology of the United States, and of the Temperate Latitudes of the North American Continent* (1857). The National Weather Service was set up within the Signal Corps of the United States Army in 1870, and in 1891 the United States Weather Bureau was established within the Department of Agriculture. By 1922, observations of meteorological variables were taken at 6000 stations, and evaporation was measured at 38 well-distributed sites. Fuller coverage of the spatial distribution of atmospheric variables at the Earth's surface led to the first (and lasting) attempts at the classification of climate — for example, by Wladimir Peter Köppen (Fig. 1.1) — and greatly improved our understanding of atmospheric processes. By plotting the data collected by weather networks on charts, the general eastwards progress of mid-latitude cyclonic storms and anticyclones was revealed. As the number of weather stations and length of records increased, so the pattern of surface conditions became more accurate and more detailed, and the surface data amassed on rainfall, temperature, and other atmospheric variables became invaluable to scientists with an interest in ecology and the physical landscape.

After about 1940, surface data were supplemented by information from balloon ascents into the upper air and the vertical structure of mid-latitude weather systems was displayed. This much further improved our understanding of atmospheric processes, and, coupled with the advent of digital computers, led during the 1950s to the building of the first numerical forecast models, the construction of which had been adumbrated 30 years earlier by Lewis Fry Richardson in his remarkable book *Weather Prediction by Numerical Process* (1922). These models, which involved the numerical solution of three-dimensional, time-dependent equations of atmospheric hydrodynamics, predicted the change in observed atmospheric temperature and pressure fields for several days ahead. They were chiefly concerned with adiabatic temperature changes resulting from air movement and pressure differences characteristic of cyclonic disturbances. Over longer periods of time, atmospheric temperature is greatly influenced by "diabatic" heating terms including the release of latent heat during condensation, atmospheric radiation processes, and convective heating from the Earth's surface. During the 1960s, these longer-term processes were combined with the equations of the weather prediction models to form the first general circulation models. In a nutshell, a general circulation model comprises a prognostic system of equations describing the physical and dynamical processes which determine climate. It usually includes also a heat and water balance model of the land surface, and a mixed-layer model of the ocean. At the heart of a general circulation model lie the governing or primitive equations: the equation of motion (conservation of momentum), the equation of continuity (conservation of mass or hydrodynamic equation), the equation of continuity for atmospheric water vapour (conservation of water vapour), and the equation of energy (thermodynamic equation derived from the first law of thermodynamics). To these equations are added the equation of state (hydrostatic equation), and, in some models, the surface pressure tendency equation. To use a general circulation model it is necessary to specify certain parameters, such as the solar constant

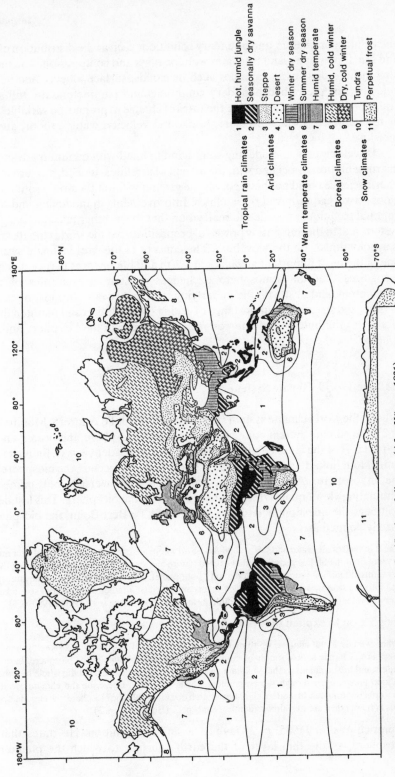

Fig. 1.1. Wladimir Peter Köppen's classic classification of climates. (After Köppen 1931)

Tropical rain climates
1 Hot, humid jungle
2 Seasonally dry savanna

Arid climates
3 Steppe
4 Desert

Warm temperate climates
5 Winter dry season
6 Summer dry season
7 Humid temperate

Boreal climates
8 Humid, cold winter
9 Dry, cold winter

Snow climates
10 Tundra
11 Perpetual frost

and orbital parameters, and boundary conditions such as the distribution of land and sea, topography, and total atmospheric mass and composition. The models also include diagnostic variables such as clouds, surface albedo, and vertical velocity. For prescribed boundary conditions and parameters, the full set of equations is solved to determine the rates of change in prognostic variables such as temperature, surface pressure, horizontal velocity, water vapour, and soil moisture.

Developments in modelling went hand in hand with quantum advances in instrumentation associated with the advent of satellites. Indeed, modern atmospheric science could not have got off the ground without the fine resolution and global coverage achieved by satellites. Improvements in modelling and observational techniques have led to realization that the weather machine is a global system, and to the immensely powerful proposition that the world climate system is synonymous with the biosphere. The concept of the world climate system is one of the most important advances in Earth and life sciences in recent years. It recognizes that the atmosphere exchanges substantial quantities of mass, momentum, and heat with the underlying land and oceans during long periods of time, and stresses the importance of seeing the evolution and maintenance of climate as the outcome of interconnected processes in the biosphere and geosphere.

1.2 The World Climate System

What is the world climate system? Opinion is divided. To Andrei S. Monin (1986, p. 2), the climate system comprises the whole atmosphere, the ocean, and the active layer of the land. Other researchers take a broader view, defining it as the sum of four linked subsystems — the atmosphere, the oceans, the biosphere, and the cryosphere. Another broad, somewhat controversial, but immensely stimulating view equates the climate system with the biosphere. This last definition begs the question: what is the biosphere? To Bert Bolin, the biosphere is strictly defined as

"the thin shell of air, water and soil around the Earth where life exists. It has developed over millions of years from the lifeless geosphere that existed before and that in turn had evolved during the first few thousand million years after the creation of our planet. The biosphere includes the atmosphere well into the stratosphere, the oceans, ice-sheets and glaciers, a rather deep layer of soil, lakes and running water on land and finally the sediments at the bottom of the sea". (Bolin 1980, p. 3)

He goes on to explain that

"when we talk about climate, we think of the physical state of essentially this same "sphere" [the biosphere]. Climate should simply be considered as the physical state of the biosphere. In order to understand the mechanisms that maintain and change the climate, we need the whole biosphere as defined above, even though the parameters which are used to describe the climate are mostly atmospheric variables. In modern climatic research one often calls the biosphere 'the climatic system', thus in reality they are two almost identical concepts". (Bolin 1980, p. 3)

Heinrich Walter (1985, p. 2) favours a similar definition. He states that the biosphere is "the thin layer of the earth's surface to which the phenomena connected with living matter are confined". He says that on land "this comprises the lowest layer of the atmosphere permanently inhabited by living organisms and into which plants extend, as well as the root-containing portion of the lithosphere, which we term the soil". Because the cycling of material is different in the watery medium of aquatic ecosystems, he divides the biosphere into the geobiosphere, comprising the terrestrial systems, and the hydrobiosphere, comprising aquatic ecosystems.

Some climatologists may balk at having the atmosphere defined as a state in the biosphere. But the term biosphere was originally used by Eduard Suess (1875) and Vladimir Ivanovich Vernadsky (1926, 1929, 1945a) to encompass life and life-supporting systems, and Bolin's and Walter's definitions stick to this eclectic view. Unfortunately, a few later ecologists, including Lamont C. Cole (1958), have polluted the term and restricted it to the totality of living systems, a definition still adopted by some climatologists (e.g. Cubasch and Cess 1990). But the interrelatedness of the components of the biosphere, and thus the world climate system, have been stressed by many modern writers (e.g. Polunin and Grinevald 1988); the following passage is typical of current thinking:

"Vegetation, as an indicator of terrestrial ecosystems on the earth's surface, responds to changes in atmospheric conditions, but it also influences atmospheric processes by affecting fluxes of energy, momentum and water. Similarly, oceans play equally interactive roles by influencing not only atmospheric conditions, but also terrestrial ecosystems. Geological processes, especially those relating to topography and soils, cause certain characteristic responses in climate and vegetation, and, in turn, geologic processes are influenced by vegetation and climate. Thus, the interactions among biospheric components are complicated, and operate over a broad array of spatial and temporal scales". (Risser et al. 1988, p. 2)

Despite the present emphasis on atmospheric interaction with other parts of the biosphere, most general circulation models describe only the physical behaviour of the biosphere; they omit chemical and biological processes. Bert Bolin (1988) pleaded for a broadening of horizons so that the dynamics of the entire biosphere, the interplay of physical, chemical, and biological processes, can be pursued (see also Rambler et al. 1989). The immeasurable worth of this integrated view should become clear in the chapters of this book.

Defining the atmospheric system as a state in the biosphere leads to a possible source of confusion as to what is meant by the term "climate". Traditionally defined, climate is the general pattern of weather in different regions and at different times. More specifically, it is the "set of weather conditions typical of a given region, together with the frequency of these conditions and their seasonal variations" (Monin 1986, p. 1); or "the composite or long-term, often spatial generalization of day-to-day meteorological conditions" (Terjung 1976, pp. 220–221). The meteorological conditions involved in traditional climatology are simple descriptors of the state of the atmosphere — air temperature, moisture content, wind speed, and so forth — in the lower layer of the atmosphere. Modern definitions of climate generally take account the all-important atmospheric energy, mass, and momentum fluxes and their interaction with other extraterrestrial and terrestrial systems. Werner H. Terjung, for instance, offers the following wording:

"the climate of an area . . . exhibits the results of meteorological conditions which are typical of a series of years . . . and are governed by solar radiation on top of the atmosphere, the composition of the atmosphere (. . . air temperature, pressure, wind, moisture content, turbidity. . .), and the structure of the earth's surface (or "controls," such as latitude, altitude, mountain barriers, ocean currents, surface conditions—the latter two not independent because of reciprocal feedbacks). In other words, the cascades of energy (radiation, conduction, convection), mass (density times volume), and momentum (mass times velocity) are linked reciprocally with the morphological components, resulting in the process-response system called 'climate'". (Terjung 1976, p. 221).

In like manner, Yeh Tuchen and Fu Congbin, stressing the interconnectedness of Earth systems, conclude that climate is more than the state of the atmosphere:

"The atmosphere is not an isolated system. It is closely connected with the sun and the other parts of the earth, such as the ocean, lithosphere, cryosphere, biosphere and so on. The connection with other parts of the earth is a function of space scale and time scale. For very limited space-scale and very short time-scale phenomena, the atmosphere may be considered approximately as being isolated. But for very large space-scale and long time-scale phenomena, the influence of the sun and other parts of the earth on the atmosphere should be taken into consideration. On the other hand, the atmosphere also has profound influences on other parts of the earth, especially on long time scales. Thus the climate can no longer be considered only the behaviour of the atmosphere, it must be defined as the behaviour of the whole climate system consisting of the atmosphere, ocean, cryosphere, lithosphere and biosphere". (Yeh and Fu 1985, p. 127).

It can be seen, therefore, that the word "climate", used in the sense of the "average state of the atmosphere", is not equivalent to the use of the word "climate" in the phrase "world climate system", where the condition of the entire biosphere is implied. It is important to be alerted to these two usages of the term climate, and to remember the while reading this book a phrase such as "soil-climate relationships" takes the more restricted definition of climate and means "the relationship between soil properties and the state of selected atmospheric variables averaged over a prescribed span of time".

1.3 Climatic Change

The state of the climate system varies with time. This fact has been appreciated since classical times: the ancients were aware that the climate in one year was not always like the climate in the previous year. During the seventeenth century, the finding of tropical fossils in temperate climes suggested that the Earth's present climate was very different from the climate in Earth's early history. In the nineteenth century, the recognition of glacial and interglacial regimes indicated that the climate had changed drastically in the relatively recent past. Theories to explain these long-term changes of climate were plentiful, but they were hard to test owing to a paucity of trustworthy climatic and palaeoclimatic data, a general uncertainty about the interpretation of climatic indicators, and the lack of a reliable timescale. Also, they were rather simplistic because many of the complex interactions and forcings of the climate system were only guessed at or remained unimagined, and because knowledge of the Solar System and Galaxy was still meagre. Over the last 20 years, the problems faced by earlier generations of theorists in trying to evaluate the worth of their conjectures have been remedied, to some extent at least. Today, dendrochronological and radiometric methods of dating provide an absolute timescale; ice cores, deep sea sediments, and palaeosols reveal the past composition of the atmosphere; a better understanding of the relation between climate and Earth surface phenomena means that climatic indicators can more confidently be used to reconstruct ancient climates; and computer models of the general circulation allow climates of the past to be simulated, and allow the effect of Earth's changing orbital motions on solar input to be calculated quickly and accurately.

Modern palaeoclimatological methods have unveiled a far more colourful spectrum of climatic changes than earlier generations of climatologists were aware of. It is now known that climate changes for a variety of reasons over timescales ranging from less than an hour, as in short-lived but severe meteorological phenomena, to over tens of millions of years, as in protracted phases of global warming and cooling. Short-term changes are present in meteorological records which disclose a number of distinct periodic components within the climate system. These include diurnal changes resulting from daily cycles of insolation receipt, diurnal and semi-diurnal tidal oscillations resulting from gravitational interaction with the Sun, Moon, and other members of the Solar System, and seasonal variations induced by the obliquity of the ecliptic. Short, irregular changes are also observed, including synoptic oscillations caused by Rossby waves which last several days over land and several weeks over the oceans; global oscillations lasting from weeks to months; and inter-annual oscillations which occur largely in the period 2 to 5 years and include the El Niño and El Niña phenomena. Very reliable palaeoclimatic indicators and climatic records suggest that climate fluctuates over periods of decades and centuries, although these variations are less intense than is the case with the shorter-period oscillations. Yet other palaeoclimatic indicators reveal very intense oscillations with periods of thousands of years (swings from glacial to non-glacial states), and tens of thousands of years (swings from glacial to interglacial states), and changes of an even more protracted nature.

The basic mechanisms which cause the climate system to change are fairly well understood in broad terms, though the relative importance of various factors at different timescales is still uncertain. The chief processes involved with the change in the world climate system over short (decades to centuries) and medium and long (thousands to millions of years) timescales are depicted in Fig. 1.2. In the short term, over years, decades, and centuries, the world climate system may change owing to external forcings or to internal dynamics. External forcing can be brought about by extraterrestrial changes in gravitational forces, variations in electromagnetic and particulate radiation receipt from the Sun and from space, the impact of planetesimals, and by terrestrial changes such as the injection of volcanic dust and gases into the stratosphere. Gravitational stresses may emanate from three sources — the Solar System, our Galaxy, and other galaxies — but only interactions within the Solar System will force the climate system in the short term when the delivery of energy is modulated by the overall motions and alignments of the planets in the Solar System and by Earth-Moon motions. Internal dynamics of the climate system, and particularly the atmos-

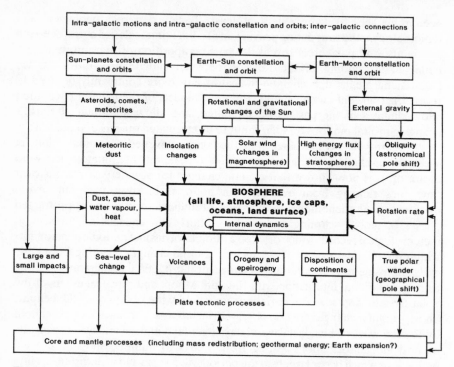

Fig. 1.2. The biosphere (world climatic system) and processes which affect it

pheric system, have been shown to involve short-term, cyclical components. Thus climatic change can occur without the aid of external forcing. External and internal forcings lead to short-term changes of climate which leave their signature in some dendrochronological, sedimentary, and geomorphological records, particularly deep-sea sediment cores, ice cores, bog and lake cores, and palaeosols.

In the long term, over thousands to millions of years, the world climate system changes owing to external forcing, both extraterrestrial and terrestrial, and to the internal dynamics of the climate system. External forcings originating from outside the Earth and acting over medium and long timescales involve orbital variations, secular changes in solar output, and, indirectly, large-body impacts which can influence endogene processes. External forcings emanating from within the Earth and acting over medium and long timescales are associated with plate tectonics processes, which in turn are driven by processes going on in the core and mantle. In the long term, plate tectonics leads to changes in palaeogeography — the redistribution of continents and ocean, the formation of mountain ranges, changes in the volume of the oceans, changes of sea level, and so on — all of which may induce change in the climate system. Redistribution of mass within the Earth can lead to true polar wander (geographical pole shift) which will have repercussions in the climate system, too. Processes in the core and mantle, which drive many crustal processes, are themselves influenced by extraterrestrial forcing: large-body impacts may force changes at the core-mantle boundary which eventually are felt in the climate system. The entire Earth is subject to gravitational forcing which may induce long-term changes of axial tilt (astronomical pole shift), again with concomitant effects on world climates. Gravitational forces also gradually change the position of the Solar System relative to the centre of the Milky Way galaxy, probably leading to very long-term changes of climate. In addition, the world climate system may change over thousands to millions of years because of processes going on inside it, such as changes in the salinity of the oceans, changes of sea-surface temperatures, the growth and decay of ice sheets and sea ice, the eustatic change of sea level, biological evolution, and changes in terrestrial biomass.

A key point to bear in mind is that, whatever cause climate to change, the change itself often involves the readjustment of the climate system as a whole. In general, climatic change is a global phenomenon because a change in any one state variable will affect, in varying degrees, the state of all other variables in the system, though the effects of global climatic change will vary from one place to another. This interconnectedness and dynamic unity of the world climate system is captured in the "Butterfly Effect", the half-serious notion that "a butterfly stirring the air today in Peking can transform storm systems next month in New York" (Gleick 1988, p. 8).

1.4 Scales of Systems

It has become clear recently that central to the elucidation of relations within the climate system is an understanding of the connections between systems of different scales within the biosphere. This finding is a development of the view that the natural world may be regarded as a nested series of systems within systems, each system being at the same time a thing in its own right and a part of a bigger, more inclusive, thing. Arthur Koestler (1967, 1978) suggested that such systems be called "holons", and a series of nested holons be termed a "holarchy". The behaviour of a holon will reflect an attempt to reconcile the forces by which the holon asserts its independence and the forces which tend to subordinate the dynamics of the holon to a larger-scale system in the holarchy. Now, within the Earth system several holarchies have been identified: the atmospheric holarchy, the tectonic holarchy, the geomorphological holarchy, the pedological holarchy, the genealogical holarchy, and the ecological holarchy. As to component holons occurring at different spatial and temporal scales in each of these holarchies, there is a lack of agreement. The loose framework depicted in Fig. 1.3, although it may not be acceptable to all readers, will be consistently applied throughout the following pages and so will help to avoid confusion with other suggested frameworks. The boxes in the diagram are meant to provide a rough guide to the size and "life-span" of Earth systems; they should not be read as hard-and-fast dividing lines between different echelons in the holarchies. Notice that for a given spatial scale, atmospheric systems survive for

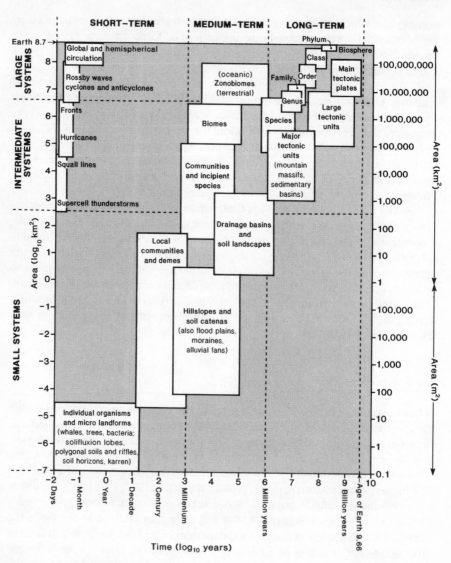

Fig. 1.3. Terrestrial systems: size and "life-span"

periods of less than a year, while tectonic systems survive for billions of years. Ecological, genealogical, geomorphological, and pedological systems generally survive for periods of years to millions of years, although the upper echelons of the genealogical holarchy and the biosphere as a whole have lasted much longer. The atmosphere is thus the most changeable of the systems of the biosphere.

Large biological and ecological systems include the entire biosphere and sizeable chunks of it. In the ecological holarchy, zonobiomes are commonly recognized as large systems; in the genealogical hierarchy, all taxa above the level of family are considered large. Large geomorphological systems include the entire surface and big crustal units such as the main tectonic plates. Nested within the large systems are systems of intermediate size which in turn comprise a myriad small systems. The terms "intermediate" and "small" are defined, somewhat arbitrarily, as follows: intermediate systems occupy an area between about 400 km^2 and $4,000,000 \text{ km}^2$. In geomorphology they range from major tectonic units, such as the Congo Basin, to large tectonic units such as the London and Paris basins; in ecology they range from biomes to communities; in genetics they range from genera to incipient species. Small systems cover an area of less than 400 km^2 down to 0.1 m^2 or even less. Their compass in geomorphology and pedology includes individual hillslopes and other landforms within drainage basins down to microrills and solution pits; in ecology and genetics it includes local communities, demes, and individual organisms.

The interplay between large and intermediate systems in the geomorphological, pedological, ecological, and genealogical holarchies on the one hand, and the atmospheric holarchy on the other, generally involves a regional response of animals, plants, soils, and landforms to climatic gradients, and especially to gradients of available radiative energy and to available moisture. In turn, climatic gradients can be influenced by conditions at the Earth's surface: there is considerable feedback at these gross scales. In general circulation models, the feedback is incorporated at the level of grid cells (system units of about $100 \times 100 \text{ km}$). The effect of Earth surface system processes at sub-grid size is included indirectly by representing the state of the grid cell parametrically, a process referred to as parameterization. A clear example of this afforded by an important sub-grid-scale element of the atmospheric system itself — clouds. Individual clouds cannot be included directly in general circulation models because no computer can yet handle small enough cell sizes. However, cloudiness can be represented by a few parameters derived empirically from cell-averaged values of temperature, winds, and humidity. In small systems, climate is often represented by a single driving variable such as mean annual effective rainfall. But it is not uncommon to consider heat and water balances of, say, soil profiles or individual organisms. The effect of small systems on climate is far too detailed to fit into general circulation models, at least in their present state of development, but local conditions at the Earth's surface do have an important influence on microclimates.

It is important to appreciate that at each echelon in a holarchy, the relations between holons will be governed by the spatial scale and time span. This means that each level of operation in the climate system is associated with virtually independent sets of state factors for which solutions can be derived by methods appropriate to that level, a fact which will be elaborated upon in the concluding chapter. In practice, therefore, the criteria involved in research projects at each level of a holarchy will differ. But holarchical levels and their state factors are not truly independent: they are linked to one another by a causal sequence involving a projection of partness and wholeness from any given level to all others (Haigh 1987). Only by establishing the scale linkages from one holarchical level to another can a comprehensive or general solution to complete climate system models be hoped for. By way of illustration, consider the question of spatial resolution in general circulation models. As was mentioned above, the smallest spatial grid cell which can be used in general circulation models is presently 100×100 km. The problem arises how to summarize biological and chemical processes operating on smaller spatial scales which influence the large-scale features of the system. Bert Bolin (1988) suggests that, as climate models basically treat the flux of energy, momentum, and water, the parameterization of sub-grid processes should bear their role in these fluxes in mind. He hopes that if large-scale models of biospherical dynamics can be formulated in this way, then the features deduced from experiments with total biosphere models can be used to work out how small-scale characteristics of the biosphere are generated and maintained. Research into scale linkage is in its infancy, but it does seem to be an exciting and potentially fruitful line of enquiry, as we shall see later in the book.

2 Air

The climate of any point on the earth's surface depends on a complex of factors, some of them due to influences arriving from outside the earth, and others purely terrestrial.

C.E.P. Brooks (1922, p. 15)

2.1 Solar Forcing

2.1.1 Cycles of Solar Activity

First spotted by Asian astronomers (with naught but their naked eyes!) in 1077, sunspots were generally ignored by western astronomers until Galileo Galilei observed them through his telescope around 1611, since soon after which time they have been surveyed more or less continuously. A German apothecary and amateur solar observer, Heinrich S. Schwabe, suspected from his observations over the years 1826 to 1837 that the number of spots on the solar disk varies regularly, and possibly periodically, with time (Schwabe 1838). Some years later he caused a stir by confirming that sunspots come and go over a cycle lasting 10 years (Schwabe 1843, 1844). He came to this conclusion simply by plotting the average number of sunspots seen each year. It seems that professional astronomers, believing the Sun to be constant, had never bothered trying this elementary exercise. Schwabe's work came to the attention of another German, J. Rudolf Wolf. On the basis 150 years of sunspot records, Wolf established an average period of 11.1 years, although he noted considerable variations in period length and amplitude (e.g. Wolf 1858).

The cycles within the Solar System influence the complex processes leading to variations in solar output which occur at many timescales. High-frequency oscillations of solar radiation, with a period of about 25 to 28 days, are due to groups of sunspots becoming aligned with the Earth disk as the Sun rotates (Smith et al. 1983). Low frequency oscillations are the integrated effect of sunspot activity through a full 11-year solar or sunspot cycle. Associated with the solar cycle is the Hale cycle, also called the double sunspot or heliomagnetic cycle, with a period of about 22 years and named after its discoverer, George Ellery Hale (1924). The Hale cycle involves a reversal of the Sun's magnetic field and sunspot variations. It has a strong influence on cycles of terrestrial magnetic activity and the production of isotopes in the atmosphere. Lower frequency oscillations in solar output seem to exist, too. Auroral evidence, for example, suggests a cycle with a period of 78 to 80 years called the Gleissberg cycle (Gleissberg 1955, 1965). Along with the solar rotation cycle, sunspot cycle, and Hale cycle, the Gleissberg cycle seems to be associated with expansions and contractions in the Sun's diameter (Gilliland 1981). Another low frequency oscillation, corresponding to the 178.73-year period of the Sun's orbital motion, was revealed by studies of the phase lag of sunspot cycles over two 178-year periods (Fairbridge and Shirley 1987). This cycle seems to accord with the recent finding of an approximate 200-year cycle in the isotopic record of tree rings (Sonett and Finney 1990; H.E. Suess and Linick 1990; D. J. Thomson 1990). A 300- to 400-year cycle has also been proposed (Link 1968). In short, solar activity, as the prime mover of the climate system, exhibits a range of short-term rhythms. These solar beats are of considerable interest to Earth scientists because they appear to have left their mark in many terrestrial phenomena.

2.1.2 Solar Signals in the Atmosphere

A link between observed sunspot activity and climatic fluctuations was made early in the nineteenth century. Monsieur Renoir (1839, 1840) opined that an increase or decrease of sunspots in the past might have affected terrestrial climate sufficiently to have caused the Ice Age, the vestiges of which had recently been discovered by Louis Agassiz. Thinking on a shorter timescale, J. Rudolf Wolf, the establisher of the sunspot cycle, believed that years in which sunspots occur are drier and more productive than years when they do not occur. Apparent proof of the sunspot period in atmospheric variables was given during the last half of the nineteenth century (see Fritz 1878). For instance, mean annual temperatures were found to parallel sunspot frequency, especially in the tropics (Köppen 1873), and famines associated with a lack of monsoon rains were connected with the sunspot cycle (Lockyer and Hunter 1877). During the opening decades of the twentieth century, the leading exponent of the solar hypothesis of climatic change was Ellsworth Huntington. Impressed by the work of Charles Julius Kullmer (1914, 1933), which showed that cyclonic storms in North America intensify within the middle part of their tracks during times with high sunspot numbers, Huntington developed a solar-cyclonic hypothesis associating high sunspot numbers with increased storminess and low global temperatures. Support for this solar-cyclonic hypothesis came from the work of a Dr. M. A. Veeder (Huntington 1917), who found that auroral displays are followed immediately by an intensification of high and low pressure cells, and a few days later by increased meridional flow. Huntington came to believe that variations in solar radiation associated with sunspots were the root cause of global climatic cycles during historical times (Huntington 1907, 1911, 1914a,b) and during Quaternary times (Huntington and Visher 1922, pp. 110–129). As to the mechanism linking solar activity to climate, Huntington was bold enough to identify variations in particulate solar radiation during solar cycles as a possible cause, and to propose that sunspot activity was itself influenced by the planets and nearby stars, suggesting that the 80-year period in sunspot phenomena was due to effects of the Alpha-Centauri system (Huntington and Visher 1922, p. 282).

The flame of interest in the solar connection, made to burn so brightly by Huntington, was all but extinguished after the mid-1930s. It was rekindled during the 1950s by several climatologists including Hermann Flohn (1951), Derek Justin Schove (1954, 1955, 1961), Franz Baur (1956, 1959), Hubert Horace Lamb (1961), and Gordon Manley (1961, 1971), and has recently flared up once more (for useful bibliographies see Pearson 1978; Sanders and Fairbridge 1987). Seeking relationships between solar activity and terrestrial phenomena is still a controversial enterprise (e.g. Runcorn and H. E. Suess 1982), but since the advent of sophisticated methods of time series analysis, which built on the pioneering work of G. Udny Yule (1927), it has gained some scientific respectability. In 1959, Helmut Landsberg and his colleagues discovered an 11-year period in a homogeneous series of temperature data from a recording station in Maryland by computing the power spectrum of the series. The effect was small but the analysis demonstrated that changes accompanying the sunspot cycle are capable of being impressed on the lower atmosphere. Weak periods of 2 and 5.6 years were also found. More recent studies using spectral techniques to tease out the harmonics within atmospheric time series have firmly established the presence of solar signals in a wide range of terrestrial phenomena. The 11-year solar cycle and its harmonics have been detected in many meteorological variables including air temperature, air pressure, and rainfall (G. M. Brown and Price 1984; Colacino and Rovelli 1983; R. G. Currie 1981b,c,d, 1987a,b, 1988; R. G. Currie and O'Brien 1988, 1990), global sea-surface temperatures (Barnett 1989), extreme hourly rainfalls in the United Kingdom (B. R. May and Hitch 1989), and also in data on the Earth's spin rate (length of day) (R. G. Currie 1980, 1981a). It has been found in the 11.6-year periodicity of the zonal index (the average pressure gradient across latitude circles) and the approximate 14-year periodicity in annual means of air temperature over central England (Gilman 1982). A 22-year signal has been detected in global and hemispherical marine temperatures (N. E. Newell et al. 1989), the incidence of tropical cyclones (Cohen and Sweetser 1975), severe droughts in the North American Great Plains (Abbot 1963), and rainfall in the broad central area of North America, from the Great Lakes to the Rockies and into southern parts of the Canadian Prairies and Ontario (Vines 1984). However, for the reasons given by Robert Guinn Currie and Douglas P. O'Brien (1990), it seems unlikely that the Hale cycle can be a factor in climatic change. The 78- to 80-year Gleissberg cycle has been found in sunspot numbers and in the proxy temperature data from the top part of the Camp Century ice core (Dansgaard et al. 1971), and appears to affect long-term temperature trends (Willett 1980, 1987). The 178.73-year cycle of solar inertial motion has been detected in the post-1715 observational record of solar activity (Fairbridge and Shirley 1987), and may account for the 170- to 200-year cycle of European temperatures (Brunt 1925) and English temperatures (Lamb 1965, 1972, 1982), for the 181-year period in isotope ratios in the top part of the Camp Century ice core (Dansgaard et al. 1971), and for a cycle of similar length in the Bai U rains of Korea (Yamamoto 1967).

    Other periods have been detected which may or may not be connected to solar cycles. In his monograph *Klimaschwankungen seit 1700* (1890), Edouard Brückner reported a cyclical change of weather lasting nearly 35 years and recorded in variations in the levels of European rivers, in the level of the Caspian

Sea, in cold winters in Europe, in the times of harvest, in the opening of rivers to navigation, and in rainfall and temperature measurements from a variety of places dotted around much of the world. Much more recently, isotope ratios in a 2000-year tree ring sequence of a cedar tree (*Cryptomeria japonica*) from Japan, as assayed by Leona Marshall Libby (1983, p. 45; 1987, p. 87), have been found to contain the following periods: for hydrogen isotope ratios — 58, 65, 86, 97, 110, and 156 years; and for oxygen isotope ratios — 55, 70, 88, 97, 124, and 156 years (see also Stuiver 1983; Schove 1987).

Convincing connections between solar activity, stratospheric winds, and surface temperatures were established by Karin Labitzke and Harry van Loon (Labitzke and van Loon 1988, 1989a,b,c, 1990; van Loon and Labitzke 1988a,b, 1990). These researchers claim that the effects they detected had previously passed unnoticed because the solar influence only operates when the wind in the stratosphere above the equator has shifted either to the west or to the east ("west" phase and "east" phase). The stratospheric wind shift occurs roughly every 26 months and is known as the quasi-biennial oscillation. When the data for both phases of the cycle are analysed together, the solar influences cancel one another out: the solar signal can only be seen if the east and west phases of the quasi-biennial oscillation are studied separately. Measuring solar activity as radio noise at a wavelength of 10.7 cm, a statistically significant relationship between the temperature at 22 km above the North Pole and the strength of the Sun during the west phase of the quasi-biennial oscillation in winter was established: warmer winters occurred when the Sun was most active and colder winters occurred when the Sun was least active (Labitzke and van Loon 1988). During the east phase, the pattern is reversed, warmer winters being associated with a least active Sun and colder winters with a most active Sun. However, no relationship could be detected when all the data, west phase and east phase, were included. Further study disclosed a significant correlation between solar activity in west phase years and pressures at sea level (van Loon and Labitzke 1988a,b). When the Sun is most active, then in west phase years pressure will be higher than usual over North America, and lower than usual over the eastern Pacific Ocean and western Atlantic Ocean. The reverse situation obtains when the Sun is least active. The pressure differences, apparently induced by solar activity, influence the flow of air and hence surface temperatures. This effect can be seen, for instance, with the quasi-biennial oscillation in west phase at Charleston and Nashville (Fig. 2.1). The solar effect also seems to influence the development of storm systems in middle latitudes, a fact first discovered by Geoffrey M. Brown and J. I. John (1979) and subsequently confirmed by Brian Tinsley (1988). During the northern summer, inter-annual variation in the stratosphere is small, but despite this, a distinct 11-year solar signal has been found in geopotential heights and temperatures (Labitzke and van Loon 1989b). This signal is present in the full time series, so no account need be taken of the east and west phase years of the quasi-biennial oscillation in summer months. The data for the Southern Hemisphere is sparse but the same signal has been found (Labitzke and van Loon 1989c). The correlations of temperature data with the solar flux have been found at 80 km (Labitzke and Chanin 1988) and as high as 170 km (Chanin et al. 1989).

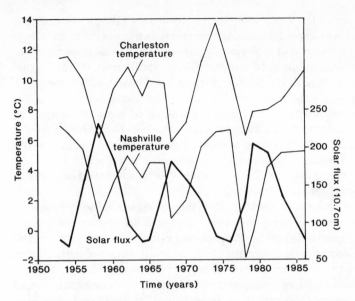

Fig. 2.1. Time series for January-February in the west years of the quasi-biennial oscillation of the 10.7 cm solar flux and air surface temperature at two stations in the United States. (After van Loon and Labitzke 1988a)

2.1.3 Mechanisms of Short-Term Solar Forcing

Changes in the energy output of the Sun over a solar cycle are probably very modest. The solar changes are magnified a millionfold in the atmosphere, an increase which is very difficult to explain and which many are still loath to accept (Kerr 1988). Nonetheless, as we saw in the previous section, there is strong statistical evidence of a connection between solar activity and climate. The big, big question is then: how can tiny changes in energy output of the Sun during a solar cycle have such a large effect on the atmospheric system? The linking mechanisms lie in ultraviolet radiation, the solar wind, or cosmic rays. Ultraviolet radiation emitted by the Sun may change the thermal structure of the stratosphere, chiefly through its influence on ozone. In turn, the thermal changes in the stratosphere may modify the vertical flux of energy in tropospheric planetary waves, or directly warm the troposphere (Lean 1984), and may generate indirect effects that could possibly account for some short-term geophysical signatures of solar activity (Gérard 1990). The solar wind interacts with the geomagnetic field through a variety of plasma populations which lie between the magnetopause and the atmosphere (see K. D. Cole 1985). It may cause a cooling of the troposphere by inducing the development of high cirrus clouds, or may cause the rate of production of ions to change so modifying thunderstorm electrification (e.g. Bucha 1984; Landscheidt 1987). The commencement of sudden geomagnetic storms, induced by solar eruptions, seems to create a highly localized warming of the stratosphere, owing to a localized ultraviolet window being created by the production of ozone-destroying molecules and ions above 30 km, and a displacement of the 7-mb polar vortex (Neubauer 1983). A mechanism linking the changing corpuscular and electromagnetic activity of the Sun through a solar cycle and 2.3-year and 11.4-year signals in mean annual river runoff in the Sahel zone of Africa has been outlined (Faure and Leroux 1990). Cosmic rays from the Sun, plus those coming in from the Galaxy, strike the polar mesosphere and stratosphere and lead to the production of nitric oxide, the catalytic destruction of ozone, and hence a cooling of the stratosphere. In addition, they create cosmogenic isotopes, such as carbon-14 and beryllium-10, recorded fluctuations of which have been identified by several workers as the firmest evidence for a link between changes in solar activity and terrestrial climate (e.g. H. E. Suess and Linick 1990; Lorius 1990; Oeschger and Beer 1990), but not everybody agrees (e.g. Raisbeck et al. 1990).

The details of how solar emissions affect the world climate system are very hazy but rapid advances in aeronomy and solar-terrestrial physics, given a big fillip by the Geosphere-Biosphere Programme, promise to cast light on the matter. Indeed, recent measurements of variations in total solar irradiance suggest a quantitative mechanism through which year-to-year changes in solar activity may influence surface temperature (Eddy 1990); and attempts to model the climatic response to solar variability, in both total irradiance and ultraviolet emissions, have met with some success (e.g. Gérard 1990). In short, the search for solar signatures is no longer so controversial as once it was and much headway has been made in recent years, as the volume edited by Jean-Claude Pecker and Stanley Keith Runcorn (1990) shows.

2.1.4 Longer-Term Variations in Solar Activity

The activity of the Sun is known to vary over centuries (e.g. Harvey 1980). There was, for example, a near absence of sunspots from A.D. 1645–1715. This fact was chanced upon by E. Walter Maunder in 1893 while poring over old books and journals concerned with the Sun, and the period of abnormally low sunspot activity has become known as the Maunder Minimum. Eighteen periods of low sunspot activity similar to the Maunder Minimum are recorded in the dendrochronological record for the past 7500 years, and include the Spörer Minimum (A.D. 1400–1510) and the Mediaeval Minimum (A.D. 1120–1280). John A. Eddy (1977a,b,c) showed that these periods of prolonged solar minima tend to be associated with periods of low temperatures on Earth. He noted, for example, the association of the Maunder Minimum with the occurrence of the Little Ice Age. The correlation between these two events is, however, debatable. Helmut Landsberg (1980) recovered historical European records which suggested to him that sunspots were in evidence during the time of the Maunder Minimum, albeit at levels lower than at present. Eddy (1983) objected to Landsberg's interpretation of the historical records of sunspot numbers and reaffirmed that the Maunder Minimum, whether it had any effect on climate or not, did occur. Landsberg also pointed out that the coldest spells during the Little Ice Age were from 1600 to 1619 and from 1800 to 1819, and were not coeval with the Maunder Minimum. However, Stephen H. Schneider and Randi Londer

(1984, p. 126) took Landsberg to task for basing his conclusions on widely scattered temperature data which are unreliable as a source of global temperature reconstruction. Analysis of a 1400-year tree-ring record from a Scots pine (*Pinus sylvestris*) in the Torneträsk region of northern Sweden has recently been used to reconstruct the mean summer temperature (April to August) in northern Fennoscandia, and the results show that the Little Ice Age at that latitude was confined to a relatively short period between 1570 and 1650, thus challenging the popular notion that the Little Ice Age occurred synchronously throughout Europe in all seasons. The matter remains unresolved, although a study made with a simple energy-balance climate model demonstrated that cool periods like the Maunder Minimum could have been brought about by a 0.22 to 0.55 percent reduction in solar irradiance (Wigley and Kelly 1990).

Solar cycles lasting millenia and millions of years, superimposed on very long-term trends in solar output, have been suggested. The record of carbon-14 in tree rings and beryllium-10 in the Camp Century ice core both contain periods in the range 2000 to 2500 years (H. E. Suess and Linick 1990; Sonett and Finney 1990). Hurd C. Willett (1961, 1962, 1980) saw changes in ultraviolet and charged particles (protons) received at the top of the atmosphere as the causal link between sunspot activity and some climatic cycles, and believed that such changes might explain all the climatic cycles which have occurred since the start of the Pleistocene epoch (Willett 1949). Thinking on a much longer timescale, Eugène Dubois (1893, 1895) contended that during his history the Sun has passed through several stages in which energy emissions have varied. A modern, more sophisticated version of these ideas was presented by Ernst Öpik (1950, 1953, 1958a,b), who attributed the glaciations of the last 500 million years to a temporary reversion of the Sun to an earlier, cooler state, and glacial-interglacial cycles to "flickerings" of a disturbed Sun. Evidence of solar cycles lasting millions of years has been uncovered by Thomas van der Hammen (1961) using several plant species, including the Monocolpites group, as indicators of palaeotemperature. The changing abundance of these plants suggests cycles of warming and cooling from the late Cretaceous period to the Upper Miocene epoch with periods of approximately 2, 6, and 60 million years. Even more protracted is the secular increase in the luminosity of the Sun during geological time. This increase may be in the order of 25 percent (M. H. Hart 1978). Energy balance models and radiative-convective models predict that if the solar constant were to fall just a few percent, then, as a result of the feedback between ice cover and albedo, the world climate system would adopt a new stable state in which the Earth would be totally covered with ice. A world-wide glaciation did not occur until the Huronian glaciation, and other factors, particularly carbon dioxide and methane levels, must have counteracted the effect of low solar luminosity during Precambrian times (Gérard 1989).

Long-term changes in solar output may arise from external causes. It has long been known that clouds of fine dust (nebulae) are draped across parts of interstellar space. In 1909, Friedrich Nölke suggested that if the Solar System were recently to have passed through a nebula, solar radiation would have been greatly reduced, thus starting the Ice Age. Later mathematical investigations of this idea showed, however, that absorption of radiation in the distance between the Sun and the Earth would be very small and a cooling effect is unlikely (Krook 1953). Other workers, rather than claiming that the passage through a dust cloud would lead to a reduction in radiation receipt, suggested that it would actually increase the solar output. The first to do so was Harlow Shapley (1921). Later, Fred Hoyle and Raymond A. Lyttleton (1939, 1950) proposed that the passage of the Solar System through a cloud of interstellar matter could result in an ice age on Earth. As the Sun swept up dust particles, it would become brighter and its output increase. Increased solar radiation receipt by the Earth would lead to increased precipitation and the growth of ice sheets (cf. G. C. Simpson 1929, 1934, 1959). This pioneering work was taken up by William H. McCrea (1975), who showed that the passage of the Solar System through a dust lane bordering a spiral arm of the Galaxy may indeed cause a temporary increase of the Sun's radiation and so lead to an ice epoch on Earth. John Gribbin (1978) developed the idea, indicating that a double-arm structure to the Galaxy would produce two ice epochs per galactic year. Evidence from lunar soil samples ranging in age from 1.7 to 0.9 billion years suggests that the Solar System did indeed thrice pass through galactic dust lanes between those dates (Lindsay and Srnka 1975).

2.2 Short-Term Gravitational Forcing

The Earth's climate is influenced by three extraterrestrial motions within the Solar system: planetary orbital motions, the Sun's barycentric orbit, and lunar tidal cycles (Fig. 1.1). All cycles in the Solar System are interlocked chronologically in a commensurable way; in other words, every fundamental cycle in the Solar System is related to every other cycle in ratios of small integers. The principal cycle is the Saturn-Jupiter lap, with a period of 19.859 years, and most other cycles are commensurable with it. For instance, the Sun's orbital symmetry progression cycle is $19.859 \times 9 = 173.73$ years. At any given time, the relationships between cycles will be approximate owing to the effects of ellipticity and other factors. The approximate commensurabilities among the cycles emphasize the dynamic unity of the Solar System (Fairbridge and Sanders 1987, p. 447). This section will consider the possible effects of lunar and planetary beats on the world climate system.

2.2.1 Lunar Signals

The first person to see a possible link between the Moon and climate was the Greek Theophrastus, who noted the tendency for the ends and beginnings of lunar months to be stormy. Isaac Newton recognized the existence of atmospheric tides, and Pierre Simon, Marquis de Laplace, sought a correlation between the tidal semi-diurnal influence of the Moon and atmospheric pressure, but without success. Not until improved instruments and longer records became available late in the nineteenth century was it possible to make the connection. Of special interest was the influence on climate of the lunar nodal precession cycle, which has a period of about 18.6 years and is tied into a nutation of the Earth's precession. N. Ekholm and Svante Arrhenius (1898) demonstrated a relation between lunar motions and the frequency of thunderstorms and the behaviour of the aurora borealis. H. E. Rawson (1907, 1908, 1909) brought forward evidence for a nodal variation in the latitude of subtropical high pressure cells in both Northern and Southern Hemispheres, and for an 18.6-year induced drought-flood in South Africa and Argentina. After Rawson's papers, interest in 18.6-year geophysical phenomena waned, although it was maintained by scientists involved with the preparation of tide tables for harbours. The subject was revived in the 1950s by scientists in the Soviet Union under the leadership of Ivan Vasil'evich Maksimov. During the 1970s, it was rekindled by Robert Guinn Currie, who was prompted by the advent of high-resolution spectrum techniques for extracting the harmonics of time series. The lunar nodal term has now been hunted and found in atmospheric variables such as air temperature (R. G. Currie 1974, 1979, 1981b,d, 1987b; Libby 1983), sea temperature (Loder and Garrett 1978), air pressure (R. G. Currie 1982, 1987b), rainfall in the west coast region of the United States (Vines 1982), rainfall in the northeastern United States (R. G. Currie and O'Brien 1988, 1989), rainfall in the American Corn Belt (R. G. Currie and O'Brien 1990), and in the 18.6-year cycle of the Great Plains drought (Kerr 1984). Also, it has been found in sea-level changes, at least within some regions (R. G. Currie 1981b,c).

Apart from the relations between changes in the seasonal and latitudinal distribution of radiation and variations in orbital parameters, which we shall explore in the next section, there is no understood association between gravitational tides and weather and climate (Bonnet 1985, p. 400). R. G. Currie and D. P. O'Brien (1990) report a model they are developing which explains how bistable switching in air pressure data may occur owing to gravitational effects caused by the Moon. Their model incorporates the acceleration of the Earth (with a period of 18.613 years), pressure gradients, and gravity into Newton's second law, which it then combines with the equations of mass conservation, state, and the first law of thermodynamics to provide a theoretical coupling mechanisms between wind systems. One wind system — a westerly flow of air induced by the 18.6-year perturbation of the Earth's rotation rate — interacts with the jet stream and wind circulation of subcontinental extent. The coupling between the two wind systems creates standing wave "cells" of subcontinental size with a period of 18.6 years. Bistable switchings of air pressure recorded at weather stations are the result of sudden changes in the velocity of the jet stream triggering the reorganization of the standing wave "cells" which adopt new, equilibrial spatial patterns.

2.2.2 Short-Term Orbital Forcing

Even if solar emissions were to stay constant, the amount of radiation received by the Earth in consecutive years would vary because gravitational interactions within the Solar System will lead to a change in the distance between the Earth and the Sun and in the solar declination angle. This can be appreciated by considering the equation describing the solar radiation incident upon a unit horizontal area at the outer limit of the atmosphere at a fixed time:

$$Q'_s = S\left(\frac{\bar{d}}{d}\right) \cos Z,$$

where Q'_s is the instantaneous flux of solar radiation, S is the irradiance of the Sun (the so-called solar constant), $\bar{d}$ is the mean distance of the Earth from the Sun, d is the instantaneous distance of the Earth from the Sun, and Z is the zenith angle of the Sun at the point of interest. Using spherical trigonometry, $\cos Z$ may be defined as

$$\cos Z = \sin \varphi \sin \delta + \cos \varphi \cos \delta \cos h,$$

where φ is the geographical latitude, δ is the geocentric declination angle of the Sun, and h is the hour angle (the angle through which the Earth must turn to bring the point of interest directly under the Sun). To find the daily flux of radiation, the equation for the instantaneous flux must be integrated from sunrise to sunset yielding

$$Q_s = \frac{S}{\pi}\left(\frac{\bar{d}}{d}\right)^2 (h_0 \sin\varphi \sin\delta + \cos \varphi \cos \delta \sin h_0),$$

where h_0 is the hour angle of the Sun at sunset. This equation, in conjunction with Kepler's equation for the eccentric anomaly and Lacaille's equation for the true anomaly, can be used to compute the daily receipt of solar radiation at the top of the atmosphere as a function of latitude and time (see Monin 1986, pp. 11–12). To find seasonal or annual radiation totals, the equation for daily insolation receipt must be integrated over an appropriate interval of time. This was done by Milutin Milankovitch (discussed below), who derived a formula for insolation receipt in summer, Q_S, and winter Q_W seasons:

$$Q_{S,W} \approx \frac{ST_0}{2\pi} \left\{ S(\varphi,\varepsilon) \pm \sin \varphi \sin \varepsilon \mp \frac{4}{\pi} e \sin \Pi \cos \varphi \right\},$$

where T_0 is the length of the tropical year (the time between successive arrivals of the Sun at the vernal equinox); $S(\varphi, \varepsilon)$ is an elliptical function describing the distribution of annual insolation along the meridian; ε is obliquity; e is eccentricity of the orbit; and Π is the longitude of perihelion. Changing gravitational fields of the Moon and the planets will cause year-to-year perturbations in ε, e, and Π which will modulate the radiation receipt at the top of the atmosphere. These perturbations are regular and will cause a systematic global redistribution of radiation which is likely to influence climate.

To estimate the changes of insolation at different latitudes for the period A.D. 1800 to 2100, Ye. P. Borisenkov and his associates (1983) used algorithms for calculating planetary ephemerides applicable to that interval of time. Their results showed that the pattern of insolation variation at high latitudes was very different from the pattern at low latitudes. Noteworthy are the cyclical components within the insolation time series. In high latitudes, at times of summer and winter solstices, an 18-year component is present; but in low latitudes the pattern is made complicated by the presence of other cyclical components of shorter period. At 20° N, five harmonics were identified corresponding to periods of 2.7, 4.0, 5.9, 11.9, and 18.6 years. The 18.6-year component represents the main nutation of the Earth's axis associated with the lunar nodal cycle; the 11.9-year term coincides with the sideral period of Jupiter; and the shorter periods can be identified with multiples of the periods of revolution of the inner planets.

2.2.3 Short-Term Changes in the Earth's Spin Rate

On a short timescale, the Earth's spin rate seems to be connected with solar and lunar variables (see Brosche and Sündermann 1990). Spin accelerations are accompanied by amplifications of zonal circulation patterns, mild winters in the Northern Hemisphere, and stronger monsoons in the tropics; whereas spin decelerations are associated with blocking circulation patterns, winters that are cooler than normal in the Northern Hemisphere, and weaker or delayed monsoons (Fairbridge and Sanders 1987, p. 447). Arguing that the Sun's gravity and rotation and the Earth's gravity and rotation are both locked into the planetary beats, Nils-Axel Mörner (1987) proposed a rotational-gravitational-oceanographic model of climatic change applicable to timescales of decades and centuries. On the basis of observations, he argued that over the last 20,000 years few major short-term climatic changes have been global; rather, they are regional, or at most, hemispherical events. These changes have been effected by the redistribution of heat by ocean currents following perturbations of the geoid, and by the interchange of momentum between the Earth's solid, viscous, fluid and gaseous layers caused by linked gravitational and rotational adjustments in response to cycles of change in planetary interactions (see also Mörner 1984a,b). This model reduces climatic change, at least over short timescales, to a simple matter of mass, momentum, and energy, and explains various terrestrial geophysical variables in terms of the interaction of the Earth, Sun, and planets.

2.3 Medium-Term Gravitational Forcing

2.3.1 Ellipticity, Precession, and Obliquity

The jostling of the planets, their satellites, and the Sun leads to medium-term orbital variations occurring with periods in the range 10,000 to 500,000 years which perturb Earth's climate. These orbital forcings do not change the total amount of solar energy received by the Earth during the course of a year, but they do modulate the seasonal and latitudinal distribution of solar energy. The variables involved in these changes were discussed in the previous section in the context of short-term changes. The same variables — obliquity, precession, and ellipticity — also change over the medium term, each having characteristic periods commensurable with the shorter-term cycles: the eccentricity cycle has two components — a short cycle of 100,000 years and a long cycle of about 400,000 years; the tilt cycle is 41,000 years long; and the precessional cycle has periodic components of length 23,000 and 19,000 years (Fig. 2.2). Cycles of approximately this length have been detected in the now fairly large collection of proxy palaeotemperature records, particularly those derived from oxygen isotope ratios in deep-sea sediment cores (e.g. Imbrie 1985; Pestiaux et al. 1987).

The idea that orbital variations might influence climate on the Earth can be traced to the seventeenth century. Monsieur de Mairan, writing in 1765, remarked on the effect of the distance of the Sun from the Earth in apogee and perigee (cited in Croll 1875, p. 528). Charles Lyell, in the first edition of his *Principles of Geology* (1830–33, vol. i, p. 110), commented on the effect of precession of the equinoxes on the receipt of "solar light and heat" in the two hemispheres. The effects of precession were explored more fully in 1842 by the French mathematician, Joseph Alphonse Adhémar, who thought that the differences in the seasons between the hemispheres brought about by precessional changes would be large enough to have caused the Ice Age. The possible effects of eccentricity on climate were raised by John Frederick William Herschel (1835), but the first detailed discussion of the matter was spelled out by James Croll in a series of papers (e.g. Croll 1864, 1867a,b) and later in his book *Climate and Time in Their Geological Relations: a Theory of Secular Changes of the*

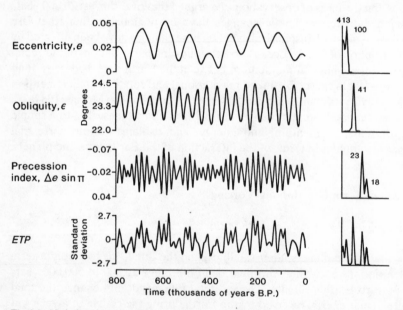

Fig. 2.2. Variations in eccentricity, *e*, obliquity, ε, and the precession index, Δ *e* sin Π, over the past 800,000 years from numerical simulations. *ETP* is a normalized and summed combination of the three other quantities. On the *right* of the figure are shown the power spectra of each curve with dominant periods in thousands of years. (After Imbrie et al. 1984)

Earth's Climate (1875). Croll argued that an ice age would occur when an elongated orbit was combined with a winter solstice that occurred near aphelion. The effects of precession, ellipticity, and obliquity on the seasonal and latitudinal distribution of radiation were studied in great detail by the Yugoslavian mathematician and engineer, Milutin Milankovitch (1920, 1930, 1938). Milankovitch's chief conclusions were that orbital eccentricity and precession produced effects large enough to cause ice sheets to expand and contract; that the climatic effects of obliquity were far greater than Croll had presumed; and that astronomical variations in eccentricity, precession, and obliquity were sufficient to produce ice ages by changing the seasonal and geographical distribution of solar radiation (Fig. 2.3). In his 1920 publication, Milankovitch suggested that a great cycle of climate, which takes roughly 100,000 years to go through one round, is produced by small variations in the orbital variables. By analogy with the annual march of the seasons, the march of the great seasons runs thus: "great winter", when the Earth is gripped in an ice age, "great spring", when there is a great thaw, "great summer", when interglacial conditions prevail, and "great autumn", when conditions start to deteriorate presaging the coming of the next "great winter".

The Croll-Milankovitch theories of climatic forcing were popular up to about 1950, after which time they were rejected by most Quaternary geologists. During the late 1960s and early 1970s, Milankovitch's cycle of great seasons was rediscovered. Evidence for a 100,000-year cycle came to light independently in loess sequences exposed in a quarry in Czechoslovakia (Kukla 1968, 1975), in

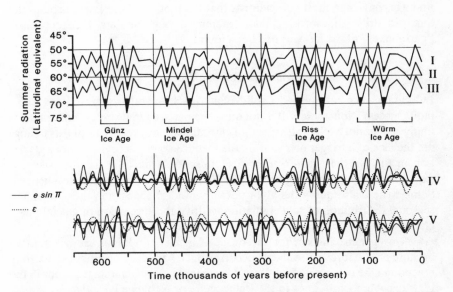

Fig. 2.3. Orbital forcing and the occurrence of ice ages according to Milutin Milankovitch. ε is the obliquity of the ecliptic; e is the eccentricity of the Earth's orbit; Π is the longitude of perihelion. (After Köppen and Wegener 1924)

sea levels (Broecker et al. 1968; Mesolella et al. 1969; Chappell 1973), and in the oxygen-isotope ratios of marine cores (Broecker and van Donk 1970; Ruddiman 1971). Moreover, both the terrestrial and marine records attested to long periods of glacial expansion (climatic cooling) abruptly ended by rapid deglaciations (climatic warming). It was realized that the 100,000-year signal might be caused by the eccentricity cycle, and James Croll's argument of 100 years earlier was resurrected: when the Earth's orbit is unusually elongate, precessional effects are amplified, producing more constrasted seasons and allowing an ice age to start. Later, it was demonstrated that climatic oscillations superimposed on the 100,000-year cycle had been produced by the precessional and tilt cycles. This demonstration required a finely calibrated calendar of Pleistocene events. Largely owing to the endeavours of the members of the CLIMAP project, a suitably detailed calendar emerged. Nicholas J. Shackleton and Neil D. Opdyke (1973), by making oxygen-isotope and magnetic measurements of a Pacific deep-sea core, and establishing that isotope Stage 19 occurs at the boundary between the Bruhnes and Matuyama epochs, gave the first accurate chronology of late Pleistocene climate. Confirmation of the Croll-Milankovitch theory was eventually forthcoming when John D. Hays found suitable cores from the Indian Ocean, which recorded climatic change over the last 450,000 years, for subjecting to spectral analysis. The results of the analysis revealed cycles of climatic change at all frequencies corresponding to orbital forcings (Hays et al. 1976). In addition to the 23,000-year precessional cycle, a 19,000-year precessional cycle component was present. This minor precessional cycle was confirmed theoretically by the Belgian astronomer André Berger (1978). The publication of these findings convinced most geoscientists that the motion of the Earth around the Sun did drive the world climatic system during the late Pleistocene. A "pacemaker" of the ice ages had been found.

2.3.2 Problems with the Croll-Milankovitch Theory

Indubitably, the refurbished Croll-Milankovitch hypothesis is enjoying enormous success. Nonetheless, it is not quite the panacea that some commentators imply: it does not explain all aspects of glaciations. Noteworthy among its failings are the lack of a trigger mechanism, the inability to predict synchronous glaciation in the Northern and Southern Hemispheres, and the powerlessness to explain the observed strong beat of the 100,000-year short cycle of eccentricity. Some writers also query the ability of very modest changes in the spatial and seasonal distribution of solar radiation to drive the succession of glacial-interglacial cycles.

A trigger mechanism for the onset of glacial conditions seems essential because orbital forcing has always occurred but it has not always induced a strong reaction in the climate system. Why, for example, did glacial stages occur in the Pleistocene epoch but not in the Palaeogene or Neogene periods? As Walter Wundt (1944) observed, it would seem that only under certain geographical and geological circumstances do orbital variations permit ice ages to make an appearance. A precondition of glaciation appears to be a cold climatic background. Now, there are many factors which could cause the atmosphere to become cooler, some of which will be discussed in the following chapters. Briefly, key factors seem to be the carbon dioxide (and possibly methane) content of the atmosphere, the arrangement of continents and oceans, and the presence of large mountain ranges. It is possible that the redistribution of continents and oceans and the growth of mountain ranges through the Cenozoic era led gradually to a cooling of the atmosphere which, by Pleistocene times, was cold enough to permit ice sheets to form during insolation minima. However, recent experiments with a general circulation model, run with orbital configurations corresponding to times of rapid ice sheet growth, raise doubts about the ability of Croll-Milankovitch forcings to trigger the growth of ice sheets (Rind et al. 1989). Under none of the orbital configurations was the model able to maintain snow cover through the summer at locations suspected of being sites where major ice sheets commenced forming, despite reduced insolation during the summer and autumn. The model also failed to preserve a layer of ice, 10 m thick, placed in localities where ice existed during the Last Glacial Maximum. Only when ocean surface temperatures were adjusted to their values at the height of the Ice Age could the model manage to preserve a smallish patch of ice in northern Baffin Island. To David Rind and his colleagues, the experiments bring out a wide discrepancy between the response of a general circulation model to orbital forcing and geophysical evidence of ice sheet initiation, and indicate that the growth of ice occurred in an extremely ablative environment. To explain how ice sheets might grow in an ablative environment requires a more complicated model or else a climatic forcing other than orbital perturbation — a reduction in carbon dioxide content is a possibility.

The strongest climatic signal in the marine sedimentary record for the last 0.8 million years corresponds to the 100,000-year short cycle of eccentricity. This is curious, and suggests a non-linear response, because eccentricity should be the weakest of the orbital forcings, simply modulating the effects of precession. Even more curious is the fact that from about 2.4 to 0.8 million years ago, the 41,000-year cycle of tilt is the dominant signal in the record (Mix 1987; Ruddiman and Raymo 1988). The changing nature of the dominant orbital signal has been underscored in other studies. Spectral analysis of magnetic-susceptibility measurements of terrigenous sediment in deep-sea cores taken from the eastern tropical Atlantic Ocean, spanning the past 3.5 million years, and from the Arabian Sea, spanning the past 3.2 million years, shows that the effect of orbital forcing changed around 2.4 million years ago (Bloemendal and deMenocal 1989). Prior to 2.4 million years ago, both records carry strong 23,000-year and 19,000-year signals of the precessional cycle, suggesting that the summer monsoons which carried the terrigenous sediments were largely modulated by insolation variations during the summer season; but after that date, the 41,000-year tilt cycle signal predominates. This switch coincides with the onset of major glaciation in the Northern Hemisphere and, in the case of the Arabian Sea site at least, is reflected in the supply of terrigenous sediment (carried by monsoonal winds) responding to a rapid increase in ice cover in Eurasia and North America. These shifts in dominant pulse suggest that forces additional to orbital variations have influenced Pleistogene climates: the Croll-Milankovitch hypothesis does not provide the whole answer. William F. Ruddiman and M. Raymo (1987) argue that the change in dominant beat must be due to a change in the configuration of land, sea, ice, and atmosphere. In particular, they conjecture that rapid tectonic uplift of the Himalayas and parts of western North America during the past few million years has led to a change in the pattern of the jet stream and the growth of cold spots over North America and Europe in the very same places that the Northern Hemisphere ice sheets were located. However, the 100,000-year eccentricity cycle has been detected in the sedimentary record before the onset of the last Ice Age (see Sect. 4.4.2), so the dominance of the tilt cycle between 2.4 and 0.8 million years ago may be anomalous, and the resumption of the 100,000-year cycle 0.8 million years ago simply a return to normal behaviour (Mix 1987).

Another puzzling aspect of orbital forcing is the apparent sensitivity of the climate system to really rather modest changes in the seasonal and latitudinal pattern of insolation receipt. The astronomer Fred Hoyle believes that, given the vast amount of heat stored in the oceans which buffers the climate system against perturbations, the changes involved are far too tiny to have any significant impact on climate. He rejects the astronomical theory of ice ages with gusto: "If I were to assert that a glacial condition could be induced in a room liberally supplied during winter with charged night-storage heaters simply by taking an ice cube into the room, the proposition would be no more likely than the Milankovitch theory" (Hoyle 1981, p. 77). Hoyle's condemnation is rather extreme, and more recent work suggests that he perhaps underestimates the degree of the seasonal insolation anomalies. Anomalies of insolation during caloric half years reach a maximum of up to about 6 kcal/cm^2 (Fig. 2.4); they decrease towards the winter poles but are still not small: an anomaly of about 4 kcal/cm^2 could melt an ice sheet 2.5 km thick in 5000 years. No, the world climate system is sensitive to the seasonal changes of climate resulting from Earth's orbital variations. The effects of orbitally induced changes of summer temperatures during the Holocene epoch, for example, are clearly recorded in melt layers in high-Arctic ice cores: the warmest summers occurred from 10,000 to 8000 years ago and the coldest 150 years ago, as would be expected on the basis of Croll-Milankovitch forcing (Koerner and Fisher 1990). The sensitivity of the climate system to orbital forcing has also been highlighted by general circulation models (Kutzbach 1981, 1987; Kutzbach and Otto-Bliesner 1982; Kutzbach and Guetter 1986; Prell and Kutzbach 1987; Kutzbach and Gallimore 1988). Studies made with general circulation models have revealed that changes in solar radiation receipt brought about by variations in the Earth's orbital characteristics elicit a different thermal response in the sea and on land, and so cause major changes in monsoons and the global water cycle, as will be seen in the next chapter. However, the matter does not rest there for, though the world climate system may indeed be sensitive to orbital forcing, it does not respond in quite the way predicted by the Croll-Milankovitch theory. Details of the last deglaciation, as encoded in the oxygen isotope record, while in broad outline consistent with the Milankovitch hypothesis in that high summer insolation in northern

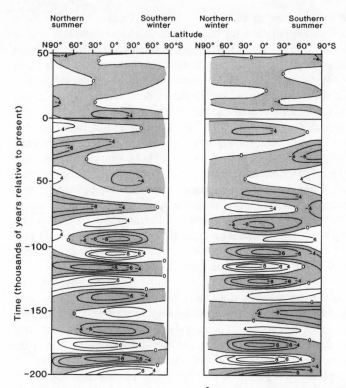

Fig. 2.4. Anomalies of insolation (kcal/cm^2) during caloric half years. 1 kcal/cm^2 = 41.9 MJ/m^2. (After Monin 1986)

latitudes caused deglaciation, indicate that the climate system responded non-linearly to orbital perturbations in the Croll-Milankovitch frequency band. For instance, individual foraminiferal oxygen-isotope data in 24 Atlantic sediment cores show three clear "steps" dated at 14,000 to 12,000 years ago, 10,000 to 9000 years ago, and 8000 to 6000 years ago (Mix and Ruddiman 1985). Similar abrupt changes of climate during the last glacial-to-interglacial transition have been detected in the marine and terrestrial records of environmental change in the high-latitude North Atlantic, Greenland, Europe, and the Caribbean and Gulf of Mexico region (Overpeck et al. 1989). In all these areas, abrupt warming took place between about 13,000 and 12,600 years ago; rapid cooling starting about 11,000 years ago to produce the cold snap known as the Younger *Dryas*; then, commencing some 10,000 years ago, a second swift warming took place to the present interglacial climate. If these "steps" in the palaeotemperature record be real, then a mechanism of climatic change additional to orbital forcing must be invoked. It has been proposed, for instance, that ice sheets might have suddenly become thinner while occupying the same area by a process of calving and unstable "downdraw" triggered by rising sea levels (Mix and Ruddiman 1985).

The formation of deep water in the North Atlantic Ocean has also been identified as a possible explanation for the pattern of climatic change during deglaciation (Broecker and Denton 1989, 1990). A deep salty current starts in the North Atlantic as northwards-flowing, normally saline surface water and then sinks and threads its way south and then east to the Indian Ocean and thence the Pacific Ocean. If the production of North Atlantic deep water cease, possibly owing to changes in seasonality brought about by astronomical forcing, then the atmosphere may switch to a glacial state. This hypothesis has been used to explain the Younger *Dryas* event, the argument running that the production of North Atlantic deep water was temporarily stopped 11,000 years ago because of the influx of glacial meltwater released from glacial Lake Agassiz (Broecker and Denton 1989, 1990). Changes in the circulation of water in the North Atlantic Ocean may have caused abrupt climatic changes in the Sahel and tropical Mexico during the last 14,000 years, too (Street-Perrott and Perrott 1990). Although the North Atlantic deep water hypothesis is highly plausible, it is the subject of lively dialogue (see W. H. Berger and Vincent 1986; Fairbanks 1989; Harvey 1989c; E. Jansen and Veum 1990; Overpeck et al. 1989; Rooth 1990).

To sum up, it would seem undeniable that the orbital forcing envisaged by Adhémar, Croll, and Milankovitch does perturb the world climate system. Whether the perturbations alone be powerful enough to start and end ice ages is doubtful. Whether they be capable of driving the more finely tuned sequence of glacial and interglacial climates during an ice age is debatable, but they do seem to play some part in this. The world climate system, being so richly endowed with feedbacks, responds in a complex way to orbital forcing and not in the relatively simple way envisaged a couple of decades ago. This complexity of response is evident from the empirical record and from theoretical models (e.g. Harvey 1988a,b,c, 1989a,b). Additional complexity lies in the fact that changes in the Earth's orbital geometry will affect several terrestrial variables at the same time, including endogene processes (Fig. 2.5). This may explain why many palaeoclimatic, palaeomagnetic, and palaeogeodetic phenomena are correlated (Mörner 1981). Indeed, it is quite possible that the driving force of climatic change in the Croll-Milankovitch frequency band is gravity acting through endogene processes, and not simply changing insolation patterns (Mörth and Schlamminger 1979; Mörner 1984a,b). Perhaps it would be prudent to conclude with Alan C. Mix (1987) that, at present, the Croll-Milankovitch hypothesis generates more questions than it provides answers. That, it might be argued, is what a healthy hypothesis should do.

2.4 Internal Dynamics of the Climate System

2.4.1 Carbon Dioxide

The possible effect of atmospheric carbon dioxide fluctuations on climate was first raised by John Tyndall (1861). Svante Arrhenius (1896, 1903) and Thomas Chrowder Chamberlin (1897, 1898, 1899) developed the idea, both believing that

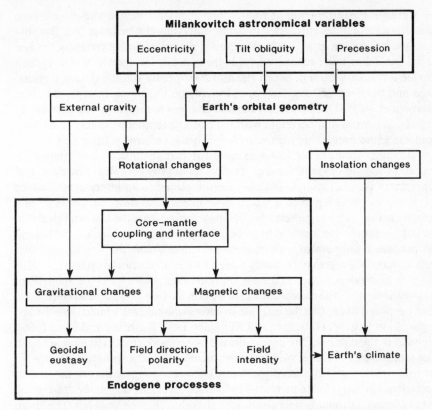

Fig. 2.5. Croll-Milankovitch forcing of terrestrial processes. (After Mörner 1984a)

changing carbon dioxide levels might have produced the Quaternary glaciations, a notion refined and championed this century by Gilbert N. Plass (1956). Later, attempts were made to assess the effect of atmospheric carbon dioxide concentrations on surface temperatures (e.g. Callendet 1938; Möller 1963; see Jones and Henderson-Sellers 1990), the most thorough being carried out with the use of general circulation models (see Bach 1984). For example, early simulations using a three-dimensional general circulation model suggested that a doubling of the present carbon dioxide concentration would raise the mean surface temperature of the atmosphere by 3.0 °C (Manabe and Wetherald 1975, 1980), and that a reduction to half the present level would reduce the mean surface temperature by 2.3 °C (Manabe and Wetherald 1967). In recent years, much effort has been put into ascertaining the effects of rising atmospheric carbon dioxide concentrations on the world climate system (e.g. Harvey 1989e,f, 1990). The results of these studies provide insights into past climatic changes. And, in pumping carbon dioxide and other greenhouse gases into the atmosphere, we are presently carrying out a massive (and possibly dangerous) experiment, the outcome of which will not be known with certainty until next century.

Reliable measurements of past carbon dioxide levels in the atmosphere come from air bubbles entrapped in Arctic and Antarctic ice cores (e.g. Barnola et al. 1987; Genthon et al. 1987). The purest record available covering the last glacial cycle has been recovered from the Vostok ice core in which carbon dioxide levels vary from between 190 and 200 ppm by volume during a glacial stage and between 260 and 280 ppm by volume during an interglacial stage (Barnola et al. 1987). The carbon dioxide changes recorded in this core are of global significance and correlate with the isotopic temperature record derived from the same core. This high correlation between temperature and carbon dioxide concentration would be expected if carbon dioxide had played an important role in climatic forcing. The correlation is complex, however, and depends on whether climate switches from a glacial to an interglacial state or vive versa: in switching from a glacial to an interglacial state, temperature and carbon dioxide move together; whereas in switching from the last interglacial to the last glaciation, the increase in carbon dioxide concentrations lagged behind the increase in temperature. There is some evidence from the Vostok core that carbon dioxide levels may be partly forced by the precessional cycle.

Several models, based on the assumption that the biological productivity of the oceans controls the concentration of carbon dioxide in the atmosphere, have been proposed to explain the increase in carbon dioxide associated with the last deglaciation (e.g. W. H. Berger and Killingley 1982; Broecker and Peng 1986). Views as to the nature of the mechanisms driving productivity changes are split. One school holds sea-level variations responsible; the other points to changes in oceanic circulation. Data from the Vostok ice core suggest that at the end of glaciations carbon dioxide increased before sea level rose and at the start of the last glaciation it probably decreased, very sharply, after sea level fell. The sharp transition from interglacial to glacial levels of carbon dioxide, and vice versa, hints at a sudden modification of the deep ocean circulation, possibly linked with sea-level decrease (Barnola et al. 1987). Changes in oceanic circulation would affect biological productivity of the seas and thus the level of carbon dioxide in the atmosphere. Models of oceanic circulation focus on changes in the formation of deep water at high latitudes and upwelling zones in low latitudes. Information on the latitudes actually involved in regulating carbon dioxide concentration in the atmosphere can be gleaned from spectral analysis of temperature and carbon dioxide time series. In marine cores, proxy temperature series carry the 41,000-year tilt signal in latitudes polewards of 45° N. The Vostok ice core temperature profile also has a strong 41,000-year component, but its entrapped carbon dioxide contains no pulse with unequivocal correspondence to the obliquity cycle. The lack of an obliquity signal in carbon dioxide in the Antarctic ice would suggest that high latitudes play but a small role in influencing carbon dioxide variations, though it is possible that they do play a greater role around the equinoxes. More plausible is the idea that the upwelling of ocean currents at low latitudes, caused by the trade winds, leads to increased biological productivity and a concomitant depletion of the atmospheric carbon dioxide reservoir. It may be more than coincidence that the concentration of aluminium in the Vostok ice core displays spikes indicative of a more vigorous atmospheric circulation at around 150,000, 70,000, and 20,000 years ago (de Angelis et al. 1987), and that these spikes correspond with very low carbon dioxide values.

Further insight into the forcing of the climate system by carbon dioxide came out of a multivariate statistical analysis of temperature and atmospheric carbon dioxide content from the Vostok ice core and several other variables including insolation in July at latitude 65° N, the total insolation at latitude 78° S during the entire year, the insolation in November at latitude 60° S, and three different estimates of ice volume changes. The chief result of the analysis was that carbon dioxide forcing makes a relative contribution to climatic change always greater than 50 percent and as high as 84 percent. This led to the conclusion that insolation changes created by relatively weak orbital forcing are magnified by carbon dioxide changes, themselves possibly orbitally induced, to cause a switch of state in the world climate system. This is essentially the idea posited by Nicklas G. Pisias and N. J. Shackleton (1984; see also Pisias and Imbrie 1986; Harvey 1989a). If carbon dioxide did act in that manner, it would nicely explain the synchroneity of major glaciations in the Northern and Southern Hemispheres, one of the main points left unexplained by the Croll-Milankovitch theory, though it has also been claimed that sea level may play a role in synchronizing climate shifts in the two hemispheres (Denton et al. 1986).

Revealingly, the inclusion of carbon dioxide in climate models leads to the finding that the pulse of climate, imputed by many to orbital forcing, may emanate from the internal dynamics of the world climatic system itself: periodicities of climatic changes may be a consequence of the inner dynamics of the climatic system and may be not due to external forcing alone (Ghil 1981). Illuminating studies on this matter have been made by Barry Saltzman, Alfonso Sutera, and Kirk A. Maasch. Saltzman and Sutera (1984, 1987), Saltzman et al. (1984), and Saltzman (1987a,b), have built a three-component, dynamic model governing global ice mass, atmospheric carbon dioxide content, and mean ocean temperature. This model possesses solutions, in a realistic parameter range, that replicate the major features of climatic variations implied by the oxygen isotope record for the last two million years. The variations replicated include a major "transition" between a low-ice (oxygen isotope), low variance mode before roughly 900,000 years ago to a high-ice, high variance mode with almost a 100,000-year period from 900,000 years ago to the present. The model only represents internal dynamics: no orbital forcing is prescribed (Saltzman 1988, p. 750). The chief features exhibited by the ice record derived from oxygen isotope data, and the atmospheric carbon dioxide record derived from the Vostok ice core — including a rapid deglaciation during which a spike of high carbon dioxide and a rapid surge in North Atlantic deep water production occurs — can be deduced as a free oscillatory solution of the model (Saltzman and Maasch 1988). The remaining variance is likely to result from external (orbital) forcing (Saltzman and Maasch 1990). Saltzman and Sutera (1987) explain that, if the physical aspects of the model be correct, then the ice ages prevailing over the last two million years are a consequence of a sensitive dynamical balance between water in its liquid and solid phases. A critical aspect of the balance is that the oceans are deep, with a large thermal and chemical capacity, and exercise some control on the carbon dioxide content of the atmosphere.

Because of positive feedbacks involving carbon dioxide, a fundamental instability is introduced that drives the ice mass variations. In an effort to arrive at a more complete theory of Pleistogene climates, Saltzman and Maasch (1990) extended their model to span the five million years of late Cenozoic climatic changes. Over that long period of time they presumed that forced and free variations in the concentration of atmospheric greenhouse gases, notably carbon dioxide, coupled with changes in ice mass and the global state of the oceans, under the additional influence of Earth-orbital forcing, would be the chief determinants of climate. The resulting model, though in a very early stage of refinement, incorporates very long-term variations of climate as well as higher frequency changes (Fig. 2.6). The model suggests that the 100,000-year cycles of climates during the Pleistocene epoch were caused by the downdraw of atmospheric carbon dioxide, possibly because of the weathering of rapidly uplifted mountains, to levels low enough for the "slow climatic system", which includes the mass of glacial ice and the state of the deep ocean, to become unstable.

2.4.2 Methane and Dust

It seems that methane, another greenhouse gas but, unlike carbon dioxide, emanating chiefly from the land, may have contributed to glacial-interglacial temperature changes (Chappellaz et al. 1990). The dust content of the atmosphere, too, may have affected glacial and interglacial climates. L. D. Danny Harvey (1988a) used an energy balance climate model to show that plausible increases in the atmospheric aerosol optical depth during the Last Glacial Maximum could have caused a 2 to 3 °C reduction of mean global temperature.

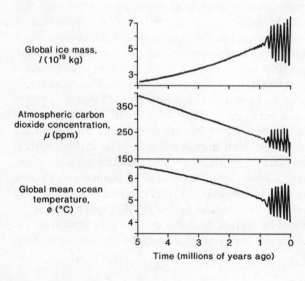

Fig. 2.6. Combined solution for Saltzman and Maasch's model for the past five million years. Low level stochastic noise is present but there is no orbital forcing. (Saltzman and Maasch 1990)

Time series from Antarctic ice cores indicate that aerosol increases occurred after significant cooling had taken place, and an increase in aerosols could not therefore have been a driving mechanism of glacial-interglacial oscillations, but rather would have served as a positive feedback mechanism, amplifying initial temperature decreases. This contrasts with carbon dioxide concentrations which appear to have led ice-volume variations, implying that the carbon dioxide cycle is a key driving mechanism of the glacial-interglacial cycles. Aerosolic dust in low latitudes, like carbon dioxide, would act to synchronize climatic change in the two hemispheres.

2.5 Geophysical Forcing

2.5.1 Volcanoes

Volcanoes are known to cause local devastation and affect immediately surrounding regions. They can also force a change in the world climate system by injecting dust and gases into the atmosphere. It seems likely that the largest eruptions, termed "supereruptions", may have global consequences, producing "volcanic winters" similar to the recently proposed "nuclear winters" (Rampino et al. 1985, 1988; Burke and Francis 1985; Kondratyev 1988; Chester 1988; see Sect. 2.5.2). The first person to attribute climatic change to volcanic eruptions appears to have been Benjamin Franklin. In 1789, Franklin proposed to the Manchester Literary and Philosophical Society that the blue haze he had observed drifting across Europe while carrying out his duties as the United States ambassador to France was produced by an eruption along the Laki fissure in Iceland and might have been responsible for the severe winter of 1783–84. A more local effect of volcanoes was described by Alexander von Humboldt after having observed "the singular meteorological process" which he gave the term "volcanic storm": the hot steam rising from a volcanic crater during the eruption spreads into the atmosphere, condenses into a cloud, and surrounds the column of fire and cinders which ascends several thousand feet and generates forked lightning (Humboldt 1849, vol. 1, p. 231). The lightning produced by the storm triggered by the eruption of the volcano of Katlagia in Iceland on 17 October 1755 killed eleven horses and two men!

Interest in the association between volcanoes and climate became avid following the massive eruption of Krakatau, Indonesia, in 1883. Spectacular sunsets which occurred for several months after the explosion, a cloud haze, and lower temperatures recorded in places far removed from Indonesia were all attributed to the explosion (Symons 1888; Wexler 1951). During the present century, many researchers have explored the relationship between violent volcanic explosions and a temporary reduction of global temperatures. They have done so for historically documented explosions, and for explosions which might have occurred in the more distant past. Charles Greeley Abbot and Frederick Eugene Fowle (1913) studied the effects of sunspots and volcanic eruptions on climate between 1880 and 1913, concluding that the combined effect of high sunspot numbers and volcanic dust injections was to depress mean annual global temperatures. A. Defant (1924) found that the atmospheric circulation was strengthened during the 2 years following the eruptions of Krakatau, St Augustin, and Bogosloff in 1883, Tarawera and Ninafu in 1886, Ritterrinsel and Bandai San in 1888, and St Maria, St Vincent, and Mont Pelée in 1902. The eruption of Mount Agung, Bali, in March 1963 led to a reduction in the temperature of the tropical middle and upper troposphere by 1 °C in late 1964 and early 1965 (R. E. Newell 1970, 1981; see also Hansen et al. 1978). On the other hand, there have been a number of great eruptions, such as that of Coseguina, Nicaragua, in 1835 and that of Mount St. Helens, United States, in 1980, which appear to have had no detectable effect on climate (Gentilli 1948; Deirmendjian 1973; Landsberg and Albert 1974; Ellsaesser 1986). The explosions of Mount St. Helens lofted roughly half the amount of material into the stratosphere compared to the Mount Agung eruption and about a tenth the amount of material of the Krakatau eruption, but despite this, they caused virtually no climatic change (Kerr 1981; Deepak 1983).

The link between volcanoes and climate was investigated in detail by Hubert Horace Lamb (1970, 1971). Part of this study involved the defining of a dust veil index, an estimate of the amount of fine volcanic dust or ash raised to the upper atmosphere by specific historical eruptions. A correlation was found between years when volcanoes erupted which had high dust veil indices and climatic cooling. However, as Lamb warned, since his dust veil index was computed chiefly on the basis of temperature variation, there is a danger of using circular reasoning when relating volcanic activity to climate by this method. Further indices were proposed by Katherine K. Hirschboeck (1980), who considered mainly the volume of an eruption, and by C. G. Newhall and Stephen Self (1982), who combined eruption volume with eruption energy in a volcanic explosivity index. However, it became clear around 1980 that key factors in understanding the relation between eruptions and climate were the composition of the ejecta, and particularly the amount of sulphur volatiles released, and the location, time of year, and prevailing climatic conditions at the time of the eruption which determine the spread and lifetime of clouds of volcanic aerosols (Rampino et al. 1988). Julie M. Palais and Haraldur Sigurdsson (1989), for instance, found that the estimated temperature decrease in the Northern Hemisphere following the eruptions of Mount Agung, Fuego, Mount St. Helens, Katmai, Krakatau, Laki, Santa Maria, and Tambora was positively correlated with the estimated yield of sulphur.

Without doubt, none of the historical eruptions was anything like as big as some of the eruptions in the geological past. Extrapolating from historical eruptions, the explosion of Toba in Sumatra some 75,000 years ago is estimated to have injected between 1000 and 5000 Mt of sulphuric acid aerosols into the atmosphere (Rampino et al. 1988). This compares with a figure of 100 Mt for the Tambora eruption of 1815. Whereas the Tambora event would have caused a dimming of the Sun, the higher estimate for the Toba event would have led to the cessation of photosynthesis. Toba-sized explosive supereruptions can be expected to have produced conditions analogous to those which would follow a major nuclear exchange, though volcanic aerosols have a longer residence time than would have smoke from fires ignited by nuclear explosions. The sheer magnitude of past eruptions makes it tempting to speculate that supereruptions (or possibly a large number of smaller eruptions) might have played a pivotal role in the initiation and timing of glacial-interglacial cycles. This idea was popular around the middle of the present century (e.g. Humphreys 1940; Gentilli 1948; Wexler 1952) and again during the 1970s (e.g. Gow and Williamson 1971; Bray 1974, 1977; Kennett and Thunell 1975, 1977; Porter 1981). J. R. Bray (1974, 1977), for instance, believed that a link exists between massive eruptions on land, especially in the Southern Hemisphere, and major advances of polar ice. The evidence is, however, far from unequivocal and lends itself to other interpretations (e.g. Ninkovich and Down 1976; Rampino et al. 1979; Kyle et al. 1981).

Volcanoes have been active throughout Earth history. Evidence suggests that bouts of increased volcanic activity have been linked with times of active tectonic plate formation and spreading. The significance of these pulses of volcanism to climate will be discussed in Section 8.3. Cyclical changes in the intensity of volcanism are superimposed on a decline of volcanic activity after the earliest phases of Earth history. Certainly, there seems little doubt that volcanic activity in the Archaean era, which ended about 2.5 billion years ago, would have been much greater than today, with carbon dioxide emissions possibly as much as 100 times their present values. The greenhouse effect produced by such enormous emissions of carbon dioxide might explain why glaciation is unknown until later in Precambrian times (Wyrwoll and McConchie 1986).

2.5.2 Vapour Plumes and Meteoritic Dust

In his book *The Evolution of Climate* (1922, p. 15), Charles Ernest Pelham Brooks made a passing reference to the fact that "the arrival of meteorites, bringing kinetic energy which is converted into heat, and introducing cosmic dust into the atmosphere" is an extraterrestrial factor of climate, though he felt it highly improbable that its effect would be appreciable. As to the recognition of the process he was right; as to the efficacy of it he erred. Dust and gases are indeed injected into the atmosphere from space. Technically speaking, such injections are extraterrestrial in origin but it is expedient to consider them under the heading of geophysical forcing, as their effect on the world climate system is similar to the effect produced by the injection of dust and gas from volcanoes. Two cases can be considered: the sudden injection of material into the atmosphere owing to the impact of a meteorite, asteroid, or comet; and the dusting of the atmosphere by micron-sized meteoritic material.

Large-body impacts would generate intense heat which may set off wildfires, so releasing soot into the atmosphere (Wolbach et al. 1985, 1988). The mechanisms by which wildfires might be ignited are somewhat debatable, but H. J. Melosh and his colleagues (1990) have come up with a plausible answer for the putative wildfires which occurred 65 million years ago at the very end of the Cretaceous period. Impacts would have produced a plume of vaporized, melted, and solid rock, and, if they should have occurred in the ocean, shock-heated steam. The thermal radiation produced by the ballistic re-entry of ejecta condensed from the vapour plume produced by a bolide with a mass of 10^{15} to 10^{16} kg could have increased the global radiation flux by up to 150 times the input from solar energy for periods of one to several hours. Thermal radiation inputs of this magnitude may well have been responsible for sparking off wildfires as well as directly damaging exposed animals and plants. The extra mass added to the atmosphere by an impact would have caused within hours a rise of mean global surface air temperature of about 1 °C, the maximum rise occurring in the vicinity of the impact. The great temperatures would have led to the formation of large quantities of the oxides of nitrogen: a very large impact could have produced up to 3×10^{18} g of nitric oxide which, in less than a year, would have spread through the atmosphere to give a worldwide, atmospheric nitrogen dioxide concentration of 100 ppm by volume, a level a thousand times higher than during the worst air pollution episodes in modern cities (Prinn and Fegley 1987). Large particles would have fallen out rapidly, but a fine cloud of dust would have spread globally. The dust cloud would have stayed in suspension for months or years, blocking out sunlight (Toon et al. 1982). Indeed, it would have been so dark that you could literally not have seen your hand in front of your face. The darkness would have brought about a drastic drop in surface temperatures with freezing conditions on continents, especially in continental interiors, and widespread and deep snowfall (Pollack et al. 1983).

The climatic effects produced by an impact are similar to the climatic effects predicted to occur after a major nuclear exchange (Crutzen and Birks 1982; Ehrlich et al. 1984; Schneider and Thompson 1988). If an impact occur in an ocean, then a period of intense cold may be followed by a longer-lasting time of much hotter conditions. The ocean water vaporized by the impact would increase the atmospheric moisture content considerably, and this would lead to the washing out of the tropospheric dust within a few weeks or months. After that, the remaining water and cloud in the stratosphere would, in the manner of a very efficient greenhouse, generate a large rise in global temperature which at the surface may exceed 10 °C: this temperature anomaly would persist for some months or years, until diffusion and photochemical processes in the stratosphere had returned the Earth to its steady-state condition (Emiliani et al. 1981). On the other hand, it is possible that a dust cloud created by even a modest impact would initiate a general and long-lasting cooling of the climate system, owing to an ice-albedo feedback mechanism: a combination of superfloods and obscuration of the upper atmosphere by dust seems a good recipe for making extensive ice sheets (Butler and Hoyle 1979; Hoyle 1981; Napier and Clube 1979). The ice-albedo feedback mechanism would be refuelled by further impacts during an impact episode, the world climate system remaining in a refrigerated state until the chance absence of impacts led to a global warming (Clube and Napier 1982).

Large-body impacts are relatively rare events. Not so the impact of meteoritic dust which takes the form of a continual rain of material falling into the atmosphere from space. Fred Hoyle and Chandra Wickramasinghe (1978) made the provocative suggestion that a close encounter between the Earth and a cometary nucleus might lead to a mass of small particles of high albedo being added to the upper air sufficient to produce an ice age. Indeed, there is now evidence that some meteoritic dust is a disintegration product of rare giant comets, 50 km or more in diameter, which, for some time after the Solar System passes through a giant molecular cloud, arrive in Earth-crossing orbits every few 100,000 years. On fragmenting, these comets produce a dense and sporadic atmospheric dust veil lasting between 1000 and 10,000 years (e.g. Clube 1986; Clube and Napier 1984, 1986). The most recent Oort cloud disturbances took place some 3 to 5 million years ago following the passage of the Solar System through the star-producing region in Orion which forms the prominent local feature known as Gould's Belt, and particularly the Scorpio-Centaurus concentration within it. The most recent injections of dust into the stratosphere are likely to have been caused by debris from a progenitor of Comet Encke which produced the Taurid-Arietid meteor stream and has dominated the terrestrial environment for the last 20,000 years. In essence, the principal climatic variations during the Holocene epoch might have been under the control of the Taurid-Arietid meteor stream (Clube and Napier 1986); for sure, strong concentrations of dust in the Taurid-Arietid stream correlate with periods of increased global rainfall. Victor Clube (1986) used Chinese, Japanese, and Korean catalogues of comets and fireballs from 200 B.C. to 1600 A.D. to show that the Northern Taurids meteor stream probably arose from a single progenitor asteroid which fragmented around 500 A.D. and might have led to 50 Mt atmospheric explosions similar to the Tunguska event of 1908, which was also likely to have resulted from a body belonging to the Taurid-Arietid stream. That the break up of a large asteroid around 500 A.D. correlates with a widespread Dark Age in Europe may not be without significance (Clube and Napier 1990, 1991).

2.5.3 Relief

A distinction between orogeny and epierogeny was made by Grove Karl Gilbert in 1890: both processes may influence the climate system. An epierogenic theory of ice ages, which posits that glaciation is caused by elevations of the Earth's crust, was first promulgated by James Dwight Dana in his presidential address to the American Association in 1855 (see Dana 1856). Dana intimated a causal relationship between late Cenozoic uplift and continental glaciation. This hypothesis of ice ages was subsequently urged by several American writers including Warren Upham. A similar view was hypothesized in the orogenic theory of glaciation as announced by Joseph le Conte (1882) and explored early in the twentieth century by Wilhelm Ramsay (1924), Rudolf Ruedemann (1939), and many others. The idea that mountain-building might be responsible for ice ages has remained popular. Fritz Albrecht (1947) reasoned that, without the Cordilleras, North America, Europe, and Siberia would have much more rain in winter, and considerably warmer temperatures, conditions which he claimed actually prevailed during Tertiary times until the cordilleran mountain ranges grew to such an extent that winter cold became so great that the Ice Age was initiated. Johannes Herman Frederik Umbgrove (1947) was won over by the view that glaciation is associated with periods of mountain-building, and identified the increase in area of land above the snow line as the causal link between the two phenomena. Whatever the mechanisms involved, mountain-building does appear to have played a role in some glaciations (Hamilton 1968; Crowell and Frakes 1970; Fairbridge 1973).

Improved stratigraphical correlation techniques developed since 1950 have shown that there is not a good relationship between orogenesis and glaciation: a far better correlation exists between epeirogeny, changes in the disposition of land masses, and palaeoclimatic trends. It is perhaps wise, therefore, to view orogeny in the context of plate tectonic mechanisms, rather than as an isolated process. In this light, virtually all tectonic events are accompanied by subsidence, uplift, and transgression or regression, and accordingly may have a major impact on regional climate. If the region undergoing tectonic change happen to be a sensitive part of the world climate system, then global climate may also be altered. This effect is illustrated by the Quaternary marine transgression in the area of southern Russia which includes the Black Sea, the Caspian Sea, and Lake Aral (Degens and Paluska 1979; Paluska and Degens 1979; Degens et al. 1981). At the outset of the Pleistogene period, the Akčagylian marine transgression led to the formation of a shallow sea, measuring some 2000 km by 1000 km, which linked the present seas and lakes of the area. About 400,000 years ago, pulses of tectonic activity led to a rapid subsidence of a basin chain running from the Caspian Sea to the lowlands of the Po Valley in Italy. At the same time, uplift occurred in parts of Russia, Anatolia, central Europe, and Scandinavia. E. T. Degens and his colleagues (1981, p. 12) think it no coincidence that three major ice ages occurred during this major tectonic pulse in Europe, and suggest that rapid tectonic change, by altering the relative area of continents and oceans, albedo, topography, orography, and bathymetry in this "weather-strategic" part of the world, may have contributed to global climatic change during the last 400,000 years.

There is now little doubt that the formation of mighty mountain ranges, particularly those with a north-south trend, can greatly influence atmospheric circulation patterns and so induce climatic change. This has been convincingly demonstrated in a series of computer experiments which test the effect of major mountain-building on climate over the past 10 to 40 million years. That orographical changes might have caused global changes of climate during this time seems feasible given the active orogenic environment: the Tibetan Plateau has been raised by up to 4 km over the last 40 million years and at least 2 km in the last 10 million years; two-thirds of the uplift of the Sierra Nevada has occurred in the past 10 million years; and similar changes have taken place (and still are taking place) in other mountainous areas of the North American west, in the Bolivian Andes, and in the New Zealand Alps (Ruddiman et al. 1989). The climatic effects of such orographical changes have been simulated using the Community Climate Model (Kutzbach et al. 1989). Experiments were run with three different "orographies", the relief being parcelled out into geographical blocks 4.4° latitude and 7.5° longitude in size: a "no mountain" case wherein present-day mountains are sliced off at 400 m; a "half mountain" case in which the added uplift is half that known to have occurred; and a "mountain" case representing present conditions. The results were given for January and July in the Northern Hemisphere. It was found that with no mountains, average wind velocity is uniform and symmetrical and planetary waves tend to be anchored at the boundaries of continents and oceans; but that with mountains, the average wind velocity is non-uniform and asymmetrical and planetary waves anchor over high plateaux. On the basis of the simulations, several features of our present climate appeared attributable to the effect of mountains (Ruddiman and Kutzbach 1989). For example, the Mediterranean climate, with its dry and hot summers on the western seaboard of the United States, is caused by the conversion of the westerly winds which would flow over the region if the Rockies were not there, into northerly winds blowing southwards from British Columbia to the Mexican border and associated with a deepened low pressure cell over the Colorado Plateau. On the other side of the Rockies, both seasons are presently wetter than earlier in the Cenozoic era, winter because the jet stream is forced south and winds which were formerly westerly are northwesterly, and summer because monsoon flows are created by the Colorado low pressure centre. Some simulated changes in climate point to the global effects of orogeny on the climate system: European winters are now colder than once they were because the Icelandic low has been displaced westwards; and Mediterranean summers are now drier than once they were owing to the development of cyclonic flow around the Tibetan Plateau and the development of a high above the subtropical Atlantic. These simulations are a great step forward in our understanding of how orogeny may affect climate, although Peter Molnar and Philip England (1990) caution that it is difficult to disentangle cause and effect — which came first, climatic change or orogeny?

2.5.4 The Arrangement of Land and Sea

The disposition of land and sea can have a profound effect upon climate. One of the first persons to recognize this fact was John Whitehurst (1713–1788), the English clockmaker and geologist. In his book *An Inquiry into the Original State and Formation of the Earth* (1778), in a discussion on the clemency of the climate in the antediluvian world, Whitehurst reasoned that seasonal contrasts in the world before the Flood were much less extreme than at present owing to the absence of large continents and mountains: the antediluvian world comprised a universal ocean with a few mountainless islands; in the postdiluvian world, which had continents and mountains, strong seasonal contrasts developed. The view that changes in palaeogeography would have a radical impact on climate was first developed fully by Charles Lyell in the first volume of his *Principles of Geology* (1830). The theme was taken up by Henry Hennessy (1859, 1860), who fancied that changes in the distribution of land masses had exercised a basic control over global climate through geological time. In essence, he showed that "if no great continents existed, but a great number of islands without any remarkable preponderance of land towards the tropical or the polar regions, the mean temperature of the earth would be increased, and the distribution of heat over its surface rendered far more uniform" (Hennessy 1860, pp. 385–386).

The role of palaeogeographical changes in climatic change was occasionally raised, for example by R. M. Deeley (1915) in connection with polar climates, during the first half of the twentieth century. But it was not until the 1970s, following the advent of the theory plate tectonics and the first modern attempts to reconstruct the geography of the geological past, that a string of papers was published in which climatic changes, and especially the contrasts between "greenhouse" and "icehouse" states, were ascribed to changes in palaeogeography (e.g. Crowell and Frakes 1970; Luyendyk et al. 1972; Frakes and Kemp 1972, 1973; Tarling 1978; Beaty 1978a,b). Each of these papers explored one of a few highly plausible ways in which geography might affect climate (Barron 1989): land provides a surface on which snow, with a high albedo, may accumulate and thus the more land there is around the poles, the greater the chances of glaciation occurring; each arrangement of land masses produces a circulation in the atmosphere and oceans different from that found today, and thus alters the polewards transport of heat (e.g. Ewing and Donn 1956, 1958; Donn and Ewing 1966, 1968); and because of the different reflective and thermal properties of land and sea, the placement and relative areas of continents and oceans will modulate climate. The often subtle effects of geography on climate are only just beginning to be appreciated (Barron 1989; Useinova 1989; V. P. Wright 1990). They will be examined a little more closely in the next chapter in the context of the causes of glaciation.

2.5.5 Geothermal Energy

For a long while it was assumed that the gradual cooling of the Earth had had a profound influence on the change of geological climates. However, Sir William Thomson (1862) calculated that the general climate of our globe cannot have been sensibly affected by conducted heat at any time more recent than 10,000 years after the commencement of the solidification of the surface, and that the present influence of internal heat upon temperature amounts to about one seventy-fifth of a degree. Likewise, Wolfgang Sartorius von Waltershausen (1865) showed that past effect of internal heat might easily have been overestimated. By 1875, James Croll could confidently conclude that "Not only is the theory of internal heat now generally abandoned, but it is admitted that we have no good geological evidence that climate was much hotter during the Palaeozoic ages than now; and much less, that it has been becoming *uniformly colder*" (Croll 1875, p. 7, emphasis in original). That, however, was not the end of the matter. In 1940, Artur Wagner examined the effect of internal heat and climate in considerable detail, basing his ideas on the fact that heat flow had been observed to affect the growth of glaciers in Greenland. He postulated that during times of tectonic stability, radioactive heat flow from the interior was at its highest and led to the warm climates of early Tertiary times; but during orogenic episodes, the heat flow was much reduced, thus promoting the growth of glaciers. Such speculations about internal heat are no longer mooted, but a possible link between internal heat production and climate has recently been brought to light. Herbert R. Shaw and James G. Moore (1988) estimated that large submarine lava flows, with apparent volumes exceeding 10 km^3, imaged on the bottom of parts of the Pacific Ocean, may produce thermal anomalies large enough to perturb the cyclical processes of the oceans, and could be a factor in the genesis of the El Niño phenomenon. They found that volumes of mid-ocean magma production would be capable of generating repeated thermal anomalies as large as 10 percent of the El Niño sea-surface anomaly at intervals of about 5 years, which happens to be the mean interval of El Niño events between 1935 and 1984; and that estimated rates of eruption, cooling of lava on the sea floor, and transfer of heat to the ocean surface could reasonably create a thermal anomaly comparable to that associated with El Niño.

2.5.6 True Polar Wander

If the globe were to move about the Earth's rotatory axis, then the world climatic zones would take up new positions. The effect of such geographical pole shift on climate was discussed by Robert Hooke (1705, 1978 edn), Johann Gottfried von Herder (1827), Isaac Newton (1729), John Lubbock (1848), William Devonshire Saull (1848), Henry James (1860), and John Evans (1866). In Britain, astronomers were set against the idea that the poles could shift by large amounts, but many German geologists — Carl, Freiherr Löffelholz von Colberg (1886), Damian Kreichgauer (1902), and Alfred Lothar Wegener (1915, 1929, 1966) — still entertained the notion well into the twentieth century. Wegener accepted that the poles had occupied very different positions in past geological ages. In collaboration with Wladimir Peter Köppen, he proposed that the sequence of Quaternary ice ages might be explained by polar displacements (Köppen and Wegener 1924). This view was endorsed by the geologist William Bourke Wright in his book on *The Quaternary Ice Age* (1937). A similar notion had, in fact, been expressed by William Morris Davis (1895–96), who suggested that the change in the limits of the wind and rain belts following a displacement of the North Pole to latitude 70° N and longitude 20° W would tend to glaciate northwestern Europe and northwestern America, would place arid trade wind climates on the northern side of the belt now occupied by tropical rain climates in Africa and South America, and would shift the tropical rain belt to the northern margin of the arid lands now found in the southern parts of Africa and South America. Alfred Wegener acknowledged that the problem of geographical pole shift required thoroughgoing mathematical treatment, and pressed for the launching of a theoretical attack on it. As it turned out, the theoretical attack did not come until 1955 when Thomas Gold, a British astronomer, presented a theory that allowed the Earth to tumble about its rotation axis, albeit slowly. Since the appearance of Gold's paper, and with the advent of palaeomagnetic data, the process of true polar wandering has been taken seriously and there is now a consensus that it does occur (see Andrews 1985; R. G. Gordon 1987). The rate at which it operates is a matter of debate, but the possibility of 5° per million years has been raised (Sabadini and Yuen 1989).

   The mutual forcing of the world climate system and true polar wander was explored by Roberto Sabadini, David A. Yuen, and Enzo Boschi (1982). These authors suggested that true polar wander, induced by the intrinsic response of a viscoelastic planet to cryospheric forcing, may be the cause underlying the termination of the last glaciation. In a subsequent paper (Sabadini et al. 1983), they floated the idea that the length of a typical ice age period (about 10 million years) is controlled by the time required for the poles to wander far enough for the geographical conditions necessary for the maintenance of a glacial state to disappear and the astronomical forcing of glacial-interglacial cycles to become inoperative. Looking at longer-term changes, William L. Donn (1982, 1987) tendered a refreshingly novel hypothesis relating geological climates to true polar wander. He based it on the assumption that the poles might have wandered greatly in the past, their position during the Jurassic period being such that most of the continental lands would have lain in tropical latitudes. This would at once account for the generally uniform distribution of temperature and the anomalous warmth of the polar regions from the Triassic period to the Eocene epoch. After the Eocene epoch, the meridional temperature gradient steepened because of increased cooling in Arctic and Antarctic latitudes until, by the late Cenozoic, glacial conditions had developed.

Against the notion of pole shift, some interpretations of environmental indicators of sediments sensitive to climate (evaporites, tillites, coal, aeolian sands, and radiolarian cherts) suggest that, while rainfall and temperature have varied at particular palaeolatitudes, marked equatorial shrinkage of subtropical high-pressure belts during glaciations, and marked polewards expansion during ice-free epochs, have not occurred: the patterns of atmospheric circulation, and thus the position of the world's chief climatic zones, have remained roughly the same for the last 500 million years (Drewry et al. 1974; see also W. A. Gordon 1975). However, not all geologists would agree with this reading of the evidence (see Sect. 3.4).

2.5.7 The Earth's Rotation Rate

It is well known that, in theory, lunar tidal friction has led to a transfer of angular momentum from the Earth to the Moon, and thus a slowing down of the Earth's rotation rate. If the Earth's rotation rate has indeed decelerated, then the length of a year and the number of days in a month will have decreased. The first evidence for these changes was detected by John W. Wells in the skeletal growth rhythms of Middle Devonian corals which lived some 390 million years ago. At this time there appear to have been about 400 days in the year (Wells 1963). Further work on fossils has revealed that the rate of decrease of spin speed has not been uniform, and indeed there have been times of spin acceleration (see Brosche and Sündermann 1990). During the Phanerozoic aeon, alternating periods of accelerating and decelerating spin velocity are separated by two types of turning point: firstly, spin maxima at transitions between acceleration and deceleration, which occurred 65 million and 375 million years ago; and secondly, spin minima at transitions from deceleration to acceleration, which occurred 235 million and 445 million years ago. These fluctuations about a general deceleration of spin velocity suggest that mechanisms other than tidal fiction are involved in the system, and a link with galactic processes seems possible (Fairbridge 1978). The relation between changes in rotation rate and climate was explored by Martin A. Whyte (1977), who found that, in general, periods of acceleration seem to have been characterized by climatic amelioration, while periods of deceleration seem to have been characterized by climatic deterioration: spin minima were preceded by ice ages with strong seasonal and zonal climatic contrasts, while spin maxima were preceded by climatic optima with warm, equable climates. However, the link between climate and spin velocity is not direct. Admittedly, a change in the length of the day would cause a small change in the diurnal temperature inequality (Lovenburg et al. 1972), but more subtle processes of climatic change appear to be involved. Spin velocity may affect climate through its effect on sea level (Whyte 1977). There is much evidence suggesting that tectono-eustasy is largely controlled by the volume of oceanic ridges and thus the spreading rate of crustal plates, and it is a distinct possibility that tidal and rotational forces would influence plate tectonic mechanisms (e.g. Mörner 1984a). Increased plate generation leads to a greater volume of oceanic ridges, an upping of sea level, and an amelioration of climate; it also appears to coincide with an acceleration of mantle spin velocity.

To study the effect of an increased rotation rate on climate, B. G. Hunt (1979) used a general circulation model with all radiative terms set at current values and with the rate of rotation boosted fivefold, a situation representative of Precambrian times (but see Williams 1980). An important result was that with a fivefold increase in rotation rate, the scale size of the meridional velocity distribution was greatly reduced and the synoptic distributions of velocity, temperature, and moisture fields became poorly correlated. The Hadley cell was reduced in extent and intensity; mean zonal winds were almost halved in intensity; the westerly jet in the troposphere was located at latitude 10°; and the stratospheric westerly jet all but disappeared as a separate entity. The upshot was a reduced polewards transport of sensible and latent heat, owing to the lack of dominant large-scale eddies, and thus warmer low latitudes and colder high latitudes than at present. The importance of the Earth's rotation rate in the evolution of climates was confirmed in the simulation studies made by William R. Kuhn and his colleagues (1989). Taking histories of solar luminosity, atmospheric carbon dioxide content, rotation rate, and continent formation, Kuhn and his colleagues generated the time evolution of Earth's surface temperature. They found that mean surface temperatures were 5 °C higher than today through much of geological time, and that the variation of mean surface temperatures has never exceeded 15 °C. But, perhaps surprisingly, the rotation rate of the Earth seems to have had a big effect on surface temperature distribution as late as 500 million years ago. For example, 3.5 billion years ago, the much faster rotation rate made little difference to equatorial temperatures relative to their present value but depressed polar temperatures by 15 °C. Indeed, the model indicated that rotation rate has wielded a major influence on surface temperature for much of Earth history, and that continental growth and albedo changes have played a secondary role.

2.5.8 Large Changes of Obliquity

The obliquity of the Earth's rotatory axis was considered by many early modern scientists to be providential. In Thomas Burnet's antediluvian paradise, it stood bolt upright, normal to the Earth's orbital plane, and consequently there were no seasons; after the Flood it leant over to its present rakish angle and the march of the seasons began (Burnet 1691, 1965 edn). Robert Hooke (1705, 1978 edn) was not ill-disposed to the possibility of large changes of obliquity, but Isaac Newton refused to countenance the idea. Calculations of obliquity changes caused by the gravitational attraction of other bodies in the Solar System were made by Pierre Simon, Marquis de Laplace, and Urbain Jean Joseph Leverrier: Laplace put the maximum variation at 1° 21', while Leverrier thought it could shift a tiny bit more, up to 1° 22' 34". George Louis Leclerc, Comte de Buffon, in his *Époques de la Nature* (1799, p. 190) was more generous, allowing a variation of up to 6°, but no more unless the orbits of the planets should change. The question of major changes in the obliquity of the ecliptic became the subject of a very lively debate between geologists and astronomers during the last century. We have already considered the effect of the small changes of tilt calculated by James Croll (1875) on the solar radiation receipt at different latitudes. Croll deemed these changes too modest to have an appreciable effect on climate. Larger changes of tilt were proposed by several people, including Alfred Wilks Drayson (1871, 1873) and Thomas Belt (1874a, 1874b), but their works now lie largely forgotten (see Huggett 1989b). In England, the debate over large changes of obliquity was brought to an abrupt end by George H. Darwin's (1880) theoretical calculations, which showed the Earth's axis of rotation to be thoroughly stable. However, changes of the Earth's axis continued to be entertained in Germany. Alfred Wegener, for instance, thought that the warmth of the poles from the Permian to Tertiary periods might be accounted for by a large change of obliquity.

Despite inexorable opposition to a large change of obliquity from many quarters, several geologists still find that it provides the most parsimonious explanation of many long-term terrestrial changes (Allard 1948; Williams 1975b; Wolfe 1978, 1980; Xu 1979, 1980; J. G. Douglas and Williams 1982), and at least two see no theoretical reason for discounting it (Williams 1972; Carey 1976). Perhaps the most contentious proposal concerning large obliquity changes has been promulgated by George E. Williams (1975b). According to Williams, the rotatory axis lay in the orbital plane 4500 million years ago and has since turned imperceptibly slowly, taking 2500 million years to complete 360° and arrive at its starting position. During the course of this very protracted obliquity cycle, the climate system would have undergone great changes. Williams conjectured that when the obliquity of the ecliptic was either 0° or 180°, which would have happened once every 1250 million years, there would have been no polar ice caps and a warm, seasonless climate would have extended to high latitudes. As the obliquity of the ecliptic increased and the seasonality of climate became stronger and stronger, so polar ice caps would have formed and girdles of glacial deposition would have pushed outwards towards the equator. When the obliquity of the ecliptic had reached either 90° or 270°, which would also have occurred once every 1250 million years, the Northern and Southern Hemispheres would have alternated between a summer of continuous heat and a winter of extreme cold, with the climate at the equator during the equinoxes being similar to the equatorial climate of today.

The problem with deciding whether obliquity has shifted by large degrees, as Williams envisions, is not so much theoretical as practical: long-term changes of obliquity are too slow to be detected by astronomical observations; they can only be inferred from the rock record. Whether large changes of obliquity would create the climatic conditions described by Williams is not clear. Simulations studies using a hemispherical general circulation model by B. G. Hunt (1982) have revealed a complex relation between mean annual climatic patterns and axial tilt. For an obliquity of 0°, where the rotatory axis stands bolt upright, the simulated climate is more vigorous than present climate, drier and colder by about 10 °C at the surface at high latitudes, slightly warmer by about 1 °C in the tropics, with an overall slight warming of the troposphere and an overall cooling of the stratosphere at all latitudes (cf. Barron 1984). Latitudinal precipitation rates are almost identical to present rates but evaporation rates reflect changes in mean temperatures, being a little higher in the tropics and lowered polewards of latitudes 45° N and S. The response to a 0° obliquity is thus complex, the results implying that the Earth's climatic zones would not change much from their present arrangement, but that the climate at very high latitudes would be much harsher in view of the greatly reduced temperatures, thus causing the habitable zone to contract. Hunt ran two experiments with an obliquity of 65°. One assumed low (that is, non-glacial) albedos at high latitudes because with such an extreme obliquity it was expected that surface temperatures would be too high to maintain snow or ice cover. The other experiment, run to provide a test of the tropical glaciation theory, assumed high (that is, glacial) albedos at low latitudes as well as high latitudes. Surprisingly, both experiments produced virtually isothermal surface and tropospheric temperatures, and much reduced latitudinal temperature gradients in the stratosphere. Temperatures at the equator were lower by about 10 °C in the experiment with high (glacial) tropical albedos, but were well above freezing at 11 °C and high enough to prevent glacial conditions from being maintained. In the case with low albedos at all latitudes, westerly jet cores were very weak, whereas low latitude easterlies in the stratosphere were strong and much extended compared with present conditions. In the case with glacial albedos in the tropics, the zonal wind distribution was even more extreme because the reversal of the low latitude temperature gradient resulted in easterly winds being dominant. The westerly jet cores were entirely removed, with the strongest winds being in the tropics where maximum wind intensity attained 34 m/s. Indeed, the conventional surface easterlies were replaced by westerlies in a reversed "Hadley cell". The difference between precipitation and evaporation, which shows considerable latitudinal imbalance at present, was locally in balance for an obliquity of 65° (Fig. 2.7). Only in the case with glacial albedos in the tropics did a significant imbalance arise owing to the reversed Hadley cell. The results of the high obliquity experiments are difficult to interpret because of the

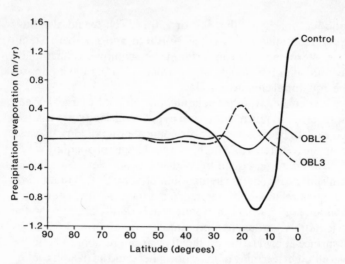

Fig. 2.7. Zonal mean difference of effective precipitation for three experiments with different obliquities. *Control* is an experiment with present obliquity (23°.5). *OBL2* is an experiment with an obliquity of 65° and low (non-glacial) surface albedos at high latitudes. *OBL 3* is an experiment with an obliquity of 65° and high surface albedos at low latitudes to allow for the possibility of glaciation in the tropics. (After Hunt 1982)

extreme seasonality which would occur in practice. But assuming the simulations be adequate representations of annual conditions, then it appears that low latitudes would be much colder and drier throughout the year, but not glaciated. Polar regions would experience long, cold winters but, because of the coldness of the winter atmosphere and its inability to hold water vapour, snowfall would be much reduced and melt during the long, hot summers. On the other hand, for current geography, sea ice should form, but, with its depth being limited by the conductivity of the ice itself, it would not grow thick enough to survive during the summer. If the ice were to persist year-round, then the pattern of climate would be drastically altered, with maximum summer temperatures in middle latitudes. As with a reduced obliquity, the habitable zone of the Earth would contract.

3 Ice and Water

When the world was still nothing but water, and the Spirit of the Lord moved upon the face of the waters, the world emerged from the water; water was the matrix of the world and all its creatures.

Philippus Aureolus Paracelsus (1951 edn, p. 13)

3.1 Terrestrial Systems and the Water Cycle

Climate is a key factor in understanding many processes which occur at the Earth's surface. Landforms and soils, animals and plants are all affected by atmospheric variables. Much of the classical work on the relations between climate and other environmental components is based on traditional, descriptive climatologies. Most of it was carried out when ideas about the dynamics of the atmospheric system and its interaction with other systems were rudimentary. Its main achievement was to establish the broad, zonal relationships between climate and Earth surface processes. In many cases, just three variables were considered — evaporation, rainfall, and temperature — global data on which were patchy. Over the last 40 years, the relationships between climate and Earth surface processes have been reformulated in terms of energy and water storage, transformation, and transfer. The heat and water balances at the Earth-atmosphere interface are the chief drivers of virtually all other forms of energy and matter transfer occurring at the Earth's surface: production and consumption in ecosystems, adaptation in animals and plants, chemical and mechanical weathering, erosion, and soil development are all strongly influenced by the supply of water and energy. Other climatic variables, such as wind speed, snow cover, lightning strike frequency, and sunshine hours, affect Earth surface processes, but temperature and available moisture can be considered the master "controlling" variables.

Because the genesis of ideas pertaining to the connection between Earth surface systems and atmospheric variables, and particularly the attempts to summarize climate in terms of precipitation and temperature, is germane to themes which will be discussed later in the book, it will be pursued a little here before proceeding to explore the relations between climatic change and the hydrosphere.

3.1.1 Descriptive Approaches

The components of the terrestrial phase of the water cycle were first quantified in the seventeenth century. Pierre Perrault (1678) and Edmé Mariotté (1686), French physicists, demonstrated that rain and snow falling on the catchment of the Seine River above a point in Burgundy were more than ample to supply the discharge of the river. In 1799, John Dalton, an English scientist, concluded from a series of experiments and observations that the rain and dew falling on England and Wales is equal to the quantity of water carried off by rivers and raised by evaporation. The first thoroughgoing effort to consider in detail the factors which influence the land phase of the water cycle was made by Nathaniel Beardmore, whose *Manual of Hydrology* (1862) was an inspiration to hydrologists and hydraulic engineers for over half a century. Not until the early twentieth century was the influence of moisture as well as temperature taken on board by ecologists, geomorphologists, and pedologists. These early seekers of relationships between climate and Earth surface processes realized that the ratio of precipitation to evaporation — the effective rainfall — is a better measure of the water available for doing work than the precipitation total alone. Temperature was often used as a surrogate measure of evaporation, largely because temperature data were available, evaporation data were not. When evaporation was used, it was generally the potential rate which was employed because, unlike actual evaporation, it is fairly readily estimated from other climatic variables. Actual evaporation is a more satisfactory measure of the moisture status of the Earth's surface, in that it reflects the simultaneous availability of water and solar energy in an environment over a given time; but, faced with a dearth of data on actual evaporation rates, temperature and potential evaporation data served early twentieth-century (and some much more recent) researchers very well.

The first person to try to find a way of summarizing effective moisture conditions at the land surface was Edgar N. Transeau. In 1905, Transeau constructed a precipitation-evaporation ratio map of the eastern United States. The value of this map is clear if the precipitation and precipitation-evaporation data for, say, St Paul, Minnesota and San Antonio, Texas be compared. These two towns have roughly the same mean annual precipitation (69.6 and 70.4 cm), but the effective moisture conditions for ecological, geomorphological, and pedological processes are very dissimilar because San Antonio's climate is actually much drier than St Paul's. This difference is borne out by the precipitation-evaporation (Transeau) ratios which are 0.51 and 1.02 respectively. In 1910, Albrecht Penck used annual precipitation and evaporation averages to classify climates. He distinguished three principal climatic regions: humid climates, in which more rain falls than is removed through evaporation, the surplus running off as rivers; nival climates, in which snowfall exceeds ablation, the excess being removed in glaciers; and arid climates, in which evaporation uses all the precipitation, thus preventing the flow of water in rivers. These three principal climates are separated by two important boundaries: the snowline, which occurs where snowfall matches ablation; and the dry boundary, which occurs where precipitation is equal to evaporation (see Penck 1973).

Whilst during the early decades of the twentieth century the importance of evaporation data was appreciated, researchers were hindered by a lack of weather stations taking evaporation measurements. Consequently, Richard Lang (1915, 1920) proposed a "rain factor" (Regenfaktor), defined as mean annual precipitation (mm) minus mean annual temperature (°C), in which temperature, readings of which were available from all weather stations, was substituted for evaporation. A world rain-factor map was prepared by Paul Hirth (1926) and rain-factor maps of other areas followed (e.g. Jenny 1930). The value of Lang's rain factor was much disputed. The main difficulty with it was that when the mean annual temperature drops below zero negative values result. This problem was overcome by Emmanuel de Martonne's (1926) indice d'aridité defined as

Indice d'aridité $= P/(T + 10)$

where P is the mean annual precipitation (mm) and T is the mean annual temperature (°C). An aridity index of less than 15 demarks the arid zone. Interestingly, there is a close correlation between monthly factors of de Martonne and the duration of precipitation expressed in minutes per month (Angström 1936). Another widely used substitute for the precipitation-evaporation ratio was Alfred Meyer's (1926) *Niederschlag Sättigungs (NS)* quotient. This is defined as the ratio of the mean annual precipitation (mm) to the absolute saturation deficit of the air (mm Hg). In the United States, the *NS* quotient map is very similar to the Transeau ratio map. *NS* quotient maps were published for Europe (Meyer 1926) and Australia (Prescott 1931), among other places.

The next major development was due to C. Warren Thornthwaite. In 1931, Thornthwaite defined a "precipitation effectiveness index", I, based on a summation of monthly moisture values:

$$I = \sum_{n=1}^{12} 115 \left(\frac{P}{T-10} \right)_n^{10/9} ,$$

where P is the monthly precipitation (inches), T is the monthly temperature (°F), and n is the number of months. In 1948, he developed this relation into a method of computing potential evaporation from air temperature. He found the following empirical relation between potential evaporation, E_0 (mm), and mean monthly temperature, T (°C):

$E_0 = 16 \, (10T/I)^a$ (mm/month),

where

$a = (492390 + 17920I - 77.1I^2 + 0.675I^3) \times 10^{-6},$

and I, an "annual heat index", is given as

$$I = \sum_{n=1}^{12} \left(\frac{T}{5} \right)^{1.514} .$$

The equation is valid only if $T < 0\ °C > 26.5\ °C$.

Rodney J. Arkley (1963a,b, 1967) refined the methods used to investigate the relation between the components of the water balance and soil processes. He defined the leaching effectiveness of the climate, Li, as the larger value of:

(1) the excess of mean monthly precipitation, P, over mean monthly potential evapotranspiration, ET_p, during the consecutive months in which P is greater than ET_p (where this condition does not occur in consecutive months, then the cumulative sum of $P\text{-}ET_p$, commencing with the first month of a moist season, is a more appropriate measure); or else (2), the average precipitation in the wettest month of the year (this is normally appropriate in arid climates where the precipitation in a single storm may so exceed the seasonal excess of P over ET_p that it is the dominant control on the depth of leaching). Furthermore, Arkley (1967) suggested that the leaching effectiveness of the climate, Li, the actual evapotranspiration, Ea, and the mean annual temperature, T, will serve to represent those climatic factors most closely circumscribing the world's zonal soil groups.

3.1.2 Energetic Approaches

Annual and monthly values of precipitation, evaporation, and temperature relate directly to two of the most important energy and mass exchange systems at the Earth-atmosphere interface — the radiation balance and the water balance. The net amount of radiative energy available to power systems at the Earth's surface is fairly well described by temperature, while precipitation and evaporation can be used to calculate the amount of water available in different environments. This is why, combined or expressed as ratios, these three variables have been so successful in capturing the zonal variation in Earth surface systems. Mikhail I. Budyko (1956, 1958) managed to justify, in physical terms, the use of air temperature as a surrogate measure of potential evaporation. In so doing, he introduced a fully energetic approach to Earth surface heat and water budgets, an approach which Thornthwaite and Arkley moved towards but never quite reached. Using a large data set, Budyko found that air temperature is related to available radiative energy in the following way:

$$R_o = 10 \sum T \text{ (ly/year)},$$

where R_o is the annual radiation balance for a wet surface with an assumed albedo of 0.18 and $\sum T$ is the sum of all daily mean temperatures in excess of 10 °C. Assuming that in the annual energy balance storage, S, and convective heat loss, H, can be neglected, it follows that

$$E_o \approx R_o / L \approx 0.2 \sum T \text{ (mm/year)}.$$

All the terms in the water balance equation can be expressed in energy terms simply by multiplying them by the specific heat of evaporation, r (Baumgartner and Reichel 1975, p. 111):

$$r(P = E + D),$$

where P is annual precipitation, E is annual evapotranspiration, and D in annual runoff. This yields

$$rP = P',$$

which is the equivalent heat of evaporation of precipitation representing the amount of heat necessary to evaporate the entire annual precipitation;

$$rE = E' \ (= L),$$

which is the latent heat of evaporation; and

$$rE = D',$$

which is the virtual heat of runoff (which reduces P'). The normalized water balance can then be written

$$(P' = E' + D') \ 1/R,$$

where R is the net radiation at the Earth's surface. The ratios P'/R', E'/R, and D'/R are very interesting because they express the dependence of evaporation and runoff on net radiation. Budyko (1956) called the reciprocal of P'/R, which he wrote as R/LP, the radiative index of dryness (Fig. 3.1). He found it very useful in delimiting vegetation zones (see Sect. 7.2.1). For mean stationary conditions, it is generally true that

$$R/P' = (1 - D/P) \ (1 + H/L),$$

where H/L is the Bowen ratio and quantifies the ratio of sensible and latent heat fluxes to or from the air (Lettau 1969, 1973). The ratio D'/R defines the relative

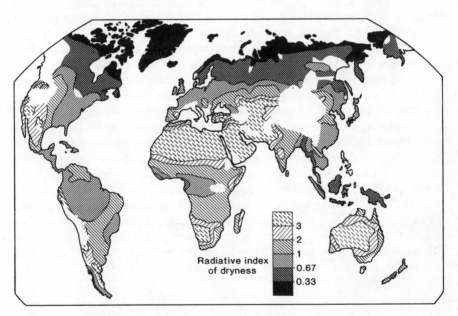

Fig. 3.1. Global distribution of Budyko's radiative index of dryness. The isolines are based on 1600 control points. *Blank areas* on land are mountains. (After Budyko 1974)

virtual heat of runoff and characterizes the partial energy of R that is withdrawn from the vaporization process over land surfaces by the runoff from land to sea (Baumgartner and Reichel 1975, p. 114).

3.1.3 Hydrological Models

The energetic approach to moisture and radiative regimes at the Earth's surface has been given a boost by the development of atmospheric general circulation models. The water balance submodel incorporated within atmospheric general circulation models is enormously valuable in providing an independent yardstick against which to match past changes in the Earth's surface hydrological cycle, and ultimately climate, as reconstructed using biological and geomorphological palaeoclimatic indicators. At present, the highest spatial resolution possible in general circulation models is about 100 × 100 km, and that can be a serious drawback in many geomorphological and pedological applications. General circulation models, for instance, are unhelpful in elucidating the mass budget of Alpine glaciers because the controlling climatic variables are determined by processes which vary at scales considerably smaller than 100 km^2. Indeed, while generally each predicting the same broad pattern of climate, such as precipitation within latitudinal belts, current general circulation models generate somewhat divergent results at a regional scale (Grotch 1988). This problem may be alleviated by the nesting of finer-scale "limited area models" within larger general circulation models (Giorgi 1989).

The pace of development on the general circulation modelling front is equalled by advances in the modelling of the terrestrial water cycle. A coupled continental water balance and water transport model was recently created as part of a study of global biogeochemistry (Vörösmarty et al. 1989; Vörösmarty and Moore 1991). The water balance component of the model predicts soil moisture, evapotranspiration, and runoff from gridded data on climate, vegetation, soils, and topography. Predicted runoff is then used in conjunction with information on fluvial topology, linear transfer through rivers, and a simple representation of floodplain inundation to predict monthly discharges from any cell within a simulated catchment. Applied to South America, the coupled model successfully predicted the timing and discharge at selected locations within the Amazon-Tocantins Basin (Vörösmarty et al. 1989); the feasibility of its being applied to the Zambesi catchment in Africa is being explored (Vörösmarty and Moore 1991). The coupled model has enormous potential: the runoff predictions, for instance, may in the future be used as a means of testing the hydrological predictions made by general circulation models. When developed more fully, it may well revolutionize many ideas on the link between the water cycle and terrestrial processes at global and regional scales.

3.2 Short-Term Changes in the Hydrosphere

3.2.1 Droughts, Floods, and the Lunar Cycle

Evidence now exists, much of it gathered by Robert Guinn Currie and Sultan Hameed, for periodic, 18.6-year lunar nodal forcing (as well as 11-year solar cycle forcing) of drought and flood in western North America during the past millenium (R.G. Currie 1981d). This drought cycle may be connected with the 20-year Kuznets cycle in the American economy (R.G. Currie 1981d). An 18.6-year drought-flood cycle has been detected in India since 1890 (Campbell et al. 1983; R.G. Currie 1984a), in northeastern China since 1470 (Hameed et al. 1983), in mid-latitude South America since 1600 (R.G. Currie 1983), and since 560 in those parts of Africa drained by the Nile River (Hameed 1984). The Nile flood records are particularly interesting. Analysis of winter and summer flood records of the Nile from A.D. 622 to 1490, using both a Fourier transform method (Hameed 1984) and maximum entropy analysis (R.G. Currie 1987a), reveals evidence for a 18.6-year cycle of flood or drought in areas within the Nile Basin. The summer flood records of the modern period (A.D. 1690 to 1962) contain the same cycle. Analysis of a drought-flood index for the environment around Beijing, China, has picked out a bistable cycle, the nodal 19-year waveform of which is shown in Fig. 3.2, where the downwards-pointing arrows mark the dates

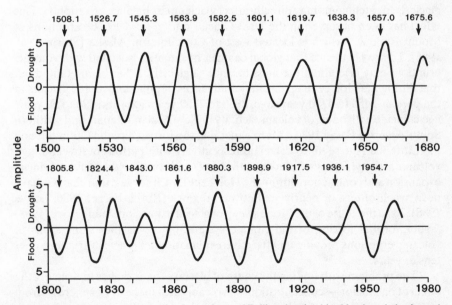

Fig. 3.2. Nodal 19-year waveform for the environs of Beijing showing bistable phase behaviour of floods and droughts with respect to epochs of the maxima in the tide, the dates of which are marked by *arrows*. This tide is the 12th largest predicted by Newton's theory. (After Currie and Fairbridge 1985)

of "epochs" of nodal tide maxima (R.G. Currie and Fairbridge 1985). From 1500 to 1680, epochs of drought and epochs of the tide were roughly in phase. The polarity must then have switched because at nodal epochs 1805.8 and 1824.4 the drought epochs and tide epochs are out of phase. The polarity had switched back by 1843.0, and then switched again at epoch 1936.1. Nodal drought has thus assumed two in-phase and two out-of-phase bistable states with respect to epochs of the tidal node over the last five centuries. These findings have implications for the changes of climate in the near future. At the epoch of the tide in 1991, and at the mid-epoch in 2001, there should be increased variability in worldwide weather with respect to drought and flood, though owing to the non-stationary behaviour of the weather system in each region, the magnitude of the floods and droughts cannot be predicted. Recent tidal epochs and mid-epochs were associated with increased flood and drought: a world food crisis occurred in the 1964.0 mid-epoch and an even more serious one in the 1973.3 epoch; increased variability of weather with respect to flood and drought was widely reported between 1982 and 1984, around the time of the 1982.6 mid-epoch.

3.2.2 Volcanoes and the Cryosphere

The reduction of global surface temperatures which are brought about by some volcanic eruptions (see Sect. 2.5.1) might be expected to be felt in the sensitive parts of the water cycle, and particularly in glaciers. Volcanoes which lie in glacierized catchments may directly affect glacier mass balance on erupting. This effect has been observed in the Drift Glacier during the 1966–68 eruptions of Mount Redoubt, which is located west of Cook Inlet in Alaska (Sturm et al. 1986). There is evidence that global changes of climate associated with volcanic eruptions may also affect the mass balance of glaciers. It has been pointed out that during the Holocene epoch, major advances of polar and alpine glaciers in the periods 5450 to 4700 years ago, 2850 to 2150 years ago, and 470 to 50 years ago coincide with bouts of volcanic activity in New Zealand, Japan, and southern South America (Bray 1974). More generally, the Little Climatic Optimum and the Little Ice Age seem to correlate respectively with periods of low and high volcanic activity. Data gathered over the last decade or so on the timing of glacial expansion and contraction during the Holocene epoch suggest that changes have been synchronous, or nearly so, between regions (Röthlisberger 1986; Grove 1988). As to the cause of the broad sweep of glacial expansion, Jean Grove (1988) favours orbital forcing of the climate system, but owns that short spells of volcanic eruptions provide an attractive explanation for glacier variations over tens of years.

The problem with finding the cause of glacial expansion lies in isolating the effect of one factor — concentrations of volcanic dust in the atmosphere — from effects of other factors (such as changes in solar activity) going on at the same time. Taking account of sunspot numbers and volcanic dust, Stephen H. Schneider and Clifford Mass (1975) used a simple climate model to predict surface global temperatures since 1600. The calculated temperatures generally agreed with historical observations, the temperature drop during the Little Ice Age, the subsequent rise to about 1800, and the fall then rise to 1880 all being predicted. A similar exercise was undertaken by Ronald L. Gilliland (1982), who modelled the effect of five external factors — atmospheric carbon dioxide content, a 76-year solar cycle (reflecting variations in the Sun's diameter), a 22-year solar cycle, a 12.4-year solar cycle, and volcanic dust variations — on world temperature. He found that each factor could influence the total temperature record, although carbon dioxide, volcanic dust, and the 76-year solar cycle had the largest influences. Indeed, with the 76-year solar cycle omitted, the model failed to reproduce with any clarity the decrease in Northern Hemisphere temperatures recorded after 1940. Given the deficiencies of the historical temperature record and the simple nature of the models it would be wrong to read too much into the results. They merely indicate that volcanic dust (and other external forcings) do seem to lead to changes of global temperature, and they lend support to the view that major volcanic eruptions, by altering the global heat balance, disrupt the mass balance of glaciers.

3.3 Medium-Term Changes in the Hydrosphere

Changes of climate over centuries and millenia, particularly those encountered during the Pleistogene period, appear to have been forced by orbital variations, although, as we have seen, there is evidence that the inner workings of the climate system may also lead to climatic pulsebeats in the Croll-Milankovitch frequencies. It has become clear over the last decade that the world climate system responds as a whole to orbital forcing, and it is therefore not surprising that orbital signals can be detected in the oceans and on land. In the oceans, orbital pulses can be detected in the oxygen-isotope record of changes in the global volume of ice, in the fossil record of changes in surface currents and surface temperatures of the Atlantic, Antarctic, Indian, and Pacific Oceans, and in the chemical record of changes in the deep water of the oceans. On land, at high and middle latitudes, orbital pulses cause changes in the size of the polar ice sheets and in the extent of mountain glaciers. At low latitudes, orbital pulses are recorded mainly as variations in rainfall and the resulting changes in the level of lakes. Over the last 18,000 years, tropical monsoon, mid-latitude, and polar climates have been greatly changed by swings in astronomical forcing functions, particularly those associated with the cycles of precession and tilt. The nature and timing of these changes have proved puzzling:

"Why were lake levels high in the American Southwest during the glacial maximum at 18 ka (thousand years ago) and low in the Northwest at the same time, and were these variations in lake levels a result of changes in seasonal precipitation or temperature? Why were lakes in the Sahara expanded during the early Holocene (12 to 6 ka), when virtually all the American deserts basins were dry? Why were summer temperatures warmest in central North America at 6 ka, when the summer radiation maximum was at 9 to 11 ka?". (COHMAP Project Members 1988, p. 1043).

An early concerted attempt to solve these riddles was made by members of the CLIMAP (Climate Mapping and Prediction) Project who, using a diverse array of marine and continental indicators, reconstructed the climatic conditions at the Earth's surface 18,000 years ago (e.g. CLIMAP Project Members 1976, 1981). This reconstruction was then used as the basis of climate models to reassemble the general circulation of the atmosphere at the time of the Last Glacial Maximum (e.g. Gates 1976a,b; Manabe and Hahn 1977). More recently, members of COHMAP (the Cooperative Holocene Mapping Project) have continued the CLIMAP strategy of using an admixture of geological data and atmospheric models to investigate the global and regional dynamics of climatic change over the last 18,000 years. COHMAP members have brought together a large body of data recording palaeoclimates from radiocarbon-dated sequences of sediments from several continents and oceans. They have then compared these data with the model simulations of past climates. The chief aim of COHMAP research is to further understanding of the physics of the world climatic system, and especially to elucidate the response of tropical monsoons and mid-latitude climates to orbitally forced changes in solar radiation and to changing Ice Age boundary conditions such as the size of ice sheets. Many other studies have revealed the nature of climatic changes during glacial-interglacial cycles and since the Last Glacial Maximum. We shall now discuss some of the advances in understanding Pleistogene climatic changes and their effects on the hydrosphere, looking at monsoons, soil moisture, lakes, and glacio-eustatic changes of sea level.

3.3.1 Monsoons

In the tropics, variations of climate at timescales spanning years to full glacial-interglacial cycles are dominated by seasonal monsoons, that is, large-scale, seasonal changes in wind and precipitation patterns such as those observed over tropical Africa and the subcontinent of India (Prell 1984; Fairbridge 1986; Prell and Kutzbach 1987). Much palaeoclimatic evidence points to significant and periodic changes in the strength of the monsoons during the Pleistogene period. The past levels of lakes, pollen profiles, and deep-sea cores all suggest that monsoons were weaker during the Last Glacial Maximum, some 18,000 years ago, and stronger during the mid-Holocene epoch, about 9000 to 10,000 years ago. General circulation models run with boundary conditions appropriate for the Last Glacial Maximum predict that the Indian summer monsoon was significantly weaker at that time (e.g. Manabe and Broccoli 1985; Kutzbach and Guetter 1986), and those run with boundary conditions appropriate to 9000 years ago (that is, with increased solar radiation in the Northern Hemisphere) predict stronger monsoons (Kutzbach and Otto-Bliesner 1982; Kutzbach and Guetter 1986). A major study of monsoon variability over the last 150,000 years, using palaeoclimatic records of the Afro-Indian monsoon and a range of general circulation models, was carried out by Warren L. Prell and John E. Kutzbach (1987). From the general circulation models were derived regional averages of climatic variables, which enabled the sensitivity of regional precipitation to changes in solar radiation and glacial boundary conditions to be gauged. The sensitivity coefficients were then used in conjunction with time series of solar radiation and glacial ice volume to simulate time series of zonal-average and regional monsoon intensity over the last 150,000 years. Finally, the simulated climatic time series were compared with the palaeoclimatic observations in the hope of gaining insight into the late Pleistogene history of monsoon variability in the African and Indian regions (Fig. 3.3). Taking all lines of evidence into consideration, several conclusions were drawn. Firstly, palaeoclimatic time series from marine sediment cores display four distinct monsoon maxima which can be widely correlated across Africa and Asia during interglacial stages. These maxima coincide with peaks of summer radiation in the Northern Hemisphere associated with the precessional cycle. During glacial stages, the time series is less easily characterized and the geological record shows regional differences, which diminishes the degree of correlation between regions. Secondly, the simulations made with general circulation models show that during interglacial stages, regional monsoon indicators (land-ocean pressure gradients, precipitation, and winds) are closely related to the departure of solar radiation from modern values, the greater the departure the greater the strength of the monsoon. Surprisingly large are the simulated changes in the water cycle, though these are consistent with the widespread geological evidence for a major change in the tropical monsoons. The greatly increased vigour of the water cycle seems to result from the non-linear relationship between temperature and saturation vapour pressure. It could be that other general circulation models, for example with interactive soil moisture, would lead to a different amplification of the water cycle; indeed, the work of Robert G. Gallimore and John E. Kutzbach (1989) shows that soil moisture alters the magnitude (but not the sign) of model sensitivity to solar radiation (see Sect. 3.3.2). Thirdly, the regional monsoon response to glacial stage boundary conditions is variable: 18,000 years ago the monsoon in southern Asia was greatly weakened but precipitation increased in the western Indian Ocean and in equatorial north Africa. But taken as zonal average values, tropical monsoon precipitation is fairly insensitive to glacial boundary conditions in comparison with its high sensitivity to orbitally induced variations in solar radiation receipt. Fourthly, agreement between palaeoclimatic data and computer predictions are quite good: four monsoon maxima for equatorial north Africa and southern Asia are present in both observed and simulated time series; and the predictions that equatorial north Africa was wetter and southern Asia was drier than today for the periods with full glacial boundary conditions, especially those prevailing between 75,000 to 15,000 years ago, are borne out by palaeoclimatic indicators. Fifthly, the general agreement (though there are some disagreements) between simulations and the geological record suggests that two cardinal factors explain major regional features of monsoon climates over the past 150,000 years: orbitally produced changes in solar radiation and the boundary conditions prevailing during glacial stages. The last of these is not routinely modelled explicitly in general circulation models.

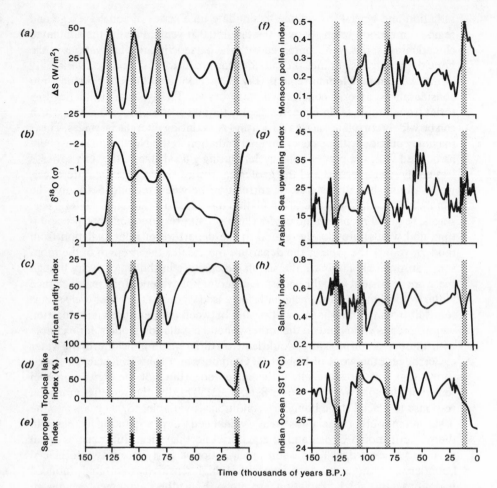

Fig. 3.3. Comparison of monsoon-related palaeoclimatic records with **a** average Northern Hemisphere summer radiation and **b** the SPECMAP composite oxygen isotope record for the last 150,000 years. (After Imbrie et al. 1984). **c** The African aridity index which is based upon the abundance of *Melosira* and indicates wet and dry periods in equatorial Africa. (After Pokras and Mix 1985). **d** The tropical lake-level record which indicates the percentage of lakes in the intertropical zone that are high or intermediate in level. (After Street-Perrott and Harrison 1984). **e** The Mediterranean sapropel index which shows the presence (*vertical bars*) or absence of sapropels. (After Rossignol-Strick 1983). **f** The record of monsoon-related pollen in the Gulf of Aden. (After van Campo 1982). **g** The Arabian Sea upwelling index which represents the faunal record of monsoon-related upwelling off Arabia. (After Prell 1984). **h** The faunal record of salinity off Arabia. (After Prell 1984). **i** The sea-surface temperature record from the western Indian Ocean. (After Prell 1984). The times of maximum solar radiation (Northern Hemisphere average for June, July, and August) are indicated by the *shaded vertical bars*. (Prell and Kutzbach 1987)

3.3.2 Soil Moisture

Given that climate, and especially the water cycle, is sensitive to orbital variations, it follows that the hydrological regime at the Earth's surface will reflect
changes in the annual pattern of insolation receipt. Until recently, soil moisture
was not allotted an interactive role in general circulation models. However, soil
moisture exercises a strong control over evaporation and so influences the water
cycle, as well as the temperature and circulation pattern over land areas.
Preliminary experiments suggest that the inclusion of soil moisture as an interactive variable in a general circulation model simulating the climatic conditions
9000 years ago produces results similar to a fixed moisture model in tropical
lands, but leads to summer warming and aridity over northern middle latitudes
(Kutzbach and Guetter 1986). Robert G. Gallimore and J.E. Kutzbach (1989)
use an interactive moisture version of a low resolution general circulation model
to examine the role of soil moisture in the climate of 9000 years ago. They are
particularly interested in determining whether soil moisture feedback in the
climate system alter the climatic response to orbital changes in radiation. The
results of the experiments, summarized in Fig. 3.4 and 3.5, show that soil moisture
and runoff increase over northern tropical lands and over eastern Asia in
mid-latitudes in response to increased precipitation in those areas. Soil moisture
is decreased over much of the northern middle latitudes, the biggest decreases
occurring in areas which in the control simulation run for present conditions are
wet — northwestern and southeastern North America and western and eastern
Eurasia. Runoff in these regions is reduced, too.

3.3.3 Lakes

A connection between orbital forcing and the fluctuations of lake levels during
the late Pleistocene and Holocene epochs has been established in some areas.
John E. Kutzbach and F. Alayne Street-Perrott (1985) used an atmospheric
general circulation model to simulate the climates of January and July at
3000-year intervals starting 18,000 years ago. The orbital changes in the seasonal
distribution of solar radiation receipt and the changing surface boundary conditions assumed to accompany deglaciation are shown in Fig. 3.6a. The simulated
changes of radiation receipt, surface temperature, and water balance in the
latitude belt from 8.9° N to 26.6° N over the last 18,000 years are shown in Fig.
3.6b, c, and d respectively. From 15,000 years ago onwards, the model predicted
a strengthening of the monsoon circulation and increased precipitation and
effective precipitation in the Northern Hemisphere tropics, which features
reached their acme in the period 9000 to 6000 years ago (Fig. 3.6d). The water
budgets predicted by the model were used to estimate the area of closed lakes
located between latitudes 8.9° and 26.6° N. Predicted lake levels matched well
actual lake levels as reconstructed from geological evidence (Fig. 3.6e, f, and g).
Recently, Françoise Gasse and her colleagues (1989) have found that water-level
fluctuations in Lake Tanganyika over the last 26,000 years are correlated with
changes in global sea level and ice volume: the lake level was intermediate

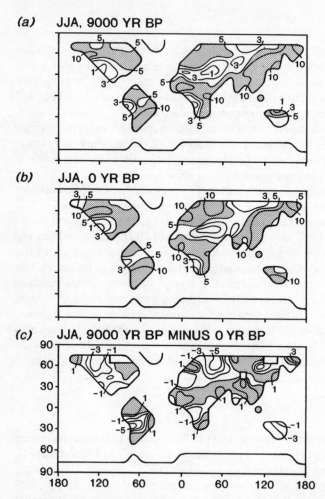

(a) JJA, 9000 YR BP

(b) JJA, 0 YR BP

(c) JJA, 9000 YR BP MINUS 0 YR BP

Fig. 3.4. The distribution of soil moisture (cm) in June, July, and August. **a** 9000 years ago. **b** At present. **c** Expressed as the difference between 9000 years ago and today. Regions with soil moisture greater than 5 cm 9000 years ago and at present are *stippled*. In the bottom diagram, **c**, regions with positive changes of soil moisture are *stippled* and *unstippled blocked out regions* over land without contours denote areas of no change in soil moisture. (Gallimore and Kutzbach 1989)

between 26,000 and 21,700 years ago, low from 21,700 to 13,000 years ago with a minimum 18,000 years ago, and high from 13,000 years ago to the present. These lake-level fluctuations were in phase with fluctuations of African lakes north of the equator which, as F. Alayne Street-Perrott and Neil Roberts (1983) have shown, are clearly linked to Croll-Milankovitch mechanisms. Gasse and her coworkers deemed this correlation puzzling, since general circulation models would lead us to expect opposing phases in Northern and Southern Hemispheres as a result of different insolation values arising from orbital motions. They

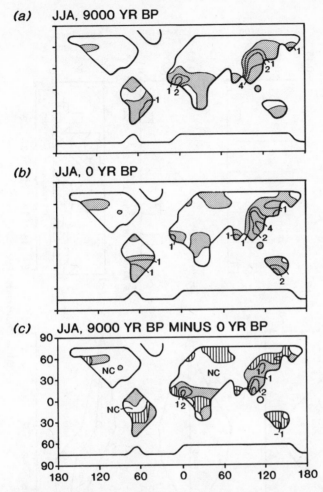

Fig. 3.5. The distribution of runoff (mm/day) in June, July, and August. **a** 9000 years ago. **b** At present. **c** Expressed as the difference between 9000 years ago and today. Regions with runoff greater than 0.01 mm/day 9000 years ago and at present are *stippled*. In the bottom diagram, **c**, regions with positive changes of runoff are *stippled*, regions with negative changes are *hatched*, and regions with no change in runoff (marked with an *NC*) are *unshaded*. (Gallimore and Kutzbach 1989)

concluded that the water balance of Lake Tanganyika reflects hydrological changes in the oceans resulting from glacial and deglacial processes and the related increased availability of global atmospheric moisture during warmer global climatic phases.

Not all changes in lake levels will necessarily reflect orbital forcing of climate. Françoise Gasse and Jean-Charles Fontes (1989), in a study of the palaeohydrology of hypersaline Lake Asal, Djibouti, discovered that, at present, infiltration of marine water is the major input to the lake. For this and other

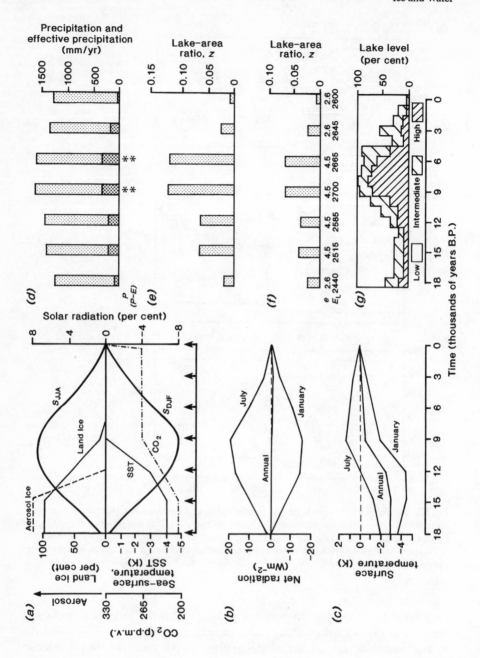

reasons they cautioned that, although the general high-stand of the lake between 8600 and 6000 years ago followed by a low level stage from 5000 years ago to the present accords well with the changes predicted by J.E. Kutzbach and F.A. Street-Perrott (1985), local conditions can cause fluctuations in lake level which are not directly related to climatic change. For instance, about 6000 years ago a sudden change of drainage occurred causing a drastic drop in lake level of about 300 m. This change was linked to an exceptional episode of rift opening and subsequent volcanism and to a decrease in continental groundwater supply owing to the sealing by argillaceous sediments of the bottom of upstream Lake Abhé.

Studies of fluctuations of individual lakes, in conjunction with many other indicators of palaeoclimates, have enabled broader studies of the changing water budget in Africa during late Pleistocene and Holocene times to be undertaken. An earlier effort in this field was made by Rhodes W. Fairbridge (1976), who dispelled the myth of the crude correlation between glacial and pluvial episodes. Recently, a continental compilation was made by Thomas Littmann (1989). By plotting palaeohydrological indicators as a function of elevation and time, Littmann was able to discern six phases of climatic and environmental change (Fig. 3.7). From 30,000 to 22,000 years ago a widespread interstadial wet phase came to an end, the tropics becoming drier, though the northern extratropics and both the tropical margins were still wetter than today. From 22,000 to 20,000 years ago the subtropical arid regions remained wetter than today, perhaps owing to the effect of mountains; the tropics themselves became arid. From 20,000 to 17,000 years ago the depression of temperature associated with the height of the glacial stage reduced the wetness of all climatic zones, save the Atlas Mountains region, which was still wetter than today. From 17,000 to 13,000 years ago, the time of the late glacial climatic instability, the subtropics rapidly became wetter over the Atlas Mountains and the central Saharan region; the rest of the continent remained drier than at present, the tropical mountains becoming even

———

Fig. 3.6. a Schematic representation of changing external forcing (Northern Hemisphere solar radiation in June, July and August, S_{JJA}, and December, January, and February, S_{DJF}, as percent different from present) and internal boundary conditions (land ice as a percentage of the ice volume 18,000 years ago; global mean sea-surface temperature as a departure from present value; excess glacial aerosol, including sea ice on an arbitrary scale; and atmospheric carbon dioxide concentration) over the last 18,000 years. *Arrows* correspond to the seven sets of simulation experiments using the National Center for Atmospheric Research's Community Climate Model. **b** Simulated net radiation, expressed as departures from the modern case, in the latitude belt 8.9° N to 26.6° N. **c** Simulated surface temperature, expressed as departures from the modern case, in the latitude belt 8.9° N to 26.6° N. **d** Simulated values of annual precipitation and effective precipitation in the latitude belt 8.9° N to 26.6° N. The *top of the columns* represent precipitation, the *lightly shaded area* evaporation, and the *darkly shaded area* effective precipitation (which is available for runoff). The *stars* indicate that the values of P and $P-E$ are significant at $p < 0.05$ in a two-sided t-test. **e** Lake-area ratios, z, assuming constant ratio of land area to internal drainage, e, and constant lake evaporation E_L. **f** Lake-area ratios, z, assuming variable values of e and E_L. **g** Temporal variations in the percentage of lakes with low, intermediate, or high levels. Maximum sample size = 73. (Kutzbach and Street-Perrott 1985)

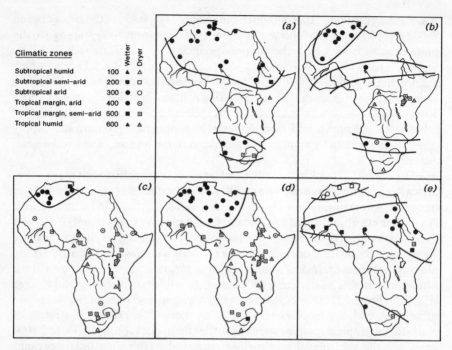

Fig. 3.7. Generalized maps of the chief water-budget tendencies in Africa from 30,000 to 10,000 years ago. **a** 30,000 to 22,000 years ago. **b** 22,000 to 20,000 years ago. **c** 20,000 to 17,000 years ago. **d** 17,000 to 13,000 years ago. **e** 12,000 to 10,000 years ago. (After Littmann 1989)

drier than they had been prior to 17,000 years ago. The period of post-glacial warming, lasting from 12,000 to 10,000 years ago, had a complicated effect on regional water budgets: the Atlas region became drier and aridity ceased in southern Africa and in the tropics, which took on modern climatic patterns. After 10,000 years ago the tropics stayed humid, although the equatorial mountains experienced another wet phase between 8000 and 4000 years ago, and changes around tropical margins were somewhat involved (see also Gasse et al. 1990; N. Roberts 1990).

It should be pointed out that all this work on past water budgets depends very much on reliable radiocarbon dates, but just what constitutes usable dated material is a matter of disagreement. For example, the reconstruction of palaeoclimatic evolution in North Africa and its interpretation in terms of changes in atmospheric circulation by Pierre Rognon (1987) has been disputed by J.-C. Fontes and F. Gasse (1989) on the grounds that evidence for a pluvial period in the northern Sahara between 40,000 and 20,000 years ago, as envisaged by Rognon, is not yet backed up by reliable data, the trustworthy evidence available to date indicating that the last major humid episode of the late Pleistocene epoch occurred some 80,000 to 150,000 years ago. Subsequent work in the Great Chotts in southern Tunisia, where uranium and thorium activity ratios were used to date material, supports Fontes and Gasse's view, two humid phases occurring at about 90,000 and 150,000 years ago with no sign of lacustrine phases in the period 17,000 to 40,000 years ago as claimed by Rognon (Causse et al. 1989).

3.3.4 Glacio-Eustatic Changes

A consequence of swings of climate from glacial to interglacial states is the waxing and waning of ice sheets. In turn, this leads to glacially controlled changes of sea level. It is to be expected that high-stands and low-stands of sea level, as recorded in marine terraces and drowned landscapes, will accord with orbital pulsebeats. With the advent of radiometric dating techniques, the relationship between terraces and past sea levels has indeed been found to match changes in orbital parameters. Wallace S. Broecker (1965) reported thorium dates of ancient coral reefs in Eniwetok atoll, the Florida Keys, and the Bahamas Islands. He noted the correspondence between three of the four maxima for Milankovitch's radiation curve for latitude 65° N and the dates of three high-stands of sea level — 120,000 years ago, 80,000 years ago, and today. In collaboration with Robley K. Matthews and others, he reported thorium dates for three coral reef terraces on Barbados which matched interglacial episodes predicted by a revised version of the Croll-Milankovitch theory (Broecker et al. 1968). The terraces were dated at 125,000, 105,000, and 82,000 years old. The inclusion of a 105,000-year-old terrace was at first a source of puzzlement as the Milankovitch radiation curve for latitude 65° N had no maximum at this date. However, examination of the radiation curve for lower latitudes, in particular 45° N, did contain peaks near all three of the Barbados terrace dates. The lack of a radiation peak 105,000 years ago at latitude 65° N results from the pulse of the tilt cycle being felt more strongly at higher latitudes. In lower latitudes, the precessional pulse is more marked. The results from Barbados were confirmed within in a few years by independent investigations on New Guinea and the Hawaiian Islands. It was these findings which rekindled interest in the astronomical theory of climate, for, if the precessional cycle were given more weight than Milankovitch had allowed, then it could explain the high-stands of sea level which took place 82,000, 105,000, and 125,000 years ago. However, a causal link between orbital variations and changes of sea level still awaited discovery. It was possible that the coincidence of the three dates of terraces with radiation maxima arose by chance. A sounder test was not forthcoming until deep-sea cores provided a "geological calendar for Pleistocene events much longer than that provided by thorium dates from ancient coral reefs" (Imbrie and Imbrie 1986, p. 146).

The relationships between dated marine terraces and orbitally driven glacio-eustatic changes is now generally well established. To take an example, Karl W. Butzer (1975), from a study of coastal stratigraphy in Mallorca in which 15 thorium/uranium dates were obtained, recognized six littoral and terrestrial "hemicycles" which accord with the 100,000 cycle of glacial and interglacial episodes (but see F.H. Fabricius et al. 1983). Similarly, the timing, frequency, and intensity of major climatic events over the last 140,000 years, as shown by the coral reef sequence from Papua New Guinea, corresponds to the predictions of the Croll-Milankovitch theory of ice ages to the extent that brief interstadial intervals of relatively high sea levels occurred roughly every 20,000 years (Aharon 1984; Bloom and Yonekura 1985).

3.4 Long-Term Changes in the Hydrosphere

During the course of Earth history the state of the hydrosphere has changed significantly. The biggest changes probably occurred during glaciations when much water was locked up in polar ice sheets and the sea level dropped considerably. Other long-term changes will have resulted from the secular decrease in rotation rate, mountain-building, continental drift, volcanism, changes in ocean circulation, and tectono-eustatic changes of sea level: as the position of the continents, the level of the oceans, and many other factors have changed over geological time, so the state of the world climate system, including the hydrosphere, will have altered. Of the many facets of long-term change in the hydrosphere which could be discussed at this juncture, just two will be singled out: the effect of palaeogeography on the siting of the world's arid and humid zones, and the question of ice ages.

3.4.1 Ancient Water Budgets

It has long been the practice of palaeoclimatologists to use supposed indicators of aridity, such as red beds and evaporites, and indicators of humidity, such as coal, to reconstruct the climates of the distant geological past (for reviews see Schwartzbach 1963; Nairn 1964; Frakes 1979; Hambrey and Harland 1981). The subject matter and literature of palaeoclimatic reassembly is vast and replete with controversy. All that will be offered here is an outline of two attempts, using general circulation models, to simulate ancient water budgets. The first predicts the climatic belts which obtained during Precambrian times when the Earth rotated several times faster than it does now; and the second considers the world climate during the Triassic period when the continents were joined, forming the supercontinent of Pangaea.

B.G. Hunt's (1979) general circulation model simulations of the world climate system with the rotation rate upped to five times its present value, as would have applied during Precambrian times, created distinct changes in the present climatic belts. The conventionally defined arid region of the globe was confined to a narrower latitudinal belt and was shifted towards the equator (Fig. 3.8). Most of the globe polewards of latitude 45° became semi-arid as the increased coldness of the troposphere led to a general decline of evaporation and precipitation in this region. The implications of these simulation results for Earth climates are very interesting. The smaller scale and less organized nature of meteorological systems in Precambrian times would have led to precipitation being less intense and more localized, which suggests that runoff and erosion

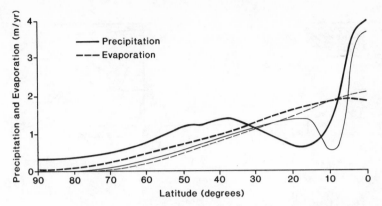

Fig. 3.8. Comparison of time-averaged zonal mean precipitation and evaporation rates for a normal rotation rate (*thick lines*) and a rotation rate five times faster than normal (*thin lines*). (Hunt 1979)

would have proceeded at lower rates. Wind stress may have been reduced by a factor of 2. In the oceans, the lower wind stress would have caused far less vertical mixing, a shallower mixed layer, a lower overall heat capacity, and a diminishment in the lateral transport of heat. All these changes would have caused the tropical atmosphere and oceans to have been warmer, and the polar regions to have been colder, than they are today. Lower wind stress would have influenced atmospheric composition: transfer of gases between the Earth's surface and the atmosphere would have been restricted, as would vertical transport in the troposphere. The colder, drier polar regions would be biologically far less active than they are today, and this could have affected the production of minor, but very important atmospheric gases, such as nitrogen oxides, ammonia, and methane.

Simulations of the climate of Triassic times, testing the effect of several different assumptions about relief, snow cover, greenhouse heating, and solar luminosity, have revealed the presence of megamonsoons on the Pangaean supercontinent (Kutzbach and Gallimore 1989). In all simulation experiments, the conjoined land masses of Laurasia and Gondwana (Fig. 3.9) exhibited extreme continentality with hot summers, cold winters (Fig. 3.9a), and large-scale summer and winter monsoon circulations. Some components of the resulting water budgets are shown in Fig. 3.9b, 3.9c, and 3.9d for the experiment with Triassic geography, no relief, modern heating, and modern snow cover. The pattern of annual precipitation (Fig. 3.9b) shows that rainfall exceeded 2 mm/day (720 mm/a) along the Tethyan coasts, which were affected by the summer monsoon, and along the eastern coast of middle and high latitudes, which received rain year round. Rainfall totals averaged 2 mm/day (720 mm/a) along the west coast in middle latitudes because westerly winds blew there during the winter, and along the west coast at the equator owing primarily to equinoctial wind convergence. The annual rainfall was less than 720 mm over almost all Pangaea. Effective rainfall (precipitation less evaporation) revealed a water

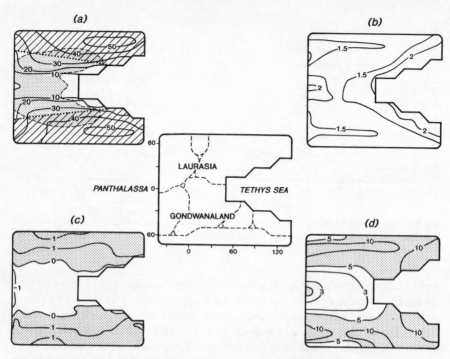

Fig. 3.9. Simulated climate on the idealized Pangaean continent. **a** Surface temperature range (K) calculated as summer temperature minus winter temperature. Seasonal temperature extremes are also indicated: the *stippled area* demarks summer temperatures greater than 30 °C; the *hatched areas* show winter temperature less than 0°C. **b** Annual precipitation (mm/day). **c** Annual effective precipitation (precipitation minus evaporation) (mm/day). The *stippled areas* are regions with an excess of precipitation over evaporation (runoff generating regions). **d** Annual soil moisture (cm). The *stippled areas* have soil moisture in excess of 5 cm. (After Kutzbach and Gallimore 1989)

deficit in low latitudes (in a band lying between latitudes 30° N and 30° S), and a water surplus in middle and high latitudes. Annual soil moisture levels were consistent with the precipitation and evaporation rates, being below 5 cm in the band circumscribed by latitudes 30° N and 30° S, and above 10 cm in areas of year-round rainfall in eastern Pangaea bordering the Tethys Sea. Only where the annual soil moisture levels exceeded 10 cm would significant amounts of runoff have been generated.

3.4.2 Ice Ages Through Earth History

Geologists have long been struck by the fact that glacial episodes have occurred at fairly regular intervals during the geological past. Early estimates put the pulsebeat of ice ages at about 210 million years. Recent work, with the advantage of a more reliable chronology of events, sets the mean period of the mean ages of glaciations at about 155 million years with a range of 140 to 170 million years, although an exception to this occurred in the Mesozoic era, which seems to have opened with a double cycle (Williams 1981a). In other words, the world climate system appears regularly to have switched from a glacial to a non-glacial mode, or what Alfred G. Fischer (1981, 1984a) called "icehouse" and "greenhouse" states. Two questions spring to mind: why should the climate system switch from one mode to another? And why should the switching be roughly periodic? We do not want for tentative answers to these fundamental questions. A few suggestions as to the causes of switching will be discussed here; the question of the pulse of glaciations will be deferred to the final chapter.

What processes might cause the climate system to change from an icehouse to a greenhouse state? The disposition of the continents is held by several geologists and palaeoclimatologists as the key to understanding the occurrence of ice ages. It has been suggested that land at high latitude is a prerequisite for the onset of an ice age. Obviously, a continental ice sheet cannot form unless there be land at or near a pole. If land lie in a polar region then it will reduce the efficiency of oceanic heat transport and provide a site where snow may be caught and held. The logic of this argument has lured several palaeoclimatologists into proposing that continentality is the chief factor in explaining the occurrence of continental glaciation through time (e.g. A. Cox 1968; Crowell and Frakes 1970; Crowell 1983; Beaty 1978a,b) and has encouraged modellers of palaeoclimates to explore the effect of palaeogeography on the simulated world climate system (e.g. A. Sellers and Meadows 1975; Donn and Shaw 1977; Barron et al. 1980, 1984; Barron and Washington 1982a,b, 1984; Barron 1985). Chester Beaty (1978a,b) believed that an essential factor in the coming of an ice age is the presence of a large land mass at a latitude high enough for sufficient snow to be caught and held. Once established, an ice age will then run its course of glacial and interglacial cycles owing to astronomical forcing until lands at high latitudes have moved far enough towards the equator for ice sheets no longer to be sustainable. According to Beaty, Eurasia and North America have been at sufficiently high latitudes since the onset of the last Ice Age and the Earth is presently locked in an icehouse state. In Beaty's hypothesis, the triggering of an ice age at high latitudes is brought about by an increase in the Earth's albedo, owing to the chance occurrence of a period of intense volcanism leading to a thick volcanic dust veil, and a slight reduction in solar output at the trough of sunspot activity. Once triggered, the ice-age state leads to a depression of tropospheric temperatures, which in turn promotes increased snowfall in critical areas. So, once up and running, an icehouse state is self-maintaining.

In broad terms, the palaeogeographical hypotheses of ice ages have been vindicated by simulations using general circulation models. Ann Sellers and A.J. Meadows (1975) showed in their simulations that glaciation occurs when land masses concentrate in polar regions, at which times global mean temperature is as much as 12 °C lower than when land masses congregate in tropical latitudes. Using an energy budget model applied to palaeocontinental reconstructions for four different times, William L. Donn and D.M. Shaw (1977) demonstrated that the general cooling of northern climates during the Cenozoic era was a direct result of an increase of land at high northern latitudes. Eric J. Barron (1983, pp. 319–320) cautioned that there are several limitations to Donn and Shaw's simulations and that their conclusion is not warranted. Specifically, Donn and Shaw used palaeocontinental positions which differed significantly from most reconstructions: they based the movement of Eurasia only on palaeomagnetic data, which gives a very large change of high-latitude land compared with reconstructions based on sea-floor spreading data and palaeomagnetic data. More damningly, their predicted global cooling appears to have resulted from their assumption that snow would cover any land surface north of latitude 60° N, this despite the fact that they allowed polewards transport of heat to mitigate the degree of global cooling. The flaws in the simulations notwithstanding, and leaving aside J. Cogley's (1979) finding in a general circulation model that the land area in the subtropics is a more important factor in producing glaciations than land at high latitudes, there is now much support for Donn and Shaw's basic conclusion that the encroachment of a land mass on either pole will lead to global cooling (Barron and Washington 1982a,b; Barron et al. 1984; Barron 1985; Crowley et al. 1986, 1987). John C. Crowell, for instance, suggested that the commencement of glaciation during late Devonian times coincided with the encroachment of Gondwana on the South Pole (Crowell 1982, 1983; Caputo and Crowell 1985; see also Powell and Veevers 1987); and Steven M. Stanley invoked an earlier such encroachment to explain the late Ordovician glaciation (Stanley 1988a, p. 343, 1988b), and the encroachment of Gondwana upon the North Pole in the late Permian period to account for the glaciation known to have occurred at that time (Stanley 1988a, p. 344, 1988b). Particularly interesting is the investigation made by Thomas J. Crowley, John G. Mengel, and David A. Short (1987) into the climatic changes associated with continental movements over the last 100 million years using an energy budget model. The model showed that the separation of Greenland from North America, and its drift to the north, could have significantly contributed to the depressing of summer temperatures and the formation of all year-round ice cover. In like manner, they found that the separation of Antarctica from Australia and its drift to the south could help to explain the growth of Antarctic ice. This study divulged an important fact: ice sheet formation may actually be inhibited when a large continent is centred over a pole because, although winter temperatures be lower, summer temperatures are too high to permit the survival of snow cover. The importance of the seasonal climatic cycle, rather than mean annual temperature, in polar ice sheet creation had been underlined in earlier studies (North et al. 1983; North and Crowley 1985; Kutzbach and Guetter 1986).

Ice sheet formation is favoured by an orbital configuration associated with cool summers which allows snow to remain unmelted from one year to the next. It has also been discovered that a supply of moisture is essential to ice sheet formation (Aksu et al. 1988), and that the best recipe for polar ice formation seems to be lots of land masses separated by seaways which moderate summer temperatures and act as a source of moisture (North and Crowley 1985). It is possible that the relation between polar climate and ice sheet formation involves thresholds and non-linearity (North and Crowley 1985). Polar climates might cool without ice sheets starting to form until a critical temperature be attained.

The smallest drop in temperature below that threshold would then lead to ice sheet growth and the establishment of a stable glacial state. Renewed warming, taking the temperature over the same threshold, might then be insufficient to return the system to the critical point. If this reasoning be correct, then it follows that an ice-free pole is not necessarily a warm pole (North and Crowley 1985).

Recent computer experiments leave little doubt that palaeogeography plays a role in the occurrence of greenhouse, as well as icehouse, episodes. A good example of this are the simulations of the climate in the mid-Cretaceous period which, according to the palaeoclimatic data, was warm and equable (Barron et al. 1981; Barron 1983, 1989; Barron and Washington 1985). The Cretaceous climate appears to have possessed the following features: warm water, greater than 12 °C, in the ocean deeps; water in the tropical oceans similar or slightly greater in warmth than it is at present; warm poles with no signs of permanent ice; less seasonality than at present in the interiors of continents; and a globally averaged mean surface temperature 6 to 12 °C higher than at present (Barron 1989). This warm and equable climate produced uncommonly large amounts of coal, evaporites, bauxite, black shale, and other rock types. Using boundary conditions appropriate to mid-Cretaceous times, a zonal energy balance model of climate produced an 1.2 percent increase in total absorbed energy, and a 1.62 °C increase in global surface temperature, compared with present conditions (Barron 1983, p. 327). This represents a significant warming of the globe, but falls far short of the temperature increase indicated by palaeoclimatic data. Much of this shortfall in temperature increase occurred in the temperate and frigid zones, the poles in particular remaining far too cold in the simulations. The point is that palaeogeography does appear to have had an effect on Cretaceous climates, but, assuming that the physical processes in the model be adequately formulated and assuming that the palaeoclimatic data be correctly interpreted, it cannot fully explain warm equable palaeoclimates (Barron 1983, p. 327). One is left to conclude, therefore, either that the assumptions of the model are faulty, or that the palaeotemperature indicators have been misinterpreted, or that some other external climatic forcing besides palaeogeography has been omitted, or perhaps a combination of the three. Factors besides palaeogeography which might have affected Cretaceous climates are the decrease in continental area resulting from high eustatic sea level, higher levels of carbon dioxide in the Cretaceous atmosphere, and amplifications of these factors through a greater polewards transfer of heat in the oceans (e.g. Barron 1983; Schneider et al. 1985; Rind 1986; Crowley 1988; Barron 1989). A variety of climate models have been employed to investigate these factors (e.g. Barron et al. 1981; Barron 1983; Barron and Washington 1985; Covey and Barron 1988; Barron 1989). Much has been learnt from these simulations. Notable points are that a combination of factors, including a fourfold to tenfold elevation in concentrations of carbon dioxide, is required to obtain the right degree of warming; that some ice may have persisted in Antarctica; and that the polewards transport of heat by the oceans, on the face of it a good candidate for explaining the warmth of polar regions, seems incapable of bringing about the changes indicated by the palaeoclimatic data (Barron 1989).

There is a tendency to think of warm, greenhouse states as the norm and cold, icehouse states as the exception. Interestingly, B.G. Hunt (1984) believes, from experiments with general circulation models, that an ice-covered Earth is the norm and it is the intervening warm periods that need explaining. This unusual idea gains some support from evidence recently found indicating that even during greenhouse stages the Earth was not totally free of ice (Frakes and Francis 1988). During the Cretaceous period — one of the warmest greenhouse excursions — cool-temperate conditions prevailed very near the poles, as we have just seen. Despite this apparent polar warmth, high-latitude ice was apparently present at sea level in central Australia: outsized erratic blocks, seemingly emplaced by ice-rafting, have been found in Lower Cretaceous mudstones of the Eromanga Basin. Strata of similar origin, ranging in age from mid-Jurassic to mid-Cretaceous, occur in other areas which were positioned between palaeolatitudes 65° and 78° N, including Alaska, Siberia, Canada, and the Spitzbergen Islands. Indeed, there is a record of high-latitude rafting throughout the Phanerozoic aeon, suggesting that ice was present on the Earth for much of its history, and that ice-free conditions could have been at most episodic over the past 600 million years (Fig. 3.10). Prior to 600 million years ago, the Earth may have been perpetually glaciated owing to the effect of a faster rate of rotation. Hunt's (1979) simulations with a faster rotation rate (Sect. 2.5.7) suggest that during most of the Precambrian the polar regions would have been about 20 °C colder than they now are, and, in spite of the very low precipitation rate at high latitudes (Fig. 3.8), conditions should have been conducive to the growth of an ice cover, providing that the land masses were suitably disposed. Hunt

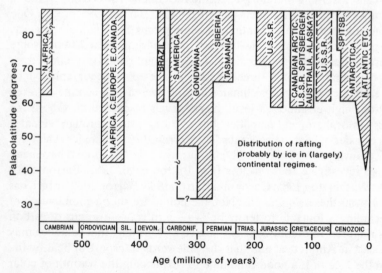

Fig. 3.10. The palaeolatitude of ice-rafted deposits through the Phanerozoic aeon. (Frakes and Francis 1988)

hypothesized that the extent of ice cover would have gradually and continuously diminished, all else being equal, as the Earth's rotation rate slowed and the circulation system began to warm high latitudes. He attributed the termination of the Precambrian ice age to the attainment of a rotational period of 20 hours, which would induce dynamical conditions less favourable to the maintenance of glaciation. Against this idea, recent studies suggest that, owing to self-regulating mechanisms within the carbonate-silicate system, global glaciation during Precambrian times would have been prevented (Kasting 1989).

4 Sediments

The relative rates of erosion in desert, tropical, and polar climates is a subject upon which there is much diversity of opinion.

Joseph Barrell (1908, p. 185)

4.1 Weathering

4.1.1 Weathering and Climate

The process of weathering is implicit in James Hutton's concept of the rock cycle. It was recognized, but not named, by the Reverend James Yates (1830–31) in a paper concerned with the formation of alluvial deposits, and in a report by John Phillips (1831) entitled *On Some Effects of the Atmosphere in Wasting the Surfaces of Buildings and Rocks*. Chemical weathering was discussed in Karl Gustav Christoph Bischof's monumental book *Elements of Chemical and Physical Geology* (1854–59). During the 1860s, students of weathering became more vocal, perhaps spurred on by the rise of fluvialism as promulgated by Joseph Beete Jukes (1862) and Colonel George Greenwood in his *Rain and Rivers; or, Hutton and Lyell Against All Comers* (1857). The term weathering now appeared regularly in the literature: G. Henry Kinahan (1866) wrote on *The Effects of Weathering on Rocks*, D.T. Ansted (1871) *On Some Phenomena of the Weathering of Rocks, Illustrating the Nature and Extent of Sub-Aerial Denudation*, and, in his report on the Geology of the Henry Mountains of 1877, Grove Karl Gilbert succinctly summarized the causes of weathering, alluding to the role played by climate. In 1880, Archibald Geikie published a paper concerned with the rate of weathering; J.G. Goodchild followed suit in 1890. A direct link between weathering and climate was forged by Jacobus Hendricus van't Hoff (1884), who found that for a 10 °C rise in temperature, the speed of chemical reactions increases by a factor of 2 to 3. Van't Hoff's temperature rule applies to many chemical reactions, especially sluggish ones, and to many biological reactions.

During the 1880s, several French scientists identified the chief types of clay mineral and their properties, and the importance of climate in determining weathering products was made clear by Ernest van den Broeck in his *Mémoire sur les Phénomènes d'Altération des Dépôts Superficiels par l'Infiltration des Eaux Météoriques Étudiés dans Leurs Rapports avec la Géologie Stratigraphique* (1881), and later by George Perkins Merrill in his famous *Rocks, Rock Weathering, and Soils* (1897) and Emil Ramann writing in his *Bodenkunde* of 1911 (see also Ramann 1928). Ramann was of the view that the hydrolytic decomposition of silicates is the most important process of weathering, and averred that the degree of dissociation of water into hydrogen ions and hydroxyl ions, which process is strongly dependent upon temperature, is a critical factor in weathering. He derived a weathering factor as the product of the annual number of days having temperatures above freezing and the relative dissociation of water. The weathering factor is greatly influenced by temperature and varies with latitude, weathering in the tropics proceeding three times faster than in temperate zones, and nine times faster than in the Arctic. Boris Borisovitch Polynov, in his book *The Cycle of Weathering* (1937), also discussed the role of climate, in conjunction with geomorphological processes and past geological conditions, on the form of the weathering crust. He did not believe that specific, dissimilar types of weathering were associated with distinct climatic conditions, which to him would imply that there exists between climate and the products of weathering an equilibrial condition and a constant relationship (Polynov 1937, p. 171). Thus, he enounced, while the allitic crust of weathering (alumina gels and their derivatives) is confined to regions of moist tropical climate, it does not follow that allitization is exclusive to that climate; rather, in the present geological epoch, allitization has reached its greatest development in countries with the most intensive weathering, and there is no reason to suppose that allitic products of weathering could not form under other conditions. This idea is similar to the view expressed by Emil Ramann that *"the chemical changes involved in the weathering of rocks and the formation of soils are really the same in different climates; the external conditions, especially the prevailing temperature, determine both the direction and rate of the reactions"* (Ramann 1928, p. 9, emphasis in original).

4.1.2 Weathering and Leaching Regimes

Modern attempts to explain the global distribution of weathering and weathering crusts usually invoke all or any of the soil-forming factors. But the primary factor, which reflects the influence of all others, is the storage and movement of water in the regolith. Climate and other soil-forming factors affect the formation of clays by weathering and neoformation mainly through their influence on the water budget of the regolith, and thus on their control on the internal microclimate of the soil (cf. Curtis 1990). Temperature influences the rate, but seldom the type, of weathering. The kind of secondary clay mineral formed in the regolith depends chiefly on two things: the balance between the rate of dissolution of primary minerals and the rate of flushing of solutes by water (e.g. Crompton 1960); and the balance between the rate of flushing of silica, which tends to build up tetrahedral layers, and the rate of flushing of cations, which fit into the voids between the crystalline layers formed from silica. Clearly the leaching regime of the regolith is crucial to these balances since it determines, in large measure, the opportunity that the weathering products have to interact. Making the same point, Ward Chesworth (1979, p. 308) writes:

"Water and its chemistry are of paramount importance to the weathering system. The availability of water, its residence time in the weathering deposit, its composition and its vertical mobility in leaching,

all in effect dictate the directions of geochemical and mineralogical evolution for the solid residua".
(Chesworth 1979, pp. 308–309)

Georges Pedro (1966, 1968, 1979, 1983) distinguished three degrees of leaching,
each of which favours the formation of a different kind of secondary mineral.
With weak leaching, a rough balance between silica and cations obtains and 2:1
phyllites such as smectite (montmorillonite and its allies) are created by the
process of bisiallitization. With moderate leaching, cations are flushed, leaving
a surplus of silica, and 1:1 phyllites, such as kaolinite and goethite, are formed
by the processes of monosiallitization. In bisiallitization and monosiallitization
hydrolysis is only partial. With intense leaching, very few bases remain un-
flushed, hydrolysis is total, and aluminium hydroxides such as gibbsite are
produced by the process of allitization, also termed soluviation (Swindale and
Jackson 1956), ferrallitization (Chesworth 1979), laterization, and latosolization
(Yatsu 1988). This association of clay minerals with leaching regimes is made
complicated by soil water charged with organic acids. Organic-acid-rich waters
lead to a complexolysis, referred to as cheluviation (Swindale and Jackson 1956)
and associated with podzolization in soils, whereby aluminium compounds,
alkaline earths, and alkaline cations are flushed out in preference to silica.

Owing to the well-established relations between clay neoformation and
leaching regime of the regolith, it is tempting, despite Polynov's caveat, to
envisage distinct types of weathering and weathering crust in each climatic zone.
At the crudest level, each zonal soil group may be characterized by a
predominant type of weathering: salinization in aridisols, calcification in mollisols, lessivage in alfisols, podzolization in spodosols, and ferrallitization in
oxisols. A more sophisticated, but still essentially climatic, scheme of weathering
is offered by Philippe Duchaufour (1982), who differentiates between three
major zones of weathering at the global scale, namely, podzolization, brunification (bisiallitization), and ferrallitization, as well as an intermediate type termed
fersiallitization (Table 4.1). To Duchaufour, climate plays a leading part in the
weathering process:

"it is involved on the one hand by the factor of *water* and on the other by the factor of *temperature*:
weathering in a hot and humid climate differs from that which characterises temperate climates, not
only in its speed by also the kind of physiochemical process"; and he is adamant that "*the type of
climatically controlled weathering is one of the essential elements of climatic soil zonation when
combined with latitude*". (Duchaufour 1982, p. 17, emphasis in original).

A. G. Chernyakhovsky and his coworkers (1976), considering processes of
chemical and mechanical weathering, group recent weathering crusts according
to the leaching (percolation) regime and the prevalence of freezing temperatures
(Table 4.2). The subtypes shown in the table are divided according to parent
rock. Each type of weathering crust is characterized by a particular combination
of weathering processes: under a non-leaching moisture regime, rocks disin-
tegrate owing to heating and cooling, and, because the small amount of water
which enters the soil adheres to grains for a long time, cations and silica become
concentrated allowing smectite to form; under a weak periodic regime, hydration by moisture in tiny cracks is important; under a strong leaching regime,

Table 4.1. Types of climatic weathering according to Philippe Duchaufour. (After Duchaufour 1982)

Climate	Weathering process	Type of clay formation and transformations	Neoformation of clays	Soil
Boreal	Complexolysis (podzolization)	Degradation and solution	Nil	Podzolic
Temperate	Complexolysis (podzolization)	Degradation and solution	Nil	Podzolic (mor)
	Gradual acid complexolysis (bisiallitization or brunification)	Transformation of phyllitic minerals to illite and vermiculite (bisiallitization)	Weak (kaolinite or halloysites)	Brown (mull)
Subtropical	Neutral hydrolysis (fersiallitization)	Inheritance of 2:1 clays; moderate transformation	Moderate (smectites)	Fersiallic
Tropical (with) dry season)	Total hydrolysis (ferrallitization)	Limited inheritance and transformation	Moderate (kaolinite)	Well-drained tropical ferruginous
	Total hydrolysis (ferrallitization)	Limited inheritance and transformation	Strong (smectites)	Poorly drained vertisols
Hot and humid (equatorial)	Total hydrolysis (ferrallitization: silica and bases moved in preference to iron and aluminium which stay in profile)	No inheritance and transformation	Strong (kaolinite; gibbsite)	Ferrallitic

chemical attack is predominant; while under a freezing regime, the freezing of water in cracks and crevices is the chief process. Figure 4.1 shows the world pattern of weathering crusts according to this classification.

Persuasive though climatic classifications of weathering be, it pays to bear in mind that soil drainage varies within major climatic zones due to the effects of topography and rock type. As Duchaufour elaborates:

"*Local* factors — those which characterise the *site* (parent rock, topography, vegetation) — also play an important part [in weathering], since in particular cases they are able to modify profoundly the climatically controlled processes as a whole. As a general rule, in a temperate climate the abundance of soluble organic products and the strong acidity speed up the weathering, but they slow down neoformation of clays or even cause the degradation of pre-existing clays. Conversely, a high amount of alkaline-earth cations and strong biological activity slow down weathering, but at the same time favour the neoformation or the conservation of clays that are richer in silica. *In addition, no matter what the climate, neoformation is favoured on basic volcanic rocks compared to acid crystalline rocks*". (Duchaufour 1982, p. 17, emphasis in original).

Because of these local factors, a wider range of clay minerals is found in some climatic zones than would be expected if climate were the sole determinant of clay formation. Take the tropics. Soils within small areas of this climatic zone may contain a range of clay minerals where two distinct leaching regimes exist in close proximity: specifically, where fast flushing, caused by high rainfall and good drainage due to high relief, removes both cations and silica then gibbsite

Table 4.2. Types of "recent" weathering crust. (After Chernyakhovsky et al. 1976)

Type and subtype (moisture regime)	Radiative index of dryness	Annual precipitation (mm)	Thickness of crust (m)	Nature of crust according to rock type[a] within leaching regime
Non-leaching regimes:				
Arid	3.5–2.5	100–400	Less than 1	• Single zone grus or silty detritus (1,2,7,9,10) • Single zone with soil horizons (3,4,5,6,8)
Semi-arid	2.0–1.0	150–900	1 to 3	• Single or double zone grus with soil and saprolitic zones (1) • Single zone grus or silty detritus (2,7,9,10) • Single zone (3,4,5,6,8)
Weak periodic leaching regimes:				
Humid boreal	1.0–0.5	500–800	Less than 1 to several	• Single or double zone grus clay (1,2,9) • Single zone with bleached upper soil horizon (3,4,5,6) • Single zone grus or detritus (7,8) • Single zone detritus or sypuchka[b] (10)
Semi-humid	1.0–0.5	500–900	1 to 10	• Double zone (single on clay rocks) with grus clay soil (bleached illuvial horizons) and grus saprolite zones (1,2,3,4,5,6,7,8,9) • Weak karst formation (10)
Strong leaching regimes:				
Humid	1.0–0.5	400–1800[+]	More than 10	• Double zone clayey or grus clay; on clay rocks often a single zone with soil horizon of tropical bleached layer and iron-rich laterite in low-lying landscape sites (1,3,5,6) • Double clay or grus clay with reddish soil zone with bleached spots in upper horizons (2) • Single or double zone with ochreous bleached spots in upper horizons (7,8)
Freezing regimes:				
Humid	0.5	100–500	1 to 10 plus	• Single zone detritus; cloddy (1,2,7,9,10) • Single zone with traces of cryogenetic deformation (3,4,5,6,8)

[a]Rock types coded as follows: 1. Acid and intermediate intrusive rocks such as crystalline schists with dioctahedral and tri-octahedral micas, chlorites, quartz, amphiboles, and feldspars. 2. Palaeotypic effusive and other altered magmatic rocks with tri-octahedral chlorite, chlorite-smectite, hydromicas, and other minerals. 3. Clastic and eluvial rocks with quartz, feldspars, hydromicas, chlorites, and smectites in the cement and the main mass. 4. Clastic and eluvial rocks with quartz, kaolinite, and iron, aluminium, and manganese hydroxides in the cement and the main mass. 5. Calcareous rocks with quartz, kaolinite, and palygorskite. 6. Calcareous marl rocks with quartz, kaolinite, palygorskite, and smectite. 7. Intermediate and basic intrusive rocks and kainotypic effusive rocks with feldspars, pyroxenes, amphiboles, and biotite. 8. Volcanic tuffs and ash with glass, feldspars, pyroxenes, amphiboles, and biotite. 9. Ultrabasic rocks with olivine, pyroxenes, serpentine, and smectite. 10. Carbonate rocks such as limestone and dolomite with calcite, dolomite, and magnesite.

[b]Loose body of fine-grained dolomite from parent rock.

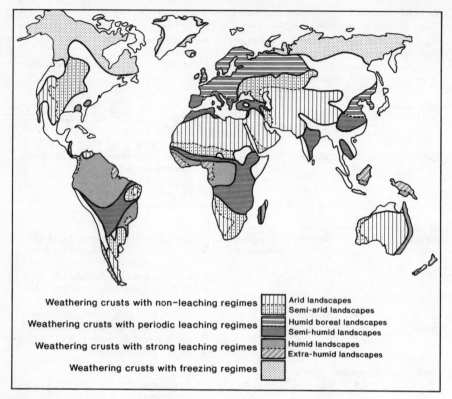

Weathering crusts with non-leaching regimes |‖‖‖| Arid landscapes
                                               |:::::| Semi-arid landscapes
Weathering crusts with periodic leaching regimes ▬▬ Humid boreal landscapes
                                               |≡≡| Semi-humid landscapes
Weathering crusts with strong leaching regimes ▓▓ Humid landscapes
                                               |///| Extra-humid landscapes
Weathering crusts with freezing regimes |⣿⣿|

Fig. 4.1. The global distribution of "recent" weathering crusts. (After Chernyakhovsky et al. 1976)

forms; and on sites where there is less rapid flushing, but still enough to remove all cations and a little silica, then kaolinite forms. G. Donald Sherman (1952), for instance, found that the type of clay formed in soils developed in basalts of Hawaii depends upon mean annual rainfall, the sequence smectite, kaolinite, and bauxite following the gradient from low to high rainfall. The same is true of clays formed on igneous rocks in California, where the peak contents of different clay minerals occur in the following order along a moisture gradient: montmoril-lonite, illite (only on acid igneous rocks), kaolinite and halloysite, vermiculite, and gibbsite (Singer 1980). And, in soils on islands of Indonesia, Edward Carl Julius Mohr and F.A. van Baren (1954) established that the clay mineral formed depends on the degree of drainage: where drainage is good, kaolinite forms; where it is poor, smectite forms. This last example serves to show the role played by landscape position, acting through its influence on drainage, on clay mineral formation. Similar effects of topography on clay formation in oxisols have been found in soils formed in basalt on the central plateau of Brazil (Curi and Franzmeier 1984). Other studies have shown that climatic factors additional to mean annual rainfall have an influence upon rock weathering. H.H. Weinert

(1961, 1965) discovered that in South Africa, weathering products are neatly packaged by isopleths of an index of "weathering climate", N, defined by the formula:

$$N = \frac{E_{Jan}}{P},$$

where E_{Jan} is the potential evaporation in January (the warmest month) and P is mean annual precipitation. In areas where N is greater than 6, hydrous mica is the dominant weathering product; where N lies in the range 2 to 5, basic rocks weather to montmorillonite (black clays) and acid rocks to kaolinite (red clays); where N is less than 2, kaolinite is more common than montmorillonite; and where N is less than 1, laterization occurs.

Time is a further factor which tends to obscure the direct impact of climate on weathering: ferrallitization results from prolonged leaching and its association with the tropics is partly attributable to the antiquity of many tropical landscapes, and is not exclusively the imprint of tropical climates, a fact intimated by Emil Ramann (cf. Pedro 1968; Chesworth 1979, 1980, 1982). This same conclusion was drawn by B.I. Kronberg and H.W. Nesbitt (1981), who argued that the extent of chemical weathering, the dominant process by which the surfaces of continents are modified, is correlated with the age of continental surfaces: in areas where chemical weathering has proceeded without interruption (though perhaps not at a constant rate) since the start of the Cenozoic era, advanced and extreme weathering products are commonly encountered; in areas where the chemical weathering "clock" has been reset by tectonic processes (which through glaciation, volcanism, and alluviation, introduce fresh rock debris into the soil landscape), young soils less than three million years old are common and display signs of incipient and intermediate weathering. In view of these complicating factors, and the changes of climate which have occurred even during the Holocene epoch, claims that weathering crusts of recent origin (recent in the sense that they are still forming and have been subject to climatic conditions similar to present climatic conditions during their formation) are related to climate, like those made by A.G. Chernyakhovsky and his colleagues (1976), must be looked at cautiously.

4.2 Denudation and Deposition

William D. Thornbury (1954, p. 28) opined that an appreciation of world climates is necessary to gain a proper understanding of the varying importance of geomorphological processes. He reckoned it self-evident that climatic factors, particularly temperature and precipitation, should influence the operation of geomorphological processes; but he was surprised at the dearth of studies which tried to ascertain the extent to which climatic variations impress themselves in the details of topography, and thought that the North Americans in particular had neglected this important aspect of geomorphology. That is not to say that relations between climate and geomorphological processes had been entirely overlooked. Joseph Barrell, in his notes on climate and terrestrial deposits, had estimated relative rates of erosion under different climates "upon average rock materials in a state of topographic maturity" (Barrell 1908, p. 188). He had maintained that arid climates had the less rapid erosion rates and rainy climates the more rapid erosion rates, and had discerned an increase in rates through arid and rainy climatic zones. Starting with the region with lowest erosion rates and moving to the region with highest erosion rates, he had suggested the following progression: warm-temperate arid climates with moderate Sun and wind action; tropical arid climates with strong Sun and wind action; cold-temperate arid climates with strong frost and wind action; temperate rainy climates with moderate chemical and mechanical disintegration; tropical rainy climates with moderate mechanical and intense chemical disintegration; and subpolar rainy climates with intense mechanical and moderate chemical disintegration. David I. Blumenstock and C. Warren Thornthwaite (1941) had indicated how major gradational processes vary in relation to temperature and effective precipitation, a theme which had been enlarged upon by Louis Peltier (1950), but this work was undertaken when very few field measurements of process rates had been taken. Measurements of the amount of sediment annually carried down the Mississippi river had been made during the 1840s, and the rates of modern denudation in some of the world's major rivers had been worked out by Archibald Geikie (1868a,b). Early measurements of the dissolved load of rivers had enabled Richard Bryant Dole and Herman Stabler (1909) and Frank Wigglesworth Clarke (1924) to estimate the chemical denudation rates early this century. But the quantitative revolution in geomorphology, started in the 1940s, was largely responsible for the measuring of processes rates in different environments.

4.2.1 Mechanical Denudation

Overall rates of denudation can be gauged from the dissolved and suspended loads of rivers, from reservoir sedimentation, and from the rates of geological sedimentation. The pattern of sediment yield from the world's major drainage basins is depicted in Fig. 4.2, and the annual discharge of sediment from the world's major rivers to the sea is shown in Fig. 4.3. It should be stressed that these figures do not measure the total rate of soil erosion since much sediment is eroded from upland areas and deposited on lowlands where it remains in store, delaying for a long time its reaching the sea (Milliman and Meade 1983). Table 4.3 shows the breakdown of chemical and mechanical denudation by continent. Robert M. Garrels and Fred T. Mackenzie (1971, p. 129) contended that the controls on mechanical denudation are so complex and the data so sparse that it was not worth attempting to assess the relative role of the variables involved. Other authors have been bolder, if less prudent, and tried to make sense of such data as are available (e.g. Fournier 1960; Strakhov 1967). Frédéric Fournier (1960), using sediment data from 78 drainage basins, correlated sediment yield with a climatic parameter, p^2/P, where p is the rainfall of the month with the highest rainfall and P is the mean annual rainfall. Although, as might be

Fig. 4.2. The sediment yield of the world's chief drainage basins. *Blank spaces* indicate essentially no discharge to the ocean. (After Milliman and Meade 1983)

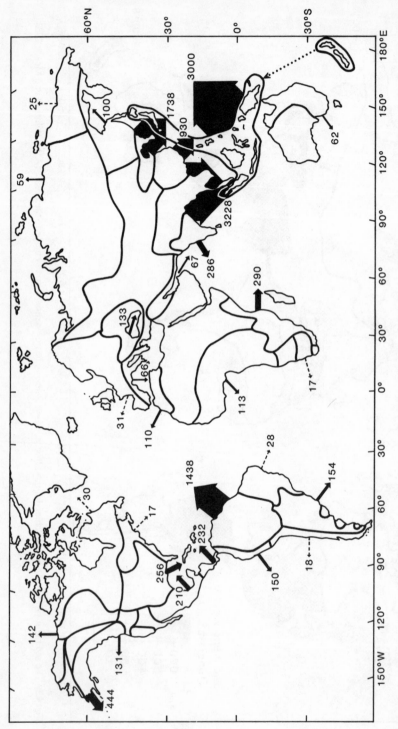

Fig. 4.3. Annual discharge of suspended sediment from large drainage basins of the world. The *width of arrows* corresponds to relative discharge. *Numbers* refer to average annual input in millions of tonnes. (After Milliman and Meade 1983)

Table 4.3. Chemical and mechanical denudation of the continents

Continent	Chemical denudation[a]		Mechanical denudation[b]		Ratio of mechanical to chemical denudation	Specific discharge (1/s.km^2)
	Drainage area (10^{6}km^2)	Solute yield (t/km^2.a)	Drainage area (10^{6}km^2)	Sediment yield (t/km^2.a)		
Africa	17.55	9.12	15.34	35	3.84	6.1
North America	21.5	33.44	17.50^c	84	2.51	8.1
South America	16.4	29.76	17.90	97	3.26	21.2
Asia	31.46	46.22	16.88	380	8.22	12.5
Europe	8.3	49.16	15.78^d	58	1.18	9.7
Oceania	4.7	54.04	5.20	1028^e	19.02	16.1

[a]Data from Meybeck (1979, annex 3).
[b]Data from Milliman and Reade (1983, Table 4).
[c]Includes Central America.
[d]Milliman and Meade separate Europe (4.61 × 10^6 km^2) and Eurasian Arctic (11.17 × 10^6 km^2).
[e]The sediment yield for Australia is a mere 28 t/km^2.a, whereas the yield for the large Pacific islands is 1028 t/km^2.a.

expected, sediment yield increased as rainfall increased, a better degree of explanation was found when basins were grouped into relief classes. Fournier fitted an empirical equation to the data:

$$\log E = -1.56 + 2.65 \log(p^2/P) + 0.46 \log \bar{H}. \tan \varphi,$$

where E is suspended sediment yield (t/km^2.a), p^2/P is the climatic factor (mm), $\bar{H}$ is mean height of a drainage basin, and $\tan \varphi$ is the tangent of the mean slope of a drainage basin. Applying this equation, Fournier mapped the distribution of world mechanical erosion. His map portrayed maximum rates in the seasonally humid tropics, declining in equatorial regions where there is no seasonal effect, and also declining in arid regions where total runoff is low. Nikolai M. Strakhov's (1967) estimates of mechanical erosion rates were lower than Fournier's but, according to Ian Douglas (1967), may be more typical of rates of erosion during much of geological time.

John D. Milliman (1980) identified several natural factors which control the suspended sediment load of rivers: drainage basin relief, drainage basin area, specific discharge, drainage basin geology, climate, and the presence of lakes. The climatic factor influences suspended sediment load through mean annual temperature, total rainfall, and the seasonality of rainfall (Berner and Berner 1987, p. 182). Heavy rainfall tends to generate high runoff, but heavy seasonal rainfall, as in the monsoon climate of southern Asia, is very efficacious in producing a big load of suspended sediment. On the other hand, in areas of high, year-round rainfall, such as the Congo Basin, sediment loads are not necessarily high. In arid regions, low rainfall produces little river discharge and low sediment yields, but, owing to the lack of water, suspended sediment concentrations may be high. This is the case for many Australian rivers. The greatest suspended sediment yields come from mountainous tropical islands, areas with active glaciers, mountainous areas near coasts, and areas draining loess soils: they are not determined directly by climate (Berner and Berner 1987, p. 183). As one might expect, sediments deposited on inner continental shelves reflect climatic differences in source basins: mud is most abundant off areas with high temperature and high rainfall; sand is everywhere abundant but especially so in areas of moderate temperature and rainfall and in all arid areas save those with extremely cold climates; gravel is most common off areas with low temperature; and rock is most common off cold areas (Hayes 1967).

Work carried out on the chemistry of river sediments has revealed patterns attributable to differing weathering regimes in the tropical zone on the one hand, and the temperate and frigid zones on the other. Paul Edwin Potter (1978) found that river sands with high quartz and high silica-to-alumina ratios occur mainly in tropical river basins of low relief where weathering is intense enough (or has proceeded uninterrupted for long enough) to eliminate any differences arising from rock type; and that river sands with low quartz content but high silica-to-alumina ratios occur chiefly in the basins located in temperate and frigid regions. A basic distinction between tropical regions, with intense weathering regimes, and temperate and frigid regions, with less intense weathering regimes, was also made by J.-M. Martin and Michel Meybeck (1979) on the basis of the composition of the particulate load of rivers. The tropical rivers studied had high concentrations of iron and aluminium relative to soluble elements because their particulate load was derived from soils in which soluble material had been thoroughly leached; the temperate and arctic rivers studied had lower concentrations of iron and aluminium in suspended matter relative to soluble elements because a smaller fraction of the soluble constituents had been removed. This broad pattern will almost certainly be distorted by the effects of relief and rock type. Indeed, J.-M. Martin and Michel Meybeck's data include exceptions to their rule: some of their tropical rivers have high calcium concentrations, probably owing to the occurrence of limestone within the basin. Moreover, in explaining the generally low concentrations of calcium in sediments of tropical rivers, it should be borne in mind that carbonate rocks are more abundant in the temperate than in the tropical zone (Meybeck 1987a, p. 414).

4.2.2 Chemical Denudation

The controls on the rates of chemical denudation are perhaps easier to ascertain than the controls on the rates of mechanical denudation. Reliable estimates of the loss of material from continents in solution have been available for several decades (e.g. Livingstone 1963), though more recent estimates overcome some of the deficiencies in the older data sets. It is clear from the data in Table 4.3 that the amount of material removed in solution from continents is not directly related to the average specific discharge. South America has the highest specific discharge but the second-lowest chemical denudation rate. Europe has a relatively low specific discharge but the second-highest chemical denudation rate. On the other hand, Africa has the lowest specific discharge and the lowest chemical denudation rate. In short, the continents show differences in resistance to being worn away which cannot be readily accounted for simply in terms of climatic differences.

The primary controls on chemical denudation of the continents can be elicited from data on the chemical composition of the world's major rivers (Table 4.4). The differences in solute composition of river water between continents result partly from differences of relief and lithology, as well as climate. Waters draining off the continents are dominated by calcium ions and bicarbonate ions. It is these chemical species that account for the dilute waters of South America and the more concentrated waters of Europe. Dissolved silica and chlorine concentrations show no consistent relationship with total dissolved solids. The reciprocal relation between calcium ion concentrations and dissolved silica concentrations suggests a degree of control by rock type: Europe and North America are underlain chiefly by sedimentary rocks, whereas Africa and South America are underlain mainly by crystalline rocks. However, it would be imprudent to read over much into these figures and to over-play this interpretation, as the continents are for the most part formed of a heterogeneous mixture of rocks.

The natural chemical composition of river water is affected by the amount and nature of rainfall and evaporation, drainage basin geology and weathering history, average temperature, relief, and biota (Berner and Berner 1987, p. 193). According to Ronald J. Gibbs (1970, 1973), who plotted total dissolved solids of some major rivers against the content of calcium plus sodium, there are three chief types of surface waters: waters with low total dissolved solid loads (~10 mg/l) but large loads of dissolved calcium and sodium, such as the Matari and Negro Rivers, which depend very much on the amount and composition of precipitation; waters with intermediate total dissolved solid loads (~10^2–10^3 mg/l) but low to medium loads of dissolved calcium and sodium, such as the Nile and Danube Rivers, which are influenced strongly by the weathering of rocks;

Table 4.4. Average composition of river waters by continent[a] (mg/l). (After Meybeck 1979)

Continent	SiO$_2$	Ca^{2+}	Mg^{2+}	Na$^+$	K$^+$	Cl$^-$	SO$_4^{2-}$	HCO$_3^-$	Σi
Africa	12.0	5.25	2.15	3.8	1.4	3.35	3.15	26.7	45.8
North America	7.2	20.1	4.9	6.45	1.5	7.0	14.9	71.4	126.3
South America	10.3	6.3	1.4	3.3	1.0	4.1	3.5	24.4	44.0
Asia	11.0	16.6	4.3	6.6	1.55	7.6	9.7	66.2	112.5
Europe	6.8	24.2	5.2	3.15	1.05	4.65	15.1	80.1	133.5
Oceania	16.3	15.0	3.8	7.0	1.05	5.9	6.5	65.1	104.3
World	10.4	13.4	3.35	5.15	1.3	5.75	8.25	52	89.2

[a]The concentrations are for exoreic runoff with anthropogenic inputs deducted.

and waters with high total dissolved solid loads ($\sim 10^4$ mg/l) and high loads of dissolved calcium and sodium which are determined primarily by evaporation and fractional crystallization and which are exemplified by the Rio Grande and Pecos Rivers. This classification has been the subject of much debate (e.g. Drever 1982; Feth 1971; Stallard and Edmond 1983; Berner and Berner 1987, pp. 197–205), but it seems undeniable that climate does have a role in determining the composition of river water, a fact borne out by the data listed in Table 4.5, which traces the origin of solutes entering the oceans. As can be seen, chemical erosion is greatest in mountainous regions of humid temperate and tropical zones. Consequently, most of the dissolved ionic load going into the oceans originates from mountainous areas, while 74 percent of silica comes from the tropical zone alone. Further work by Michel Meybeck (1987b) has clarified the

Table 4.5. The geographical origins of dissolved loads to the oceans and variations of chemical erosion according to morphoclimatic regions. (After Meybeck 1979)

Climatic region	Area of exoreic runoff	Exoreic runoff	Transport of silica		Transport of ions		Chemical erosion (t/km².a)
			Dissolved load to oceans (10^6 t/a)	%	Dissolved load to oceans (10^6 t/a)	%	
Tundra and taiga	20.0	10.7	15.0	3.9	466	13.1	14
Humid taiga	3.15	3.4	5.0	1.3	74	2.1	15.5
Very humid taiga	0.2	0.6	1.1	0.2	9	0.25	32
Pluvial temperate	4.5	15.3	45	11.8	540	15.4	80
Humid temperate	7.45	7.75	17.5	4.6	407	12.0	35
Temperate	6.7	3.35	9.4	2.5	301	8.8	28
Semi-arid temperate	3.35	1.05	2.7	1.0	130	3.7	24
Seasonal tropical	13.25	5.85	31.1	8.2	119	3.4	6.4
Humid tropical	9.2	8.85	38.2	10.2	239	8.0	15.5
Very humid tropical (plains)	6.9	18.45	78.6	20.8	165	4.8	22
Very humid tropical (mountains)	7.95	24.05	130	34.4	908	25.6	67
Arid	17.2	0.65	3.9	1.0	3457	2.8	3
Total of tropical zone	37.3	57.2	278	73.6	1431	41	
Pluvial regions of strong relief	12.65	40	176	47	1457	42	≈ 74

association between chemical weathering, mechanical weathering, lithology, and climate. Chemical transport, measured as the sum of major ions plus dissolved silica, increases with increasing specific runoff but the load for a given runoff depends on underlying rock type (Fig. 4.4). Individual solutes show a similar pattern. Dissolved silica is interesting because, though the rate of increase with increasing specific discharge be roughly the same in all climates, the actual amount of dissolved silica increases with increasing temperature (Fig. 4.5). Meybeck concludes that, although lithology, distance to the ocean, and climate all affect solute concentration in rivers, transport rates, especially in the major rivers, depend first and foremost on specific river runoff (itself related to climatic factors) and then on lithology.

4.2.3 Slope Processes

Extensive field measurements since about 1960 show that slope processes appear to vary considerably with climate (A. Young 1974; Saunders and Young 1983). Soil creep in temperate maritime climates shifts about 0.5 to 2.0 mm/a of material in the upper 20 to 25 cm of regolith; in temperate continental climates rates run in places a little higher at 2 to 10 mm/a. Generalizations about the rates of soil creep in other climatic zones are as yet not forthcoming owing to the paucity of data. In Mediterranean, semi-arid, and savanna climates, creep is probably far less important than surface wash as a denuder of the landscape. Such studies as have been made in tropical sites indicate a rate of around 4 to 5 mm/a. Solifluxion, which includes frost creep caused by heaving and gelifluxion, occurs 10 to 100 times more rapidly than soil creep and affects material down to about 50 cm, typical rates falling within the range 10 to 100 mm/a. The rate of surface wash,

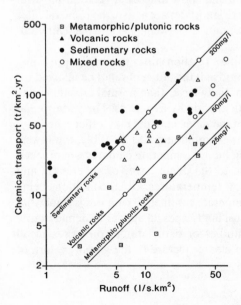

Fig. 4.4. Chemical transport of all major ions plus dissolved silica versus runoff (specific discharge) for various major drainage basins underlain by sedimentary, volcanic, and metamorphic and plutonic rocks. (After Meybeck 1979, 1987b)

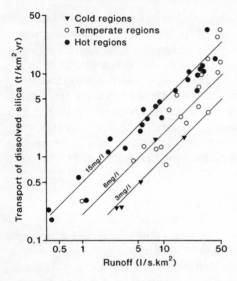

Fig. 4.5. Evolution of the specific transport of dissolved silica versus runoff (specific discharge) for cold, temperate, and hot regions. (After Meybeck 1987b)

which comprises rainsplash and surface flow, is determined very much by the degree of vegetation cover, and its relation to climate it not clear. Solution (chemical denudation) probably removes as much material from slopes as all other processes combined. Rates are not as well documented as for other slope processes, but typical values, expressed as surface lowering rates, are as follows: in temperate climates on siliceous rocks, 2 to 100 mm/millenium, and on limestones, 2 to 500 mm/millenium. In other climates data are fragmentary, but often fall in the range 2 to 20 mm/millenium and show little clear relationship with temperature or rainfall. On slopes where landslides are active, the removal rates are very high irrespective of climate running at between 500 and 5000 mm/millenium.

Overall rates of denudation do show a relationship with climate, providing infrequent but extreme values be ignored and the effect of relief be allowed for (Table 4.6). Valley glaciation is substantially faster than normal erosion in any climate, though not necessarily so with erosion by ice sheets. The wide spread of denudation rates in polar and montane environments may reflect the large range of rainfall encountered. The lowest minimum and, possibly, the lowest maximum rates of denudation occur in humid temperate climates where creep rates are slow, wash is very slow due to the dense cover of vegetation, and solution is fairly slow because of the low temperatures. Other conditions being the same, the rate of denudation in temperate continental climates is somewhat brisker. Semi-arid, savanna, and tropical landscapes all appear to denude fairly rapidly. Clearly, further long-term studies of denudational processes in all climatic zones are needed to obtain a clearer picture of the global pattern of denudation.

Table 4.6. Rates of denudation in climatic zones. (Saunders and Young 1983)

Climate	Relief	Typical range for rate of denudation (mm/millenium)	
		Minimum	Maximum
Glacial	Normal (= ice sheets)	50	200
	Steep (= valley glaciers)	1000	5000
Polar and montane	Mostly steep	10	1000
Temperate maritime	Mostly normal	5	100
Temperate continental	Normal	10	100
	Steep	100	200+
Mediterranean	—	10	?
Semi-arid	Normal	100	1000
Arid	—	10	?
Subtropical	—	10?	1000?
Savanna	—	100	500
Tropical rain forest	Normal	10	100
	Steep	100	1000
Any climate	Badlands	1000	1000000

4.2.4 Regional and Global Patterns of Denudation

Enormous variations in sediment and solute loads of rivers occur within particular regions owing to the local effects of rock type, vegetation cover, and so forth. Attempts to account for regional variations of denudation have met with more success than attempts to explain global patterns, largely because coverage of measuring stations is better and it is easier to take factors other than climate into consideration. Positive correlations between suspended sediment yields and mean annual rainfall and mean annual runoff have been established for drainage basins in all parts of the world, and simply demonstrate the fact that the more water that enters the system, the greater the erosivity (cf. Walling 1987, p. 110). Like suspended sediment loads, solute loads exhibit striking local variations about the global trend. The effects of rock type in particular become far more pronounced in smaller regions. D.E. Walling and B.W. Webb (1986) reported that dissolved loads in Great Britain range between 10 to more than 200 tonnes/km^2/a, and that the national pattern is influenced far more by lithology than by the amount of annual runoff. Very high solute loads are associated with outcrops of soluble rocks. An exceedingly high solute load of 6000 tonnes/km^2/a has been recorded in the River Cana which drains an area of halite deposits in Amazonia; and a load of 750 tonnes/km^2/a has been measured in an area draining karst terrain in Papua New Guinea.

All the general and detailed summaries of global and regional sediment yield (e.g. Fournier 1960; Corbel 1964; Strakhov 1967; Holeman 1968; Ahnert 1970; J.M.L. Jansen and Painter 1974; Jansson 1982, 1988; Milliman and Meade 1983) split into two camps of opinion concerning the chief determinants of erosion at large scales: one camp sees relief as the prime factor influencing denudation rates, with climate playing a secondary role; the other camp casts climate in the leading role and relegates relief to a supporting part. Everybody seems to agree that either relief or climate (as measured by surrogates of rainfall erosivity) are the major controls of erosion rates on a global scale: the problem is deciding on the relative contribution made by each factor. Jonathan D. Phillips (1990) has set about the task of solving this problem by considering three questions: whether indeed relief and climate be major determinants of soil loss; if so, whether relief or climate be the more important determinant at the global scale; and whether other factors known to influence soil loss at a local scale have a significant effect at the global scale. Phillips's results showed that slope gradient (the relief factor) is the main determinant of soil loss accounting for about 70 percent of the maximum expected variation within global erosion rates. Climate measured as rainfall erosivity was less important but with relief (slope gradient) and a runoff factor accounted for 99 percent of the maximum expected variation. The importance of a runoff factor, represented by a variable describing retention of precipitation (which is independent of climatic influences on runoff) was surprising. It was more important than the precipitation factors. Given Phillips's findings, it may pay to probe more carefully the pattern of sediment yields disclosed by Margareta B. Jansson (1988), wherein variation in sediment yield within climatic zones is greater than the variation between climatic zones.

4.3 Short-Term Forcing of Denudation-Deposition Systems

4.3.1 Solar Signals in Sediments

The thickness of laminae in the sediments known as varves is generally thought to be a function of climate. It is generally true to say that the thin laminae found in varve sequences register seasonal deposition, although exceptions are known. Varve thickness thus provides a rough guide to precipitation and temperature. Indeed, in some cases it is possible, by pollen analysis of microtome sections of individual laminae, to discover the season of greatest runoff and the type of precipitation. Given that varves are sensitive to climate, it is reasonable to suppose that if short-term changes in solar activity do affect climatic patterns, then solar signals should be present in varve deposits. A survey of periodicity reported in varve series made by Roger Y. Anderson (1961) revealed an impressive number of occurrences of an 11-year period, and less clear but discernible periods of about 2 and 5 years. Anderson reported the power spectra of several varve series computed by F.W. Ward Jr. A series for the Jurassic Todilto limestone of New Mexico, in which varve thickness changes were thought chiefly to record the effect of temperature, contained power peaks corresponding to periods of 2, 5, 10, and 13 years, the last two being near the sunspot period (though none of the signals was impressively strong). Clearer signals were extracted from the Upper Devonian Ireton shale varves of the Woodbend Formation, Alberta, Canada, which consist of alternating bands of green-grey calcareous shale and brown organic-rich calcareous shale. The laminations were thought to be influenced by temperature, which would affect chemical precipitation, and oceanic circulation or storminess and precipitation. A 22-year signal was very pronounced in this series. In an attempt to detect some overall pattern of spectral peaks, Anderson plotted the prominent peaks derived from ten varve series. The peaks formed clusters corresponding to periods of 8.3, 13.1, 22.1, and 40.2 years. An 11-year cluster of peaks, corresponding to the sunspot period, is noticeable by its absence. Anderson suggested that this might be because its effect is too diluted by terrestrial factors to be apparent, or that the 22-year heliomagnetic cycle may exert a disproportionately large control over weather, though this seems very unlikely (see R.G. Currie and O'Brien 1990). Anderson (1961) also showed that, by smoothing the varve series using a 31-year moving average, longer-term cycles were also present. In the Jurassic Todilto limestone, the average time between maxima of the smoothed series was 187 years, and smaller maxima contained within the larger one had periods of 68 to 88 years. These quasi-periodic trends may result from long-term persistence in temperature, precipitation, and oceanic circulation (Anderson 1961). A correspondence with the Sun's orbital cycle and the Gleissberg cycle, respectively, seems possible.

Solar signals are present in more recent sediments, too. The thermoluminescence profile of a young marine sediment core, which provides information on climatic change over the last 1800 years, records a stable and well-defined 11.4-year cycle and an 82.6-year cycle, the amplitude of both of which is modulated by a 206-year wave (Castagnoli et al. 1990). A weak solar signal detected in recent varves of glacial Lake Skilak, southern Alaska, may arise from periods of high solar activity resulting in increased temperatures, greater annual meltwater runoff, and the deposition of thicker varves (Sonnett and Williams 1985).

4.3.2 Lunar Signals in Sediments

Lunar signals have recently been detected in ancient "varves". George E. Williams interpreted narrow laminae in the Precambrian Elatina Formation as annual varves, and believed that these varves contained periodicities corresponding to the solar cycle, the Hale cycle, and a longer sunspot period of about 100 years (Williams 1981b, 1985; Williams and Sonett 1985; Sonett and Williams 1987). However, after having made a study of sedimentary rhythmites of siltstone and fine sandstone from late Precambrian glaciogenic formations in South Australia (the Reynella Siltstone), ranging in age from about 650 to 800 million years, Williams (1988, 1989) now thinks that the laminae are ebb-tidal deposits encoding a full spectrum of tidal cycles ranging from semi-diurnal, through diurnal, fortnightly, and monthly cycles to the lunar nodal cycle. If this interpretation be correct, then the rhythmites provide information on palaeotidal periods and the Earth's palaeorotation with an accuracy hitherto unattainable for Precambrian times. And they also furnish direct evidence of the sensitivity of sedimentary process to the Earth-Moon coupling (see also Sonett et al. 1988; Archer et al. 1990; Williams 1990a,b).

4.4 Medium-Term Forcing of Denudation-Deposition Systems

In 1895, Grove Karl Gilbert announced that the cyclical changes in the carbonate content of hemipelagic beds in the Cretaceous rocks of Colorado reflected the response of sensitive sedimentary facies to climatic rhythms driven by precession (see also Gilbert 1900; Blytt 1899). This illuminating and perspicacious sugges-tion was deemed too far-fetched, and the effects of orbital cycles on climates and sediments decried for being too nebulous, for "cyclostratigraphy" to emerge as an important branch of stratigraphy (cf. Fischer 1988). With the recent renewed interest in the effects of astronomical forcing on climate, the realization that the state of the world climate system seems very sensitive to extraterrestrial forcing, and the advances in sedimentological investigation such as the development of instrumental scanning methods, cyclostratigraphy is now emerging as a produc-tive field of study (see Einsele et al. 1990). As Alfred G. Fischer proclaims:

"Allocyclic sedimentary rhythms — driven by orbital reaction of the Earth with its moon, its sibling planets, and the Sun, and transmitted through climate — are real. They are more complex than imagined by Gilbert, but they pervade the sediments, constituting a frontier for exploring the functioning and history of the Earth and the solar system". (Fischer 1986, p. 373).

4.4.1 Terrestrial Deposits

Orbital signals have been detected in sediments and soils from a range of terrestrial and semi-terrestrial environments. George J. Kukla (1968) discovered a 100,000-year climatic pulse in loess deposits of Czechoslovakia (Sect. 2.3.1). From an examination of the nature of the fossil content of the loess deposits, he discerned that the cooling phase of the climatic cycle was much longer than the warming phase. Indeed, the transitions from dusty, polar desert to deciduous forest were so abrupt that they appeared as lines in quarry sections. Investiga-tions of magnetic susceptibility variations in a Pleistocene loess section spanning 250,000 years at Halfway House, central Alaska (Fig. 4.6) were carried out by James E. Begét and Daniel B. Hawkins (1989). The Arctic loess contains thick palaeosols formed during early parts of interglacial stages, and thinner palaeosols formed during interstadial events. Cross correlations between the palaeosols and marine oxygen-isotope curves were made: the thicker palaeosols correspond to isotope Stages 7 and 5; the thinner palaeosols correspond to isotope early Stage 7 and Stage 3. Autocorrelation and time-series analysis of the loess data exposed spectral peaks at 125,000, 41,600, and 23,000 years, showing for the first time that orbital pulses influence, and are registered in, proxy climatic data from terrestrial aeolian deposits at high latitudes. Soil-sedimentological records from the eastern Mediterranean and the south part of the North Sea also reveal several superimposed cycles of change during the Pleistocene epoch (Paepe et al. 1986). There are four "maxi-cycles", each

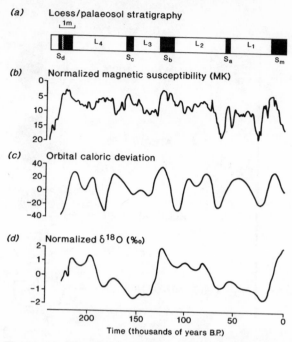

Fig. 4.6. a Stratigraphy of soil and loess at the Halfway House site, central Alaska: S_m modern soil; S_a, S_b, S_c, S_d palaeosols; L_1, L_2, L_3, L_4 massive loess beds. Old Crow tephra is shown by a *stippled pattern within S_d*. **b** Normalized magnetic susceptibility variations through the loess at Halfway House site. The sampling interval was 10 cm. **c** Retrodicted insolation values at latitude 65° N. (After Berger 1978). **d** Normalized and orbitally tuned marine oxygen isotope variations. (After Naeser et al. 1982). (Begét and Hawkins 1989)

600,000 years long, consisting of "cryomeres" lasting 250,000 years and "thermomeres" lasting 350,000 years, each comprising a number of warm-to-cold cycles with periods of 100,000 years. Precessional and obliquity forcing in soils of the region is evident as far south as the southern Peloponesos.

Some fluvial and deltaic sediments have yielded signs of orbital forcing. In the Nile delta, sediments record changes in climate affecting East Africa. Heavy minerals in radiocarbon-dated cores collected within the delta were employed by Alain Foucault and Daniel Jean Stanley (1989) as distinct markers of climatic shifts during the late Pleistocene and Holocene epochs. The fact that the White Nile, Blue Nile, and Atbara Rivers all contribute different amounts of pyroxenes to the sediments in the main Nile River enabled Foucault and Stanley to use an index expressing the frequency of amphibole to the frequency of amphibole plus pyroxene in sand fractions, IAmph, to reconstruct the relative sediment load contributions from the major tributaries of the Nile over the last 25,000 years (Fig. 4.7). It would seem that climatic belts migrated considerably during the late Pleistogene period, modifying Nile discharge and sediment load. The results agree with oscillations in African palaeoclimate established by other, inde-

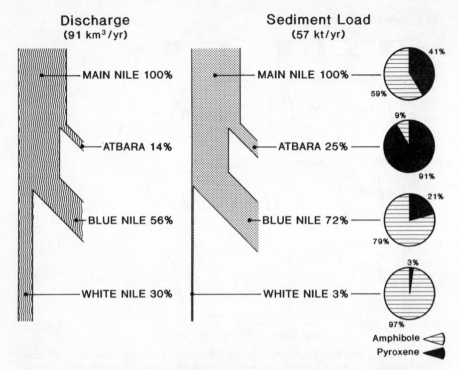

Fig. 4.7. Diagrams showing (*left*) the total discharge of the River Nile and the relative proportions of discharge supplied by its three chief tributaries, and (*right*) the total sediment load of the Nile River and relative sediment load contributions of its tributaries. The *pie diagrams* show the relative proportions of amphibole and pyroxene in the present bed-load of the Main Nile and in its tributaries. (Foucault and Stanley 1989)

pendent methods such as studies of lake levels (Fig. 4.8). Similarly, Françoise Gasse and her colleagues (1989) showed that freshwater diatoms carried from west African rivers into the Atlantic Ocean, there to accumulate, may be used as a tool for reconstructing palaeoclimates and for establishing correlations between environmental changes on land and in the sea. Much older fluvial and deltaic sediments also bear orbital signals: cycles in sandstone-shale alternations in the Devonian pro-delta facies of the Catskills (van Tassel 1987) and in Miocene rocks of California (Clifton 1981) have been attributed to orbital effects.

4.4.2 Marine Systems

Signs of orbital forcing have been found in the full range of marine sedimentary systems: epicratonic basins, marine carbonate platforms, and pelagic and hemipelagic systems. A fine example of sedimentary cycles in epicratonic basins comes from the middle of the North American continent, where Middle and

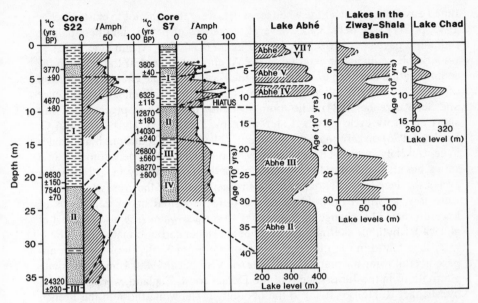

Fig. 4.8. *I*Amph index patterns plotted alongside radiocarbon-dated lithostratigraphic sections of cores S22 and S7. I to IV are lithostatigraphic units. *I*Amph is defined as the ratio of the frequency of amphibole to the joint frequency of amphibole and pyroxene. Time correlation is indicated by *dashed lines*. *I*Amph patterns are correlated with lake level variations, including those of Lake Abhé, lakes in the Ziway-Shala Basin, and Lake Chad. (Foucault and Stanley 1989)

Upper Pennsylvanian and Permian rocks exhibit the classic alternations of marine to non-marine sediments produced by transgressive cycles and designated "cyclothems" (Wanless and Weller 1932). In Illinois, a typical cycle commences with alluvial channel sandstones; these are followed by an underclay, then coal, then marine shales containing limestone, and finally, but not always, a brackish or limnic shale. Each cycle lasts about 400,000 years, the approximate length of the long eccentricity cycle, and may be related to glacio-eustatic variations effected by the rhythmic growth and decay of Carboniferous ice sheets (Wanless and Shepard 1936; Crowell 1978). The cyclothems are best developed in Kansas, where they show a fourfold substructure, presumably related to the 100,000-year short cycle of eccentricity. It is interesting that in these sequences the shorter rhythms appear to have been suppressed in favour of the long eccentricity rhythm. This "red shift" is more noticeable than in Pleistocene deposits, and suggests that Carboniferous climates reacted with greater inertia than did Pleistocene climates (Fischer 1986, p. 371).

Cycles of sedimentation on marine platforms have come largely from studies of Triassic rocks in the Alps and Hungary (Schwartzacher 1975). A number of workers in these areas have found cycles suggestive of glacial control forced by cycles of precession and ellipticity (Goldhammer et al. 1987; Hardie et al. 1986; Fischer 1986). Tantalizingly, these sequences suggest a Croll-Milankovitchtimed eustatic cycle at a time when ice sheets were thought not to exist (Fischer 1986, p. 371) (see Sect. 3.4.2). Several sequences of pelagic and hemipelagic sediments have been investigated (e.g. Fischer et al. 1985; Arthur et al. 1986). A remarkable sequence is exposed in the Apennines of central Italy. Rhythmic sedimentation has been investigated in the Early to Middle Cretaceous portion of the sequence by Walther Schwartzacher and Alfred G. Fischer (1982; Fischer and Schwartzacher 1984), who found evidence indicative of precessional and short ellipticity cycle signals. Later work by Timothy D. Herbert and A.G. Fischer (1986) on carbonate production in pelagic mid-Cretaceous black shales in central Italy, quantified by calcium carbonate and optical densitometry time series, has made manifest the influence of the short eccentricity and precessional cycles: the deep sea was anoxic when both eccentricity and the precessional index were low, at which times black shales were deposited under conditions of lowered rates of exchange of deep-water to surface-water (see also Herbert et al. 1986). Rhythmic alternations of carbonate-richer and carbonate-poorer beds were also found in the Pliocene Trubi formation of Sicily, the beds here apparently following the precessional cycle (de Visser et al. 1989). Limestone-shale alternations in the Jurassic beds of the Dorset coast, England, were attributed by Michael R. House (1985) to the obliquity cycle, while those in the French Fosse Vocontienne and the Maritime Alps were linked, via the supply of detrital material, to precessional or obliquity cycles (Fischer 1986). Evidence of the obliquity and precessional cycles has been unveiled in the English chalk, too (M.G. Hart 1987).

Aeolian components of more recent pelagic sediments reveal indirectly that changes occurring on land are locked into orbital variations. T.R. Janecek and D.K. Rea (1984), using mass accumulation rate and grain size of the total aeolian component isolated from pelagic clay from two North Pacific cores, one under the westerlies and one under the tradewinds, gauged changes in intensity of the atmospheric circulation over the last 700,000 years. Aeolian gain size, which is a good indicator of wind intensity, fluctuates at periodicities roughly the same as the Croll-Milankovitch variables. This lends support to the view that the intensity of the atmospheric circulation responds to changing solar insolation receipt resulting from changes in the Earth's orbital parameters. Warren L. Prell (1984), by studying planktonic formaminifers in the Arabian Sea, showed that the strength of the Southwest Indian Monsoon has varied greatly during the late Pleistogene period. Relatively warm sea-surface temperatures, recorded by oxygen isotope ratios, are taken to represent a decrease in coastal upwelling in the Sea owing to weaker, low-level winds during the summer monsoon. The 164,000-year record of monsoonal upwelling studied is consistent with changes in solar radiation induced by precessional forcing. Interestingly, there is a lag between changes in summer radiation and changes in upwelling, suggesting to Prell that other processes, such as seasonal snow cover, modulate the direct response of the monsoons to solar heating.

A number of tentative conclusions may be drawn from studies of cyclical sedimentation in pelagic and hemipelagic systems. In general, bundles of bedding couplets (e.g. carbonate-richer and carbonate-poorer beds) are distinct and mirror the precessional index. Additionally, some sequences are dominated by the simple obliquity cycle. Deposits of the same age march to the precessional beat in some areas, chiefly in low latitudes, but to the obliquity beat in other areas, mainly in high-latitude sequences. The situation is not always clear-cut, however, some sequences marching to a mixture of both beats, and others displaying an irregular succession of beds (cf. Fischer 1986).

The pervasiveness of basin-wide, shallowing-upwards cycles of sedimentation in units 1 to 5 m thick bounded by surfaces produced by geologically instantaneous base-level rises has led Peter W. Goodwin and E.J. Anderson (1985) to submit "punctuated aggradational cycles" as a new general hypothesis of episodic stratigraphic accumulation. Glacial eustasy driven by astronomical forcings in the Croll-Milankovitch frequency band is pinpointed as the most likely root cause of punctuated aggradational cycles. By superimposing 20,000-year and 100,000-year cyclicities on varying amounts of continuous subsidence, Goodwin and Anderson showed how total local base level, which equates with the amount of room available for stratigraphical accumulation, increases. They found that base level rises by steps, rapid rises being followed by periods of stability (Fig. 4.9). With high subsidence rates, the stepwise creation of stratigraphical room would allow five punctuated aggradational cycles per 100,000-year cycle of ellipticity; in regions of low subsidence, just one or two punctuated aggradational cycles might be preserved owing to non-deposition or erosion (or both) which would occur at times of sea-level fall during the second half of the 100,000-year cycle.

A word of caution is in order at this juncture. Thomas J. Algeo and Bruce H. Wilkinson (1988) entreat us to beware reading too much into the calculated periods of sedimentary cycles. By analysing data on more than 200 mesoscale sedimentary cycles in Phanerozoic rocks, they show that cycle periods are randomly distributed with respect to the four major Croll-Milankovitch orbital parameters, save for the cycles which ran their courses in late Mississippian through to late Pennsylvanian times and show a positive clustering around the long eccentricity period. Cycle period is largely a function of average cycle thickness and periods in the Croll-Milankovitch frequency band will tend to be extracted for thicknesses of between 1 and 20 m regardless of the actual cause of the cyclicity. Algeo and Wilkinson also warn us that, on average, deposition occurs for just one-thirtieth of the average period of a cycle: long, unconstrained intervals of non-deposition predominate in most cratonic and continental-margin cyclical sequences. They do not deny that sedimentary cycles may bespeak astronomical forcing of climate; they merely point out the pitfalls which arise in establishing periodicity in some sedimentary environments, and that autocyclic mechanisms might also generate cycles with periods in the Croll-Milankovitch frequency band. The best place to look for unequivocal signs of orbital forcing, they advise, is tectonically stable and climatically sensitive lacustrine and barred basinal systems in which varves are preserved and provide an in-built geochronometer. It is to such sedimentary environments that discussion will now move.

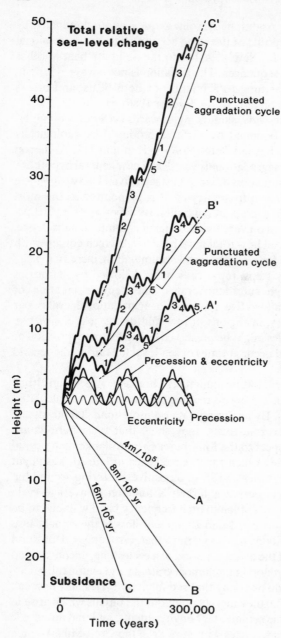

Fig. 4.9. A general model for episodic stratigraphic accumulation created by eustatic responses to orbital perturbations in the Croll-Milankovitch frequency band superimposed upon continuous subsidence. The precessional cycle and short eccentricity cycle are represented by sine waves and are summed in a third curve representing the general form of eustatic response to the selected simple pattern of orbital variations. This total eustatic response is then added to each of the three subsidence curves (A,B, and C) to give three sample patterns of total relative sea-level rise (lines A', B', and C'). The resulting patterns could produce punctuated aggradation cycles and sequences. The sequences might be truncated by erosion in regions of low subsidence (line A') or be more complete in areas of high subsidence (line C'). (Goodwin and Anderson 1985)

4.4.3 Lake and Evaporite Deposits

The record of astronomical forcing in sediments is probably best expressed in lacustrine sediments, which register variations in lake level, and in evaporite sediments, which record variations in salinity. These are essentially closed systems in which varving provides a time base. Studies of long varve sequences in lakes have revealed strong pulses in the Croll-Milankovitch frequency bands — strong precessional signals grouped into 100,000-year bundles corresponding to the short ellipticity cycle. In the Eocene Green River Formation (a lake-playa complex which filled synorogenic intramontane basins in the middle Rocky Mountains in Colorado, Utah, and Wyoming), 2 to 3 m couplets of varved and kerogen-rich carbonate beds (oil shales) and marlstone betoken oscillations of lake levels. Extrapolating annual varve thicknesses within these couplets, a mean rhythm of 21,630 years is obtained which may be identified with the precessional cycle (Bradley 1929, 1931). In other parts of the Green River Formation, the couplets seem bundled into groups of three to six, which correspond to the short eccentricity cycle. These bundles are themselves bunched into 400,000-year superbundles, which fit tidily into the long ellipticity cycle.

Another example is afforded by cycles of sedimentation in the Newark rift valley during Triassic times, produced by climatically controlled rise and fall of the Lake Lockatong complex, first described by F.B. van Houten (1964), and re-examined by Paul E. Olsen (1984, 1986). The lakes fluctuated between expanses of water over 7000 km^2 in extent and in excess of 100 m deep, and restricted playas or completely dry lake basins. Annual laminations (varves) show that short sedimentary cycles involving about 5 m of sediment were produced by lakes whose depth oscillated with a period of about 21,800 years. Longer cycles, involving about 25 m and 100 m of sediment, echo periodic changes in the magnitude of the 21,800-year lake-level cycles. They are characteristically 101,000, 400,000, and 418,000 years in length. During the Triassic period the Newark Basin lay at palaeolatitude 15° N, or thereabouts. At that latitude, solar insolation changes should reflect the precession and eccentricity changes, but not changes of tilt. This theory seems to hold, for there is no sign of a 41,000-year obliquity cycle in the Lockatong sediments. In another study, Roger Y. Anderson (1984) found that, in the varved Castile evaporites formed at the close of the Permian period in the Delaware Basin of western Texas and southeastern New Mexico, the start and finish of deposition, the trends in annual thickness, and the symmetry of response are consistent with control by a 100,000-year eccentricity cycle. Nearly symmetrical changes in the rate of calcium sulphate deposition in the same deposits accord with a cycle of precessional changes of around 20,000 years. There is no apparent sign of forcing by the tilt cycle, but, as with the Lockatong sediments, that may be due to the low palaeolatitude of the area. Though limited in number, studies of medium-term rhythms in lake varves prompt two important conclusions (Fischer 1986, p. 363): the precessional and eccentricity cycles have existed for at least 200 million years; and during that time, the cycles have stuck to the same relative and absolute timing (but see A. Berger et al. 1989).

4.5 Long-Term Changes in Sediments

The long-term history of weathering, denudation, and deposition is intimately linked with the changing states of the atmosphere and oceans. In turn, atmospheric and oceanic conditions have been greatly influenced by the development of life, and in particular by biological innovations such as oxygen-liberating photosynthesis, various chemolithotrophic activities of bacteria (nitrogen fixation, sulphate reduction, and the like), the rise of metazoans, the invention of burrowing, the evolution of calcareous and siliceous skeletons, the colonization of lands, the appearance of calcareous skeletons in plankton, the appearance of siliceous skeletons in the phytoplankton, the invention of flowers, and the rise of Man (Fischer 1965, 1972, 1984b). Indeed, if the Gaia hypothesis be accepted, then since life first appeared on Earth it has actively regulated the state of the air, ocean waters, and surficial sediments (Lovelock 1972, 1979, 1988, 1989; Lovelock and Margulis 1974a,b). Even if the Gaia hypothesis be disallowed, there is little doubt that many sedimentary cycles are effectively controlled by biological processes (e.g. Schidlowski 1987, 1988), and great strides are being made in understanding the dynamical stability of the atmosphere and oceans as maintained by biogeochemical cycles over billions of years (e.g. Kump 1988a,b, 1989; Kump and Garrels 1986). Conditions at the land surface, too, must have altered in response to the changing state of the atmosphere and, later, to the various stages in the colonization of the land (Ollier 1979, 1981). A full account of the evolution of weathering and denudation regimes and sedimentary environments would be out of place here. All that will be presented are two topics: the value of palaeosols in climatic reconstruction; and the enigma of glacial sediments at low latitudes.

4.5.1 Palaeoclimatic Significance of Palaeosols

Some features of sediments and soils reflect the state of the atmosphere at the time they were formed and may thus be used in reconstructing palaeoclimates. It is impossible here to enter into a lengthy discussion of sediments and soils as indicators of past climates. To give the flavour of ideas in this field, the following phenomena and their relations with climate will be briefly considered: duricrusts, carbon isotopes, and major oxides in palaeosols.

It is a common and generally acceptable practice to take duricrusts as strongly suggestive of past water regimes. Calcretes are presently formed chiefly in semi-arid areas where the annual rainfall is about 200 to 500 mm (Watkins 1967; Goudie 1985). It is not unreasonable, therefore, to use their presence in ancient sediments to infer desert-like conditions with an annual water deficit at the time of their formation (e.g. V.P. Wright 1990). Ferricretes and alcretes are widely accepted as being products of humid climates, ferricretes requiring alternations of wet and dry seasons. There is, however, some evidence that ferricrete may form under permanently moist tropical climates (McFarlane 1976, p. 45). Less clear are the climatic conditions under which silcretes form. Polar views on the matter are represented by Erich Kaiser (1926), who calls for extremely arid climates, and K.W. Whitehouse (1940), who invokes humid tropical conditions. Michael A. Summerfield (1983) opines that silcretes may actually form under two distinct climatic regimes, distinguishing between "non-weathering profile" silcretes produced by the localized mobility and high concentration of silica in more alkaline weathering environments under a mainly arid to semi-arid climate, and "weathering profile" silcretes, which are created under a humid climate in a highly acidic weathering environment with poor drainage. Thus, while there is some justification for using duricrusts as broad indicators of past climatic regimes, it is advisable to be cognizant of the problems entailed in doing so.

Carbon isotopes from soil carbonates and soil organic matter provide information on palaeoclimatology and palaeoecology because the carbon in soil carbonate forms in isotopic equilibrium with carbon dioxide in the soil, the isotopic composition of which is determined by local plant cover (Magaritz et al. 1981; Cerling 1984; Cerling and Hey 1986; Quade et al. 1989b), inheritance from parent material being insignificant (Cerling 1984; Quade et al. 1989a). The carbon isotopic composition is determined by the proportion of C_3 to C_4 plants in the local biomass. In general, C_3 plants (which include nearly all trees irrespective of climate, nearly all shrubs and herbs, and grasses which favour a cool growing season) have an average carbon isotope ratio, $\delta^{13}C$, of -27 permille with a range of -35 to -20 permille depending on genus, plant longevity, moisture stress, light intensity, and other factors (Ehleringer 1989). C_4 plants (which include a few shrubs in the families Euphorbiaceae and Chenopodiaceae and grasses favouring a warm growing season) have an average carbon isotope ratio of -13 permille. Soil carbonate formed in the presence of pure C_3 biomass at 25 °C would have a carbon isotope ratio of about -12 permille; soil carbonate formed in the presence of pure C_4 biomass at 25 °C would have a carbon isotope ratio of about +2 permille (Quade et al. 1989b).

As an example of inferring palaeoclimatology and palaeoecology from carbon isotopes in soils, we shall mention the study made by Jay Quade, Thure E. Cerling, and John R. Bowman on the Siwalik Group sediments in the Potwar Plateau region of northern Pakistan (Quade et al. 1989b). These sediments, fine-grained floodplain facies of large river systems, contain a well-exposed record of palaeosols spanning the last 18 million years. Calcium carbonate nodules were taken from the calcic palaeosol horizons, roughly one palaeosol being sampled, as it were, every 130,000 years. The carbon isotope ratios in the nodules, when plotted against time, display three distinct phases (Fig. 4.10a). From 18 to about 7.2 million years ago, the carbon isotope ratios fall in the range -13 to -9 permille. These values suggest that a pure, or almost pure, C_3 biomass, probably a mosaic of closed canopy forest and grassland, dominated floodplains during that period. From about 7.2 to 5 million years ago, during the late Miocene epoch, the carbon isotope ratios shifted towards more positive values, marking the gradual expansion of C_4 plants (essentially grassland) on the floodplain. By 5 million years ago, C_4 grasses constituted more than 90 percent of the biomass. From about 5 to 0.4 million years ago the carbon isotope ratios stayed within the range -2 to +2 permille, indicating that the C_4 grasslands remained dominant

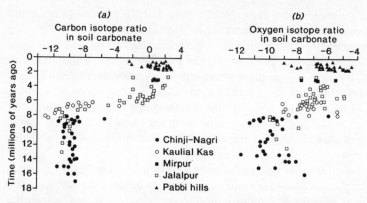

Fig. 4.10. The Siwalik sequence, northern Pakistan. **a** The carbon isotope ratio in palaeosol carbonate nodules plotted against age. The negative values before 7.4 million years ago indicate that a C_3 biomass dominated the floodplain, whereas the more positive values in the Plio-Pleistocene palaeosols are suggestive of C_4 grasslands being dominant. **b** The oxygen isotope ratio in palaeosol carbonate nodules. Oxygen isotopes ratios shift towards more positive values in the late Miocene epoch, indicating a change in the average isotopic composition of soil water, probably owing to regional climatic change (Quade et al. 1989)

throughout that time. The oxygen isotope ratios for the same palaeosols also shift towards more positive values in the late Miocene epoch (Fig. 4.10b). The ecological shift in late Miocene times might have occurred for one of two reasons: either the climate might have changed, allowing the invasion of C_4 grasses from outside the region; or it might have resulted from the evolution of C_4 plants, and thus a floral invasion unrelated to climate. In every section studied, the oxygen isotope values shift before the carbon isotope ratios. As the oxygen isotope value of soil carbonates has been shown to relate to the isotopic content of local rainfall (Cerling 1984), which in turn depends on temperature, the isotopic composition of regional rainfall, the seasonality of precipitation, and other factors, it is tempting to opt for the climatic, rather than the evolutionary, explanation. To be sure, a climatic explanation accords with other evidence. For instance, a number of soil features change after about 7 million years ago: depth of leaching decreases; the colour of the leached zone changes from dominantly strong reds and oranges to yellower hues; humic soil horizons, possibly owing to the appearance of grassland, become commoner. A major faunal turnover also occurred 7.4 million years ago (see Sect. 7.4.1). As to the cause of the climatic change, Jay Quade and his colleagues (1989b) identify the appearance (or strong intensification) of a monsoon system in the region between about 7.4 and 7.0 million years ago. The modern monsoonal climate, with its moderate rainfall, strongly seasonal rainfall distribution, as well as the prevalence of natural fires, favours grasslands over forests, and its seems fair to assume that the arrival of a strong monsoonal system in the late Miocene epoch would have enabled grasslands to oust the extensive patches of closed canopy forests.

Palaeozoic and Precambrian palaeosols have recently been used to assess the composition of ancient atmospheres (Holland and Zbinden 1988). An investigation of a palaeosol developed in basaltic rock and exposed along the north bank of the Sturgeon River, some 25 km southwest of Baraga, Michigan, has shown that free oxygen must have been present in the atmosphere 1.1 billion years ago, though with a partial pressure possibly as much as 1000 times lower than at present (Zbinden et al. 1988). The Flin Flon palaeosol, formed 1.8 billion years ago in the Amisk Group volcanics, shows many characteristic features of modern well-drained soils: corestones of spheroidally weathered pillow lavas occur at the base of the C horizon, and decrease in size upwards eventually to disappear in the hematite-rich horizon at the top of the palaeosol; oxides of calcium and magnesium and ferrous iron decrease up the profile while oxides of aluminium and titanium increase in the same direction (Holland et al. 1989). The ferrous iron was apparently oxidized to ferric iron and held within the palaeosol profile during weathering, indicating the presence of oxygen in the atmosphere, albeit at a much lower partial pressure than at present. In the 2.2 billion year-old Hekpoort palaeosol of the Transvaal, South Africa, the profile is depleted in iron (Retallack 1986). The difference between the Flin Flon and Hekpoort palaeosols suggests that the oxygen content of the atmosphere rose appreciably between 2.2 and 1.8 billion years ago. Indeed, all palaeosols formed after about 2.0 billion years ago indicate that the atmosphere contained sufficient oxygen to oxidize all the iron in soils developed on igneous rocks. For instance, palaeosols formed in late Ordovician andesite flows exposed along the coastline near Arisaig, Nova Scotia, display chemical variations similar to modern soils developed on mafic volcanic rocks and contain virtually all the iron which had been oxidized during weathering (Feakes et al. 1989). In two of the three palaeosols, a small amount of ferrous iron escaped oxidation and was precipitated at the base of the palaeosol profiles, perhaps owing to the activity of non-vascular plants.

Lithified palaeosols have been identified in Mesozoic strata. Fifteen of them, occurring in the top 90 m of the late Albian Boulder Creek Formation in the foothills of northeastern British Columbia, Canada, were studied by Dale Leckie, Catherine Fox, and Charles Tarnocai (1989). The palaeosols have well-developed profiles some 0.5 to 1.5 m thick and include A, B, and C horizons. They developed in unconsolidated floodplain alluvium during a time when one or more basin-wide unconformities resulted from eustatic lowering of sea level or local tectonic events. Careful analysis of the palaeosols enabled some aspects of the environment in which they formed to be reconstructed. Overall, it would seem that the environment in which the soils developed had low relief, was subject to little erosion, and enjoyed a humid to subhumid climate. The water table was high for part of the year, which could have arisen from either a low-lying topographic position or else a highland setting with a perched water table, and did not necessarily betoken high rainfall. Pointers to the prevalence of a humid to subhumid climate include clay translocation features, complete alteration and removal of weatherable minerals such as biotite, and the absence of caliche, calcrete, and calcareous horizons. There is some evidence of periodic drying when dessication cracks opened up, later to be filled with other sediments, and organic debris was oxidized.

The study of very old palaeosols is in its infancy and not without its teething problems, the difficulty of establishing whether certain parts of stratigraphical sections represent past soils being a prominent one (e.g. Leckie et al. 1989; Holland and Feakes 1989; Palmer et al. 1989); but it is shaping up as a fruitful field of research, as the case studies brought together under the editorship of V.P. Wright (1986) and J. Reinhardt and W.R. Sigleo (1988), and the general review penned by Greg J. Retallack (1990), show.

4.5.2 Glacial Deposits in the Tropics

The occurrence of ice sheets and periglacial landscapes near sea level in tropical latitudes during late Precambrian times poses a big problem for palaeoclimatologists. One possible explanation would be that, owing to much reduced solar output, the entire globe was gripped by a very severe glaciation. Much evidence invalidates this idea, not the least of which is the lack of glacial deposits on land which occupied high latitudes at the time. At least two other explanations come to mind (Embleton and Williams 1986): the axial geocentric dipole model of the Earth's magnetic field is invalid for the late Precambrian era; and the obliquity of the ecliptic was greatly increased during late Precambrian times, the Earth's rotatory axis standing at more than 54°. As there is little reason for supposing that the axial geocentric dipole model be wrong, we are left with the possibility that the obliquity has changed by large degrees. George E. Williams (1975b, 1981b, 1986) believed that this explanation is consistent with the puzzling features of the late Precambrian glaciation: it appears to have occurred mainly in low to middle latitudes; it shows strong seasonal changes, even at low latitudes; and it involves a curious mixture of glacial rocks (tillites) with rocks normally indicative of warm conditions such as dolostones, evaporites, and banded iron formations. All these enigmatic features would make sense if the tilt of the Earth were increased to 54° at the time of the glaciation, some 750 million years ago, since low latitudes would receive less radiation than high latitudes, seasonal contrasts would be very intense, and climatic zonation would be much weakened (see also Kröner 1977; Embleton and Williams 1986). Similarly, Williams (1981b) invoked increased tilt to explain some rather odd features of the late Ordovician glaciation as recorded by Rhodes W. Fairbridge (1970, 1971) in northwest Africa: sand is the predominant sediment in the rocks; there are vast outwash sheets of sand covering thousands of square kilometres with sandstone units up to 20 m thick and hundreds of kilometres in extent showing evidence of very high velocity currents which might have resulted from "a catastrophic decantation of meltwaters" (Fairbridge 1970, p. 878); long, parallel grooves are cut into outwash sandstones and extend for hundreds of kilometers; there is evidence that loose sands were frozen into temporary "bedrock" upon which tills were deposited; the basement of Precambrian rocks is weathered and bleached to depths of 3 to 4 m, and capped by a residual hematitic crust or fossil soil; and finally, traces of trilobites and other marine life are present through almost the entire sequence of rocks. These features, including the consociation of glacial and tropical deposits, may be explained as resulting from a much increased obliquity: in winter, unconsolidated sands would have frozen and, with marine sediments, been overridden by advancing ice. In summer, rapid melting of ice sheets would have led to catastrophic floods, marine life would have spread far into polar waters, intense weathering would have produced sand in abundance and affected the underlying rocks, and large blocks of ice might have calved from the melting ice sheets and have slid down the outwash plains to have scored the frozen sands for many kilometres.

5 Landforms and Soils

We shall arrive at a clearer understanding of the processes of the formation of soil if we take *Climate* as our starting point.

Emil Ramann (1928, p. 2)

Any one who has observed land forms produced under widely different climatic conditions can hardly fail to note significant differences in the landscapes and a certain degree of harmony between climate and landscapes, particularly with respect to the lesser topographic features.

William D. Thornbury (1954, p. 54)

5.1 Soils and Climate

5.1.1 The Climatic Factor in Soil Formation

In 1786, Horace Bénédict de Saussure wrote in his *Voyages dans les Alpes* (Vol. iv, p. 208) that climatic factors are responsible for the existence of differing organic matter levels in soils. A fuller appreciation of the climatic factor in soil formation has been attributed to Johann Christian Hundeshagen who, in his *Forstliche Berichte und Miscellen* (1830–32), mentioned that the climatic forces — water, heat, light, and oxygen — must act in unison to create a productive soil (Joffe 1949, p. 12). G.C.L. Krause, in his *Bodenkunde und Klassifikation des Bodens* (1832), correlated soil productivity with climatic factors. The importance of climate in understanding rock weathering and soil formation was clearly appreciated by Karl Sprengel. Writing in his *Die Bodenkunde oder Lehre vom Boden, nebst einer vollständigen Anleitung zur chemischen Analyse der Ackererde* (1844), Sprengel explained how native rocks are decomposed by water, oxygen, carbon dioxide of the air, heat and cold, vegetation, and electricity. He realized as well that soil properties vary under different climates: warm climates produce soils better suited to plant growth than do cold climates because warmer environments favour a greater production of ammonia and nitrates in the process of organic decay; all organic bodies decay more slowly in cold climates than in warm climates and so manure remains in the soils of cold environments much longer; and although cold climate soils contain more organic matter, they yield less well than warm climate soils, chiefly owing to the relative brevity of the growing season.

Despite all these advances in the appreciation of soil formation and its relation to climate, it was not until the second half of the nineteenth century that the nature of soil as an independent body was acknowledged. The first to advocate that soil is a formation in its own right, and not a mere rock formation, was Friedrich Albert Fallou (1855, 1862, 1875). But it took many more decades for this germ of an idea to grow to fruition. Ferdinand Paul Wilhelm, Freiherr von Richthofen, was a follower of Fallou but stuck to a "geological" view of the nature of soils. He journeyed widely and was one of the first to lay down the geographical concept of the distribution of soils: he saw that rock decomposition and neoformation was correlated with the environment, especially with climate (Richthofen 1886). As late as 1918, Emil Ramann was still clinging to the "geological" view of soils, even though he used climatic factors as the basis of his soil classification. A change of emphasis did not emerge until the genetic school of pedology identified the facts and factors of soil formation as distinct from weathering. The role of climate as a soil-forming factor was recognized independently in the United States and Russia in the last decade on the nineteenth century. In America, Eugene Woldemar Hilgard carried out extensive studies of soils which led him to appreciate that different soils tend to be associated with different environmental conditions. In his monograph on *The Relation of Soils to Climate* he wrote of a more or less intimate relation between the soils of a region and the prevailing climatic conditions (Hilgard 1892), and in his book *Soils* (1906) he recorded the tendency of climate materially to influence the character of soils formed from the same rocks. Hilgard's views were based on a large body of data he had collected on the acid-soluble constituents of soils in the arid and humid subtropical lands of the southern United States. Taking hundreds of samples, he found that the mean percentage of total soluble material in arid soils was 30.84, almost twice as much as in humid soils where the figure was 15.83 percent. These findings prompted to him to propose a lixivation or leaching theory which attributed the relative richness of acid-soluble constituents in arid soils to the lack of leaching under an arid climate.

In Russia, Vasilii Vasielevich Dokuchaev, a geologist by training, came to view soil as an independent body after having surveyed vast stretches of the chernozems underlying the Russian steppes. In 1879, he expressed the view that soil is the surface mineral and organic formations produced by the combined activity of animals and plants, parent material, climate, and relief. Here was the first explicit statement of the factors of soil formation (see Joffe 1949, p. 17). From it followed the principle of geographical soil types: each soil formation is associated with a definite climatic belt, unlike the underlying rock formations which are unrelated to climate. Dokuchaev's pioneering thesis was taken up by his collaborator and disciple, Nikolai Mikhailovich Sibirtsev. The achievement of Sibirtsev was to integrate Dokuchaev's factors of soil formation and establish their differential role in soil development. For Sibirtsev, moisture was the primary climatic factor in soil formation, and advanced the idea that the zonal distribution of soils is based on climatic zonality (see Joffe 1949, pp. 18–19). Dokuchaev's ideas were carried into Germany and the United States through the writings of Konstantin Dimitrievich Glinka, especially his *Die Typen der Bodenbildung* (1914, 1927) and a lecture delivered at the First International Congress of Soil Science (1928) which looked at the close connection between climate and soils in Russia. Emil Ramann, in his *Bodenkunde* (1914), considered the same intimate connection in the European setting. In North America, Curtis Fletcher Marbut was perhaps the most prominent pedologist in the early decades of the twentieth century. Greatly impressed by Glinka's pronouncements, Marbut was instrumental in spreading the word about soil-forming factors in the United States. His soil classification, sketched out in 1927 and published in the *Atlas of American Agriculture* (1935), included a major (albeit misleading) distinction between "normal" soils in which calcium carbonates accumulate (pedocals) and "normal" soils in which aluminium and iron accumulate (pedalfers). The credit for outlining the relations between soil and climate on a global scale must be given to Richard Lang (1920), although a cursory treatment had been offered by Hermann L.F. Meyer (1916). According to Lang, climatic soil belts, each circumscribed by definite numerical values of his "rain factor", succeed one another from pole to equator.

5.1.2 Soil Types and Climate

An outcome of all the early work on soil-climate relations was a number of climatic classifications of soils (see Clayden 1982). Successful though these systems were, they suffered from using ratios of annual precipitation to temperature, evaporation, or humidity, or simply annual precipitation and temperature indices, which can be empirically related to soil patterns but which cannot be related directly to soil processes. The direct relationship between climate and processes which operate within the soil system was not investigated until C. Warren Thornthwaite developed his water-balance approach to climatic classification. Russian researchers attempted to relate soil types to the radiation balance and water balance of the Earth's surface. Mikhail I. Budyko (1974), for instance, found that the world's major soil types are closely correlated with the radiative index of dryness and the net radiation available at the surface. As the radiative index of dryness increases, so the following sequence of soils occurs: in the tundra zone, tundra soils; in the forest zone, podzols, brown forest soils, yellow Earths, red earths, and lateritic soils, depending on the value of net radiation, R; in the steppe zone, black earths (chernozems) and black savanna soils, depending on the value of net radiation, R; and in the semi-desert, chestnut soils; and in the desert, grey earths (sierozems).

The first person to inquire into the relations between the water balance and soil processes as a possible guide to the zonal soils one would expect to find under different climates was Rodney J. Arkley (Arkley 1963a,b, 1967; Arkley and Ulrich 1962). In an investigation of the distribution of soils in the western United States in relation to climatic parameters, Arkley (1967) undertook water-balance analyses of more than 1000 climatic stations. From the climatic records, he computed the leaching effectiveness of the climate, Li (Sect. 3.1.1), and extracted the actual evapotranspiration and the mean annual temperature. He then plotted the great soils groups associated with each climatic station against the three climatic variables. This exercise showed that some great soils groups — red soils, desert soils, reddish brown soils, and reddish chestnut soils, for instance — have distinct limits set by climatic factors (Fig. 5.1). There is some overlapping of great soil groups owing to the influence of other factors on soil distribution, but the broad relations between soil type and climate are abundantly clear. Arkley drew several general conclusions about the relations between leaching effectiveness and soil type and actual evapotranspiration and soil type. The amount of water available for leaching has a large influence upon soil formation. Podzols, brown podzolic soils, grey-brown podzolic soils, sols bruns

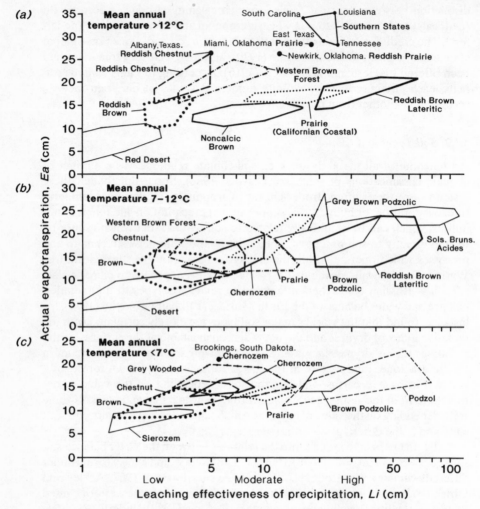

Fig. 5.1. The climatic range of great soil groups in the United States. **a** Warm regions. **b** Temperate regions **c** Cold regions. (After Arkley 1967)

acides, and reddish brown lateritic soils, all of which have low base saturation or strong acid reaction or both, are found where the leaching effectiveness of the climate exceeds 46 cm, or in cold climates, or where local conditions such as coarse texture enhance leaching and the leaching effectiveness exceeds 30 cm. Prairie soils, non-calcic brown soils, brown forest soils, and grey wooded soils, which are generally leached free of carbonates and retain a fair amount of exchangeable bases within the solum, are associated with moderate leaching (a leaching effectiveness of 15 to 40 cm). Red desert soils, desert soils, sierozems, reddish chestnut soils, chestnut soils, chernozems, and parts of the western brown forest and grey wooded soils, all of which retain bases in all or a large part of the solum, occur where leaching is slight (leaching effectiveness less than 15 cm) and no water can be expected to penetrate below the rooting depth of ordinary plants. Soil formation also depends upon actual evapotranspiration rates. Red desert soils, desert soils, and sierozems, all low in organic matter, are found in regions of low actual evapotranspiration (6Ea less than 25 cm). Reddish brown soils, non-calcic brown soils, and brown soils, also generally low in organic matter, occur in cooler climates where actual evapotranspiration is moderate (6Ea lies in the range 25 to 38 cm). All soils formed under climates where actual evapotranspiration is high (6Ea in excess of 33 to 38 cm) tend to be rich in organic matter, save for some reddish brown lateritic soils where high leaching, intense organic matter decomposition, and low fertility lead to low organic matter content.

Arkley's work, and similar work carried out by others (e.g. Fränzle 1965), provides some justification for including climatic factors in classifications of soils. The soil taxonomy of the United States recognizes ten soil orders (Soil Survey Staff 1975). Nine of the ten orders developed during the Holocene epoch and may be interpreted as stages in a developmental sequence (Ruhe 1983), while the remaining order, the Ultisols, is essentially a relict of pre-Holocene times. Climate will have surely made an impression on soils in each of the orders. It seems acceptable, therefore, that the Soil Survey Staff introduced soil temperature and soil moisture regimes at the level of the suborder. But there is a danger here and with all other schemes which portray major soil types neatly delimited by climatic boundaries: climatic parameters built into soil-classification systems are derived from data collected over the last 40 to 80 years which represent less than 1 percent of Holocene time (Ruhe 1983). As the soil groups formed during, or in one case, before, the Holocene epoch, it is perilous to read too much into the present coincidence between climate and soil types.

5.1.3 Soil Processes and Climate

Relating soil processes to climate, especially fast-acting processes, should be less besets with problems than is relating soil type to climate. Certainly, Hans Jenny's classic work on the *Factors of Soil Formation* (1941) and its update *The Soil Resource* (1980) make good reading and seem to establish beyond a shadow of a doubt the efficacy of climate as a determinant of soil processes. Jenny's purpose was to see how the soil system is influenced by a single climatic factor, to establish univariant "climofunctions". In fact, relatively few climofunctions have been established over the last 50 years or so. Most of them relate soil properties to mean annual temperature or mean annual rainfall, or to some measure of effective rainfall. For instance, the dependency of total organic nitrogen content of loamy surface soils (0 to 20 cm) of former grasslands of the Great Plains of North America upon moisture and temperature gradients was established by Jenny (1941, p. 171). Soil nitrogen was found to relate to moisture, as measured by Meyer's *NS* quotient, and mean annual temperature in the following way:

$$N = 0.55e^{-0.08T} \left(1 - e^{-0.005m}\right),$$

where N is nitrogen content of surface soil (percent), T is mean annual temperature (°C), and m is the NS quotient. The soil nitrogen "surface" described by this equation is depicted in Fig. 5.2. For a fixed temperature, soil nitrogen content increases along gradients of increasing humidity from deserts near the Mexican border to more humid regions near the Canadian border. Notice that the rate of increase is steeper at lower mean annual temperatures, and that the soil nitrogen content of soils in the northern part of the Great Plains is greater than the nitrogen content in the southern part. The increase is exponential: nitrogen contents rise faster per unit increase in moisture as the Canadian border is approached. For a fixed moisture level, soil nitrogen content decreases from north to south. The decline is exponential, the nitrogen level falling a bit faster in northern regions than in southern regions. Jenny (1935) also derived a clay-climate "surface" for soils derived from a mixture of granites and gneisses in the eastern United States (Fig. 5.3), the equation for which is

$$c = 0.0114m \bullet e^{0.0140T},$$

where c is average clay content to a depth of 40 inches (percent), T is mean annual temperature (°C), and m is Meyer's NS quotient. Interesting features of this surface are the paucity of clays under conditions of limited moisture and high temperatures (deserts and semi-deserts) and under conditions of abundant moisture and low temperatures (Arctic and cold regions); at constant temperature, a general increase in clay content with increasing humidity, the rate of increase being strongly temperature-dependent (least in the north and greatest in the south); and, at constant moisture, an exponential increase in clay content, the rate of increase being very sensitive to temperature (least in arid regions and greatest in humid regions).

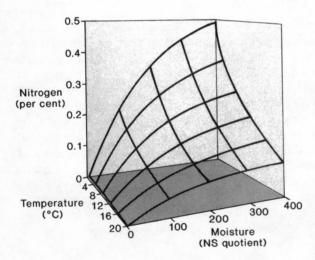

Fig. 5.2. An idealized nitrogen-climate surface for grassland soils of the Great Plains area of the United States. (After Jenny 1941)

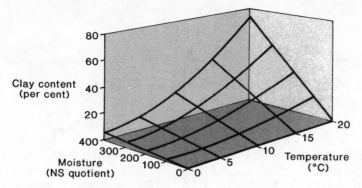

Fig. 5.3. An idealized clay-climate surface, showing the variation of "climatic clay" in soils derived from granites and gneisses as a function of moisture and temperature. (After Jenny 1935)

A very convincing climofunction relates the depth of carbonate accumulation in loessal soils formed on broad ridges to mean annual rainfall along a transect following the 11 °C isotherm running from the semi-arid parts of Colorado, through Kansas, to the humid areas of Missouri (Jenny and Leonard 1934). The data (Fig. 5.4a) are described by the following equation:

$$d = -30.5 + 2.5P,$$

where d is depth to the lime horizon (cm) and P is mean annual rainfall (cm). Rodney J. Arkley (1963b) established a similar climofunction for soils developed on alluvial terraces in Nevada and California. He found the relation (Fig. 5.4b):

$$d = 4.41 + 1.39P \ (n = 27; r = 0.761),$$

where the variables are as previously defined. Notice that both regression coefficients are smaller than they were in the case of the Great Plains soils. This suggests that the cold winter rains which occur in California are more effective displacers of calcium carbonate than are the warm summer showers in Colorado, Kansas, and Missouri (Jenny 1980, p. 326). More recent studies have successfully related soil properties to climatic variables. Giles M. Marion (1989) correlated the long-term rate of calcium carbonate formation with mean annual precipitation in desert soils of the North American Southwest. Peter W. Birkeland and his colleagues (1989) found that accumulation indices for pedogentically significant aluminium and iron, and a depletion index for phosphorus, calculated for soils in chronosequences from Baffin Island in the Canadian Arctic, the alpine Sierra Nevada and Wind River Range in the western United States, the alpine Khumbu Glacier area of Mount Everest in Nepal, and the alpine Southern Alps in New Zealand, when ranked according to the degree of soil development, provided a sequence related to climate. The greatest accumulation and depletion has occurred in the warmest and wettest environment, and the least in the coldest and driest environment.

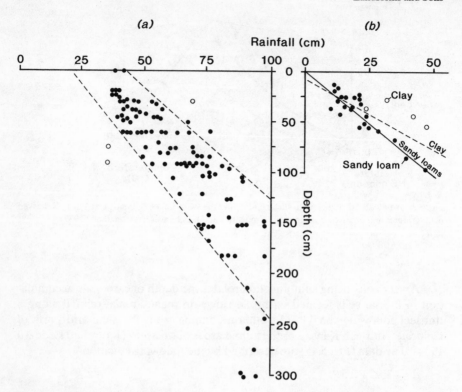

Fig. 5.4. a Depth to carbonate horizon or thickness of carbonate-free soil related to mean annual rainfall in soils of the Great Plains area, United States. (After Jenny and Leonard 1934). **b** Depth to carbonate horizon of alluvial terraces in Nevada and California. (After Jenny 1980)

The climofunctions described thus far apply to soils in areas large enough to incorporate gradients of climatic variables such as temperature and moisture. Where these gradients are caused by latitudinal and longitudinal changes of climate, then fairly large areas are required to pick up the climatic imprint in the soil. At a small scale, it becomes difficult to winnow the effect of climate from the effect of topography and soil history (e.g. Vreeken 1975a). Where the gradients are a response to altitudinal variations, much more local differences in soil properties can be attributed to climate. For instance, C.J. Chartres and C.F. Pain (1984) investigated three soils developed in volcanic ash at altitudes of 1040 m, 1720 m, and 2350 m in Papua New Guinea's Enga Province. The sample points were located within about 40 km of each other. With increasing altitude, greater proportions of silt-sized, unweathered and partially altered primary minerals, and increasing molar ratios of calcium, magnesium, sodium, and potassium to aluminium, were discerned. Chartres and Pain attributed these changes in soil properties to the decline in temperature, and associated changes in evaporation and leaching, with altitude.

There really are very few rigorously evaluated and reliable climosequences (Yaalon 1975; Huggett 1982; Catt 1988). The chief reason for this seems to lie in the difficulties of gauging the effect of climate on a soil property whilst holding all the other soil-forming factors constant. It is usually the case that many soil-forming factors are not truly independent variables (climate, for example, is affected by topography), that some factors are not amenable to quantification, and that the constancy of most of the factors is very difficult to establish. These difficulties are illustrated by the case of the carbonate-climate function (Catt 1988, p. 541). A prerequisite to establishing the relation between depth to carbonates and mean annual rainfall is that a number of sites can be located where rainfall has been consistently different but other climatic factors have been the same, and the soils at each site have been developing for the same length of time, in very similar parent materials, in similar geomorphological situations, and have all experienced the same changes of vegetation and the same disturbance by animals. Some of these prerequisites, such as geomorphological situation and vegetational history, can be met by field observations and laboratory work on the pollen content of the soils or, failing that, of nearby lakes. The similarity of the influence of animals at all the sites may, perhaps, reasonably be assumed. But other preconditions are more elusive. Except in exceptional cases such as soils formed on volcanic deposits, the age of a soil has to be assessed indirectly using stratigraphical evidence or using internal soil evidence. Both these lines of evidence for soil age are beset with problems (Vreeken 1975b). Decalcification is a slow process, its effect being measurable over thousands to hundreds of thousands of years. Over time intervals of that duration, annual rainfall is hardly likely to have held constant. If all sites had experienced the same climatic changes, this inconstancy of annual rainfall rate might not be too serious a drawback, but normally the sites will be in different regions, each of which will have had a different climatic history. In the light of these problems, it is not surprising that trustworthy climofunctions have been unforthcoming except in a few cases.

Recently, the validity of one of the classic climofunctions — the changing sequence of soil properties from eastern Iowa to western Kansas across the prairies in the midwestern United States (Fig. 5.5) — has been questioned by Robert V. Ruhe (1984) on the grounds that soil-landscape systems behave in a more complicated way than simple soil-climate relationships would have us believe. The soils are all Mollisols. They are divided at about longitude 96° to 97° W into Udolls, lying to the east, and Ustolls, lying to the west. The relationships between soils and climate are complicated by soil stratigraphy and climatal and vegetational change. The Udolls are formed in Peoria loess of late Pleistocene age and the maximum duration of weathering in them, as established by radiocarbon dating, is 14,000 years. This contrasts with the Udolls, formed in Bignell loess of Holocene age, where weathering can have proceeded for a maximum of 9000 years. The 5000-year head start by the Udolls directly affected the climatic impact on the soils. Before the Ustolls formed, the Udolls supported coniferous and then deciduous forests under a moist climate. The deposition of the Bignell loess in which the Ustolls formed appears to have taken place under

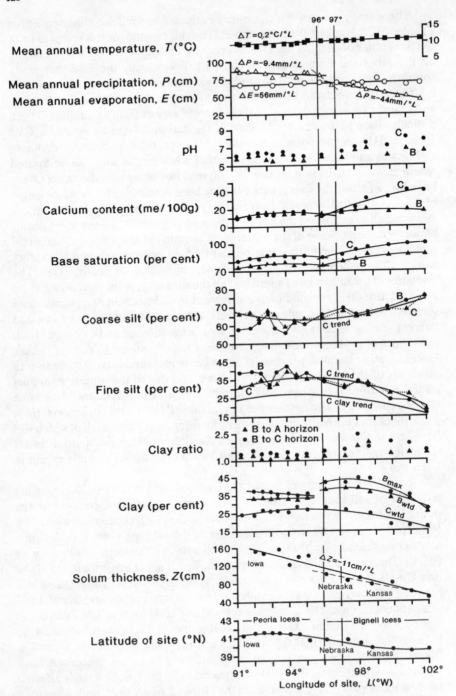

Fig. 5.5. Longitudinal distributions of soil properties and climate along a traverse across the prairies in the mid-western United States. (After Ruhe 1984)

a severe climate with hot, arid summers followed by cold, dry winters. For 5000 years, the Udolls and Ustolls formed under a climate some 40 to 50 percent drier than today, with moisture deficits increasing in magnitude and annual duration from east to west. The weathering record in the soils is chiefly contained in their base status, which varies inversely with age and climate. The base status of the Udolls is lower than that of the Ustolls and varies less with longitude. This is probably the result of 5000 years' extra weathering and two additional spells of water surplus. The base status of the Ustolls is higher than that of the Udolls and varies more with longitude. Indeed, the base status changes across the Udoll soil sequence may be a true climofunction developed during the Holocene epoch. But the major change in the loessal sediments and the chemical and physical properties of soils formed in them is a discordance at about longitude 96° to 97° W at the junction of the Peoria and Bignell loesses. This discordance is not climatically determined.

5.2 Landscape Morphometry and Climate

Two great methodological traditions of climatic geomorphology run throughout the mainstream literature on the subject: the application of the concept of landform zonality, based on climatological and ecological principles; and the "inductive definition of climate-process provinces based on assumed general relationships between the efficacy of selected geomorphic processes and standard climatic means" (Derbyshire 1976, p. 1). In this section, we shall peruse three variations on the morphoclimatic theme — morphological regions, morphometry and climate, and the origin of asymmetrical valleys — about all of which prosecutors from both great traditions of climatic geomorphology have had something to say.

5.2.1 Morphoclimatic Regions

A cardinal tenet of climatic geomorphology is that the "relief sphere" (Büdel 1982, p. 3), or configuration of the ground surface, is essentially a product of exogenic processes. That is not to deny that topography is produced by endogenic processes as well, but the "raw form" produced by endogenic processes is acted upon by exogenic processes to produce a distinctive "actual form" (Büdel 1982, pp. 18–19). The energy which powers the exogenic relief-forming processes comes from the Sun. Solar energy acts directly on the land surface, speeding up chemical and mechanical reactions in sediments and soils. It also powers the water cycle, the land portion of which involves exogenic processes. The key to understanding the spatial differences in Earth processes is the basic transfer of energy and matter which results from the inequality of radiation receipt between land, sea, and the atmosphere, and between the tropics and the poles. The circulation patterns so set up lead to spatial regularities in the world climate system which, when modified by the effect of relief, produce regional differences of climate. These contrasts between climatic zones lead to regional differences in the stores and fluxes within Earth surface systems. This is a fairly uncontentious point. A hotly debated matter is whether each exogenic process lead to a characteristic relief product, and what causes even more disquiet among geomorphologists is the suggestion that, as exogenic processes vary from one climatic type to another, it is possible to define areas (morphoclimatic zones) wherein relief development is governed by a particular set of exogenic processes.

The notion that a specific climate might produce a distinctive suite of landforms can be traced to Louis Agassiz and the birth of the glacial theory in 1840 (Derbyshire 1973). But climatic geomorphology really began with William Morris Davis's essays on the geographical cycle. In his works, Davis considered a normal (temperate) fluvial cycle (Davis 1899), an arid cycle (Davis 1905), and a glacial cycle (Davis 1900, 1906). His assumption was that distinct assemblages of geomorphological processes and landforms occur under different climates, an idea later ardently espoused by Carl Ortwin Sauer (1925). It is curious that, though Davis journeyed in tropical areas, he did not distinguish an equatorial or humid tropical landform assemblage (Stoddart 1969a, p. 161). However, Charles Andrew Cotton added a savanna cycle to Davis's list in 1942, and again in a much revised form in 1961. After Davis, climatic geomorphology was pursued with rigour in France and Germany, and it is in those countries that the subject is still treated in depth, English-speaking geomorphologists generally taking a somewhat dim view of it.

The term climatic geomorphology seems first to have been used by the Frenchman, Emmanuel de Martonne, in 1913. In his *Traité de Géomorphologie Physique* (1909), de Martonne discussed, as Davis had done, temperate, arid, and glacial cycles of erosion, but he undertook much field work in Brazil and in 1940 published a classic paper on humid tropical processes and landforms. Pierre Birot, a pupil of de Martonne's, worked in tropical and Mediterranean areas and published several statements on climatic geomorphology, including his *Essai sur quelques Problèmes de Morphologie Générale* (1949) and *The Cycle of Erosion in Different Climates* (1968). The stoutest and most prolific proponents of the French school of climatic geomorphology have been Jean Tricart and André Cailleux. Their *Traité de Géomorphologie* includes volumes on periglacial (1967), glacial (1962), arid (1969), and hot landforms (1965a), as well as a general introduction to climatic geomorphology (1965b, 1972). In Germany, the explorations of Ferdinand Paul Wilhelm, Freiherr von Richthofen, in China (1886) and of Siegfried Passarge (1926, 1973) and Franz Thorbecke (1927a, 1973) in Africa, fostered an early awareness of geomorphologists in the role of climate on landform development. Albrecht Penck, for instance, distinguished several landform assemblages associated with different climates (1905, 1910, 1914a,b, 1973). These early developments of climatic geomorphology in Germany led to a discussion meeting of the Düsseldorfer Geographentag on the *Morphologie der Klimazonen*, the proceedings of which were edited by Franz Thorbecke (1927b), and are a landmark in the development of the subject. Later, a periglacial cycle of erosion was suggested by Carl Troll (1944). Work continued in low latitudes, significant contributions coming from Otto Jessen (1936) in Africa and Karl Sapper (1935) in Middle America and Melanesia, and much later from

Herbert Louis (1957, 1973), Herbert Lehmann (1957), and Herbert Wilhelmy (1958). But the leader of the German school was undoubtedly Julius Büdel, who published a long stream of papers on the relations between climate and landforms, as well as morphoclimatic schemes (e.g. Büdel 1982). His morphoclimatic regions are depicted in Fig. 5.6. Climatic geomorphology is still a major theme in German geomorphology, witness Hanna Bremer's *Allgemeine Geomorphologie* (1988).

The work of the French and German schools of climatic geomorphology was not without its followers in North America. Stephen Sargent Visher (1941,

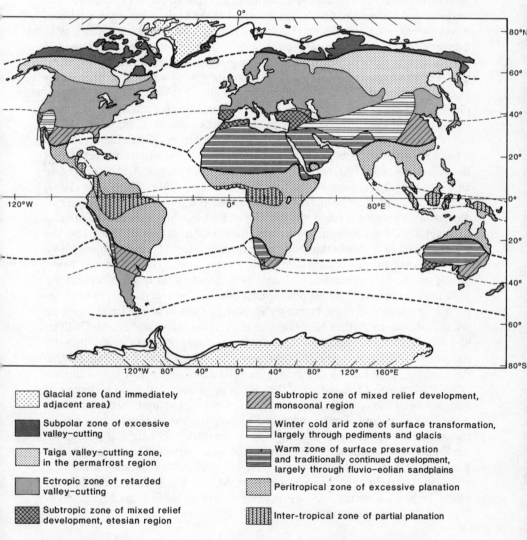

Glacial zone (and immediately adjacent area)

Subpolar zone of excessive valley–cutting

Taiga valley–cutting zone, in the permafrost region

Ectropic zone of retarded valley–cutting

Subtropic zone of mixed relief development, etesian region

Subtropic zone of mixed relief development, monsoonal region

Winter cold arid zone of surface transformation, largely through pediments and glacis

Warm zone of surface preservation and traditionally continued development, largely through fluvio–eolian sandplains

Peritropical zone of excessive planation

Inter-tropical zone of partial planation

Fig. 5.6. Morphoclimatic regions of the world according to Julius Büdel. (After Büdel 1982)

1945) considered the relations between climate and geomorphology, listing and mapping a wide range of climatic variables which might be expected to influence geomorphological processes. The later attempt by Louis Peltier (1950) to define morphogenetic regions inductively by considering the nature of climatic control unfortunately shunned Visher's long list of climatic factors and simply tried to elucidate the effect of mean annual rainfall and mean annual temperature on the effectiveness of individual processes of weathering and erosion, and then delimited morphogenetic types as combinations of individual processes within limits set by temperature and precipitation. Several criticisms have been levelled at Peltier's scheme: temperature and rainfall provide too gross a picture of the relationships between rainfall, soil moisture, and runoff; and the magnitude and frequency of storms and floods, so important to the understanding of landform development, are not included (Stoddart 1969a; see also Stoddart 1969b). W.F. Tanner (1961) tried to upgrade the notion of morphogenetic regions by adopting evaporation in place of temperature on the grounds that precipitation-evaporation relations more realistically define the availability of water. Nonetheless, his characterization of climate was still crude, as a glance at world maps of geomorphologically important precipitation and temperature characteristics will show (e.g. Common 1966). The latest work on morphoclimatic regions does take on board the details of the climatic mosaic. For example, in 1975 Peltier proffered a more all-inclusive theory of climatic geomorphology which overcame some of the objections to his system of a quarter of a century earlier. And, in setting up climatic landscape regions of the world's mountains, Will F. Thompson (1990) used several criteria: the height of timberline, the number and character of altitudinal vegetational zones, the amount and seasonality of moisture available to vegetation, physiographic processes, topographic effects of frost, and the relative levels of the timberline and permafrost limit (see also K.R. Young 1989).

The work on morphoclimatic regions prior to the late 1940s was based chiefly on the field observation of land form, and to a far lesser extent on the field measurement of geomorphological process. The rise of the physical approach to geomorphology, begun by Robert E. Horton in 1945 and taken up enthusiastically by Arthur N. Strahler in the United States and André Cailleux and Jean Tricart in France, prompted some geomorphologists to question the uniqueness of landforms in different regions: as physical laws are immutable, so the thinking ran, it follows that landforms will be the same regardless of climate (Pitty 1982, p. 82). (The notion of thresholds in geomorphological systems had not then been thought of). The consequence was that many British and North American geomorphologists, who had always tended to remain aloof from climatic geomorphology, regarding it as an eccentric and harmless pastime of the French and Germans, came round to the view expressed by Lester Charles King (1957) that all hill forms occur in all geographical and climatic zones (see also Frye 1959). In 1964, Luna B. Leopold, M. Gordon Wolman, and John P. Miller, in their influential *Fluvial Processes in Geomorphology*, could confidently pen the following statements:

"Climate affects hillslope forms through the action of such factors as precipitation and temperature on weathering and removal. Specific conditions do not produce unique landforms; there are innumerable combinations of climatic factors and lithologies and structural and stratigraphic relations which will produce the same forms". (Leopold et al. 1964, p. 384).

"All kinds of hill forms are found in all kinds of climates. The inselbergs of the tropics are also found in the semiarid desert. The cliff face, debris face, talus-veneered rock surface, and concave footslope are found in the Arctic regions of snow and ice, the tropical forests of Hawaii, and the semiarid regions of the Sudan and South Africa". (Leopold et al. 1964, p. 383).

This idea was taken up by Robert V. Ruhe (1975), who averred that "fully developed slopes" (slopes with profiles consisting of summit, shoulder, back-slope, footslope, and toeslope) will, owing to the action of subaerial processes, have the same form in all environments, regardless of climatic differences. Nicholas J. Cox (1977) allowed that Ruhe's basic thesis may be correct, but took issue with the evidence used to support the case (see also Ruhe 1977). The argument is still unsettled.

A problem that tends to have been overlooked in studies which try to associate geomorphological form variables with climate is that the fluvial system acts as a whole, as Stanley A. Schumm (1977) so clearly and cogently argued. All drainage basins have a perimeter, an outlet, a channel network, and a suite of valley-side slopes: a change in the geometry or position of any of these elements will necessarily cause alterations in the rest, even though a time-lag may be involved in the adjustment. For this reason, one cannot realistically investigate the relation between climate and individual form variables: all form components of a drainage basin must be included (Kennedy 1976, p. 171). Furthermore, variations in individual variables may not respond directly to climate, but may adjust to climatically induced modifications of some other feature of a drainage basin, as would be the case if slopes were to change following a relocation of the outlet after a post-glacial rise of sea level (Kennedy 1976, p. 171). Add to these considerations the orientation of a drainage basin, the complexity of climatic variables themselves, and the rather unclear association between single geomorphological processes and slope form, then the difficulty of probing into climate-landform relationships becomes apparent. However, the would-be climatic geomorphologist should not be too disheartened, for, although unambiguous links between climatic and topographic features are likely to prove very rare, it is possible to identify some associations between them, as we shall now see.

5.2.2 Morphometry and Climate

Even if the concept of morphoclimatic regions be invalid, that would not discount the possibility that some morphometric properties of a landscape are adjusted to the prevailing climatic conditions. As long ago as 1913, Emmanuel de Martonne suggested that the forms of fluvial erosion are sensitive to, and so are indicators of, certain climatic elements. It was not until the late 1950s that this idea was tested. Richard J. Chorley (1957) investigated the morphometry of drainage basins on three areas of well-dissected "mature" relief which had different climates: Exmoor, England, and Pennsylvania and Alabama in the United States. He characterized climate and vegetation by an index, I_c, defined as

$$I_c = \frac{I \cdot v}{P \cdot r},$$

where I is Thorthwaite's precipitation effectiveness index, v is vegetation cover, P is amount of precipitation, and r is intensity of precipitation. Drainage basin area and stream length were found to increase, and drainage density to decrease, with increasing I_c. In a more detailed investigation of drainage basins in Arizona, Colorado, New Mexico, and Utah, Mark A. Melton (1957) found that 93.2 percent of the variation in drainage density was explained in terms of Thornthwaite's precipitation effectiveness index, percent bare area, infiltration, and rainfall intensity (defined as the amount of rain of 1 hour's duration in a 5-year period), and that the three climatic variables (precipitation effectiveness, infiltration, and rainfall intensity) alone explained 92 percent of the variation. Despite the significance of his findings, Melton rightly stressed the difficulty of disentangling the effects of climate, lithology, and structure on drainage basin form, admonishing us that "geologic and climatic elements act in complex ways, through the agencies of vegetal growth, soil formation, run-off and erosion, infiltration, and soil creep" (Melton 1957, p. 37).

From a re-examination of Melton's data set, which is still one of the most comprehensive sources of material for the investigation of drainage basin features in general and hillslope form in particular, Barbara A. Kennedy (1976) raised two points of interest. Firstly, she reaffirmed the weak overall correlation between the precipitation-effectiveness index and the values of maximum valley-side slope angles. Coupled with other evidence for the non-uniformity of response of angles to increasing values of drainage density, this finding underscores how difficult, if not misleading, it is to try to establish broad regional associations between selected climatic and morphometric variables. Secondly, she highlighted one of the major features of climatic geomorphology: although the direct association between climate and landform is never clear and evidence for it somewhat thin, the indirect evidence for an element of climatic influence is far more substantial. Relative relief and drainage density, factors which are more immediately linked to changes in maximum slope angles than is the precipitation-effectiveness index, are themselves measures of the landscape which reflect climatic, as well as structural and lithological, influences, as witnessed by Melton's discovery that drainage density has a strong negative correlation with the level of precipitation effectiveness. In short,

"Melton's data provide a clear picture of the pattern of success that may be expected in any wide-ranging discussion of the influence of climate upon drainage basin forms in general and slope forms in particular. At the regional scale, the patterns of association will appear blurred, for non-climatic forces will intrude upon and modify the action of climatic elements. Only when structural considerations, in particular, are held relatively constant will it be possible to identify co-variation in climatic and morphological indices and it is to be expected that the nature of such relationships will vary with the precise location and composition of the basins concerned". (Kennedy 1976, pp. 179–180).

There is really little later work which has added very much to the clarification of climatic impact on drainage basin form at the regional scale. The study made by H.S. Sharma (1987) in Rajasthan, if not so extensive as Melton's study, at least used a large sample of drainage basins. Morphometric properties were measured for 116 drainage basins, 54 third-order basins, and 62 random-order basins from various climatic zones in the area. The morphometric properties measured were stream number, stream length, drainage density, steam frequency, texture ratio, dissection index, and stream sinuosity index. These were correlated with mean annual rainfall (Fig. 5.7). Several correlations between form variables and climate were uncovered (Fig. 5.8): drainage density and stream frequency increase from arid to semi-arid regions, and decrease in humid regions; landscape dissection is maximal where rejuvenation has occurred, or in areas of higher rainfall; the texture ratio increases with increasing rainfall from arid to semi-arid zones, and decreases in the humid zones; and stream sinuosity is maximal in semi-arid regions.

A drawback with much of the work on morphometry and climate has been the over-reliance on mean values, mean frequencies, and mean extremes (cf. Eybergen and Imeson 1989). An attempt to rectify this problem was made by Frank Ahnert (1987). Using suitable meteorological data, Ahnert carried out a magnitude-frequency analysis of daily rainfall, short duration rainfall, the duration of rainless periods, the severity of frosts, freeze-thaw cycles, and wind velocities. The method was to regress the magnitudes of the meteorological events against the decadal logarithm of their recurrence interval, expressed in years. Doing this, the constant of the regression equation (the first regression coefficient or intercept value) defines the magnitude of the 1-year event; the sum

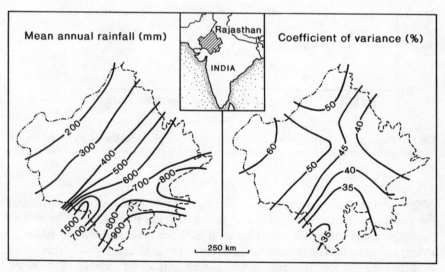

Fig. 5.7. Mean annual rainfall and rainfall variability in Rajasthan. (After B. L. Gupta and Singh 1981)

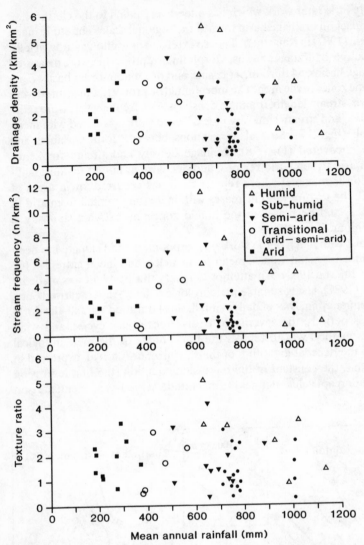

Fig. 5.8. Drainage density, stream frequency, and texture ratio plotted against mean annual rainfall for a sample of 54 third-order drainage basin in five climatic zones in Rajasthan. (After Sharma 1987)

of the constant and the second regression coefficient (the slope of the regression line) defines the magnitude of the 10-year event; the sum of the constant and twice the second regression coefficient defines the magnitude of the 100-year event, and so forth. The first and second regression coefficients can be combined to form a magnitude-frequency index, as well as mapped individually to show the spatial variation of magnitudes and frequencies (Fig. 5.9). As Ahnert asserted, this kind of study might serve as the core of process-orientated mor-

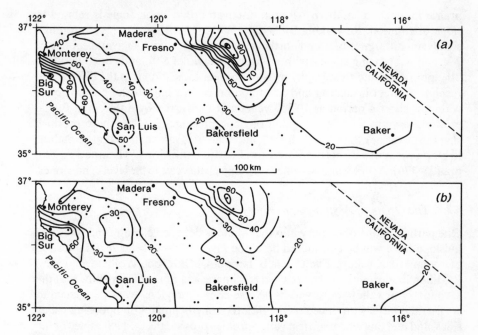

Fig. 5.9. Isoline maps of **a** the regression constant, a, and **b** the regression coefficient, b, of the equation used for estimating magnitude and frequency in a belt between 35° and 37° N latitude across the State of California. The equation takes the form $P_{24} = a + b \log_{10} RI$, where P_{24} is the daily amount of precipitation and RI is the recurrence interval expressed in years. (After Ahnert 1987)

phoclimatology. To be sure, because of its focus on process, it stands more of a chance of elucidating the relations between landforms and climate than does a purely morphological approach.

As was mentioned earlier, some geomorphologists, Robert V. Ruhe (1975) in particular, claim that slope form is similar in kind, if not in detail of expression, in all environments. The question is, do the "details of expression" attributable to individual processes vary under different climates? A problem in elucidating the effects of a single process on slope form is that very few cases are known where a slope has been produced by the action of just one process. Furthermore, few, if any, slope processes can be linked in a simple way to specific climatic or meteorological variables, and in most cases the temperature and moisture conditions at the soil surface or in the soil will be all-important, and not the temperature and moisture measured in a Stevenson screen. Thus, "in the overwhelming majority of situations we have to deal with slopes exhibiting a mixture of processes, each one of which is related in a complex fashion to the magnitude, frequency and timing of the incidence of sun, snow, rain and wind" (Kennedy 1976, p. 172). This should not, however, be seen as cause for despair. There are situations where conclusive evidence for climatic control of microlandforms has been found. M.J. Bik (1968), for instance, in a study of prairie mounds in southern Alberta, discovered that slope angle is related to slope orientation. Reworking Bik's data, Barbara A. Kennedy (1976) found a significant difference between northeast-facing slopes, with an average angle of 7.3° and west-facing slopes, with an average angle of 6.0°, and concluded that "the prairie mound topography provides a clear indication that the differential receipt of direct insolation and precipitation by low-angled slopes may, even within as short a period as 10,500 years and unassisted by differential fluvial erosion, serve to modify the action of slope processes sufficiently to produce significant variations in profile form", and that "it is safe to infer that variations in slope form are directly related to climatic conditions" (Kennedy 1976, pp. 175–176). A problem arises when we bump up the scale to the level of hillslopes.

5.2.3 The Origin of Asymmetrical Valleys

One particular line of enquiry into the relations between climatic variables and hillslope form will be considered here: the investigation of the vexed question of asymmetrical valleys. Even Luna B. Leopold, M. Gordon Wolman, and John P. Miller, that triumvirate so opposed to climatic geomorphology, allow that the asymmetry of valleys observed in diverse climatic regions from periglacial to semi-arid is largely produced by the different microclimates on opposing valley sides, and feel that asymmetrical valley profiles provide the best examples of the effect of different climates on rock types. Be that as it may, demonstrating the primacy of climate in the creation of asymmetrical valleys is no easy task. Much depends on finding a suitable definition of asymmetry. In the past, geomorphologists have been rather lax in carefully defining the phenomena they purport to explain and have been less than zealous in considering the possible role of non-climatic factors. Barbara A. Kennedy (1976, p. 181) traces the roots of this imprecision to a curious desire to identify a global pattern of asymmetrical valley forms linked to the major climatic zones, a desire that has led to the overshadowing of the more basic discussion of the manner in which individual cases of valley asymmetry can be linked to climatic factors. To tighten up the definition of asymmetry, Kennedy proposed that the Strahler maximum angle (the angle of the "constant slope" sector, or "backslope" in Ruhe's terminology) should be employed. If everybody adhered to this, then comparison between different studies would be greatly facilitated. According to Kennedy, there are four forms of valley asymmetry, all of which may be considered as arising from an imbalance in the volumes of material transported down opposing hillslopes to the valley floor, to variation in the rates of removal of debris and solid rock from the base of the slopes, or to some combination of these two. None of these causes is wholly non-climatic in origin, but neither is it totally climatic. The four forms are: firstly, "valley-wide" assymetry where one valley is prone to either greater overall denudation or to greater basal erosion along the length of a basin; secondly, "localized" asymmetry, where either valley side experiences greater denudation or more severe basal attack at a limited number of points within a basin; thirdly, an asymmetry resulting from the superimposition of valley-wide and localized asymmetry, where a general tendency for greater or less erosion along the length of one valley side is coupled with localized steepening or reduction at certain points; and lastly, an asymmetry resulting from a complex combination of valley-wide and localized asymmetry, where, though there be no general tendency for one valley side to be steepened or reduced in angle, there is a tendency for localized erosion to vary in intensity with the orientation of the profile concerned. After having established the main forms of asymmetrical valley, Kennedy goes on to identify eight possible causes of them: the Coriolis force (which she feels can safely be eliminated), differences in insolation and precipitation receipts, differences in slope dimensions, variable lithology, geological structure, warping, evolution of the drainage net, and glaciation. Clearly, neither the forms of asymmetrical valleys nor their causative agents are simple. This is why attempts to clarify the role of climatic factors in asymmetrical valley creation have not been too successful. Particularly problematic is the role of structure, a factor whose variation is almost as widespread as climatic variation but whose manner of operation remains somewhat obscure.

Despite these problems, the majority of geomorphologists have held firm to the belief that many asymmetrical valleys are produced by climatically controlled variations in slope processes. Opinions differ as to the climatic regime which created the asymmetrical valleys: in general, American workers consider that the valleys have been, and are being, produced by forces at work now in the landscape; European geomorphologists favour the view that the valleys are "fossil" forms, survivals from a "periglacial" climate. In an effort to resolve some of these problems, Kennedy examined the asymmetry of selected sets of small east-west-trending valleys, and sets of small north-south-trending valleys, in North America. She concluded that "climatic" elements in valley asymmetry should be seen

"not simply as the products of differential fluvial attack, but rather as the outcome of microclimatically controlled variations in the responses of opposing valley sides to basal corrasion. The development of any valley-wide asymmetry depends upon the creation, first of relatively steep slopes by stream activity and second, upon the existence of microclimatic variations that are sufficiently strong to allow some degree of difference in the nature or rate of operation of the subaerial processes which then set to work upon those steepened slopes". (Kennedy 1976, p. 197).

The valleys she studied in North America showed that the second criterion appears to be rarer in north-south-trending valleys than in those that run east-west; and that sufficiently strong microclimatic differences are less common in areas of semi-arid climate; which leads us to the thorny question of the link between valley asymmetry and the major climatic zones.

The old idea that many asymmetrical valleys in Europe were produced under periglacial conditions has been shown to be an oversimplification. Steeper north-facing and east-facing valley profiles have been described in past and present areas of periglacial environments, and steeper south-facing and west-facing profiles have been discovered beyond the reach of periglacial zones, past and present. To explain the first of these discoveries, Jean Tricart (1963) proposed two forms of periglacial activity: "warm" or "marine", characterized by steeper south-facing and west-facing slopes; and "cold" or "continental", creating steeper north-facing or east-facing slopes. By looking at the latitudinal zonation of these two types, H. Karrasch (1972) attempted to resolve the problem of asymmetrical valley distribution in Europe. Although the Tricart-Karrasch classification may hold in certain areas of Europe, it breaks down in Britain and North America (Kennedy 1976, p. 198). The basic questions are these: is it reasonable to assume that there should be a distinction (in the Northern Hemisphere) between asymmetrical valleys with steeper north-facing or east-facing profiles on the one hand, and those with steeper south-facing or west profiles on the other? And, even allowing that such a distinction be justified, is it possible to educe a mechanism that will create clear-cut zones of each asymmetrical valley type? Until more be understood about the processes of symmetrical valley creation, the first question is unanswerable, and until it be answered, the second question is redundant. In short, schemes of global patterns of asymmetry are a nonsense when understanding of the basic mechanisms at work is so precarious (Kennedy 1976, p. 198). That does not mean the matter can never be resolved. The main need is an agreed definition of asymmetry and fresh data, particularly on the control of moisture conditions within the soil mantle and the relationship between present climatic events and subaerial processes. An improved understanding of the structural control of hillslope form would also be an asset.

5.3 Landforms and Soils During the Pleistogene Period

The effect of climatic change on soils and landforms will be to change process rates or to alter the dominant process. In both cases, the landscape will be forced into disequilibrium and geomorphological and pedological activity will increase for a while. This will be especially so with a change in process regime because the landscape will necessarily be in disequilibrium with the new processes. A phase of intense activity will ensue, involving the reshaping of hillslopes, the reworking of soil profiles, and the changing of sediment stores in valley bottoms. However, many landforms and soil bodies are relatively long-lived, resilient elements of the physical soil landscape. For this reason it is common for a cliff, a floodplain, a cirque, some soil horizons, and many other soil-landscape features to survive longer than the climatic regime which created them: seldom does the erosion promoted by a new climatic regime renew all the landforms and soil bodies in a landscape; far more commonly, remnants of past landforms and soils are preserved. Consequently, most soil landscapes are a complex collection of landforms and soils inherited from several generations of soil-landscape development, a palimpsest of past forms. In some soil landscapes, the inherited forms have been formed by processes similar to those now operating there, but it is not uncommon to find polygenetic soil landscapes in which the processes responsible for a particular landform or soil body no longer operate. The clearest and least equivocal example of this is the glacial and periglacial landforms left as a vestige of the Ice Age in mid-latitudes. But polygenetic soil landscapes are common. In deserts, ancient river systems, old archaeological sites, fossil karst phenomena, high lake strandlines, and deep weathering profiles are relict elements which attest to past humid phases (Goudie 1983, 1985); while stabilized fossil dune fields on desert margins are relics of more arid phases. In the humid tropics, a surprising number of soil-landscape features are relict. For example, in the central Amazonian Basin (Tricart 1985) and in Sierra Leone (Thomas and Thorp 1985) vestiges of fluvial dissection which occurred under dry conditions between about 20,000 and 12,500 years ago have been discovered.

It is now known that soil landscapes change owing to climatic and endogenetic forcing and to internal adjustment in landform and soil systems. However, the links between climatic forcing and geomorphological and pedological change are not well understood in theoretical terms. Most geomorphologists working on Pleistocene and Holocene timescales adopt a rather spongy paradigm involving the concepts of thresholds, feedbacks, complex response, and episodic activity (Chorley et al. 1984, pp. 1–42). But despite the lack of a tight theoretical base and the complex dynamics of landform and soil systems, soil-landscape changes over periods of 1000 to 100,000 years display consistent patterns which are largely forced by climatic, eustatic, or tectonic conditions. In this section, we shall examine these changes in the context of glacial-interglacial cycles during the late Pleistocene and Holocene epochs, and then look at attempts to model the response of landforms and soils to medium-term climatic change.

5.3.1 Soil-Landscape Change and Glacial-Interglacial Cycles

The switch from a glacial to an interglacial climate is generally thought to entail a change from cold and dry conditions to warm and moist conditions. Although this appears to be a crude characterization of the climate system in glacial and interglacial states, it does seem to capture the broad changes involved, despite regional anomalies. A study by J.R. Petit and his colleagues (1990) of dust deposits in the Vostok ice core, recovered by Soviet Antarctic expeditions, shows that changes in dust concentration vary globally during glacial-interglacial cycles, and, as James E. Begét and Daniel B. Hawkins (1989) suggested, are linked to orbitally forced fluctuations of climate. Eighty percent of the dust in the Antarctic ice comes from continental sources, the remaining 20 percent being volcanic aerosols. The continental material would have been carried from arid and semi-arid parts of the Southern Hemisphere, the unglaciated parts of Antarctica, and the continental shelves, which were progressively exposed to wind as sea level fell during a glacial stage. The amount of dust deposition would have depended on aridity, the wind speed over source areas, and the ability of the atmosphere to transport material meridionally. In general, the high dust concentrations occurred during cold stages and are indicative of arid conditions.

The alternation of cold, dry and warm, moist climates during the Pleistogene period would have affected denudational, depositional, and pedogenetic regimes, instigating changes in both the type and rate of processes encountered. Broadly speaking, processes during warm and wet stages would have been mainly chemical (such as leaching and piping) and would have led to the formation of soils and deep regoliths; processes during cold and dry stages would have been associated with the existence of permafrost, ice sheets, and cold deserts (Starkel 1987). The landforms and soils produced by glacial and by interglacial process regimes are generally distinctive, and are normally separated in time by erosional forms created in the shortish transition period from one climatic regime to another. When the climate is in transition, both glacial and interglacial processes run at levels exceeding thresholds in the slope and river systems (Fig. 5.10). Leslek Starkel (1987) summarized the changes in a temperate soil landscape during a glacial-interglacial cycle (Fig. 5.11). During a cold stage, erosion is dominant on the upper part of valley-side slopes, while in the lower reaches of valleys abundant sediment supply leads to overloading of the river, to deposition, and to braiding. During a warm stage, erosion thresholds are not normally exceeded, most of the slopes are stable, and soil formation proceeds (cf. Kukla 1977; Catt 1988). Meandering channels tend to aggrade, and erosion is appreciable only in the lowest parts of undercut valley-side slopes and in headwater areas. All these changes create distinct sequences of sediments in different parts of the fluvial system. Similar changes of process regimes occurred in arid and semi-arid environments. For instance, talus deposits at the foot of escarpments in hot deserts register changing climates during the Pleistogene period. Rand Gerson and Sari Grossman (1987) showed that taluses formed during prolonged mildly arid to semi-arid climatic regimes (pluvial modes), and were eroded by gullying leaving talus flatiron relicts during arid to extremely arid climatic regimes (interpluvial modes). Similarly, in northwestern Texas and eastern New Mexico, a vast sheet of Quaternary loess covering more than 100,000 km^2 and up to 27 m thick, known as the Blackwater Draw Formation, records more than 1.4 million years of aeolian sedimentation. Vance T. Holliday (1988, 1989a) uncovered six buried soils in the formation, analysis of which suggests environmental oscillations. Stable landscapes obtained under sub-humid to semi-arid conditions, similar to those of the past several tens of thousands of years, whereas regional wind deflation and aeolian deposition prevailed during periods of prolonged drought.

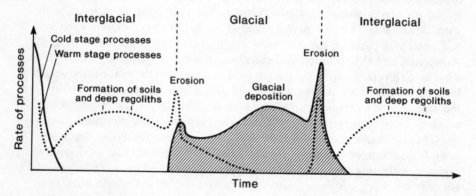

Fig. 5.10. Changes in geomorphological and soil systems during a glacial-interglacial cycle. (After Starkel 1987)

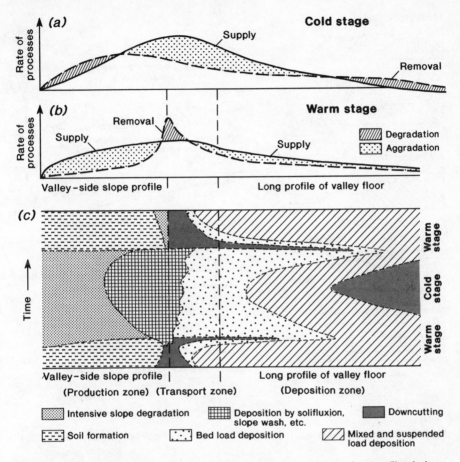

Fig. 5.11. The typical sequence of changes in valley-side slopes and valley long-profiles during a glacial-interglacial cycle. **a** The relationship between removal and supply in the valley long-profile during a cold stage. **b** The relationships between removal and supply in the valley long-profile during a warm stage. **c** A glacial-interglacial transect showing the changing slope and river processes as the climate shifts from interglacial to glacial and back to interglacial states. (After Starkel 1987)

The effects of climatic change on soils and landscapes during the Pleistogene period are commonly most clearly displayed in tectonically active areas. Where uplift has occurred, parallel alluvial fills give way to a staircase of terraces or pediment levels; where subsidence has taken place, normal aggradation sequences occur, climatic changes being echoed in changing grain size and changing facies. Figure 5.12 portrays five basic interactions between tectonic activity and medium-term climatic change. Changes of this kind, in which climate and tectonism act simultaneously on soil landscapes, have been modelled with a degree of success by J. Boll and his colleagues (1988) who simulated the development of river terraces in a tectonically active landscape during glacial-interglacial cycles.

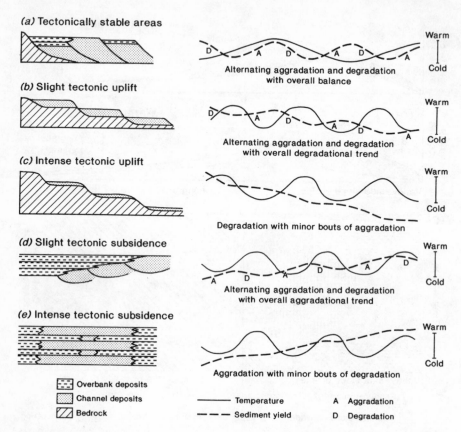

(a) Tectonically stable areas

Alternating aggradation and degradation
with overall balance

(b) Slight tectonic uplift

Alternating aggradation and degradation
with overall degradational trend

(c) Intense tectonic uplift

Degradation with minor bouts of aggradation

(d) Slight tectonic subsidence

Alternating aggradation and degradation
with overall aggradational trend

(e) Intense tectonic subsidence

Aggradation with minor bouts of degradation

Overbank deposits
Channel deposits
Bedrock

——— Temperature   A Aggradation
– – – Sediment yield   D Degradation

Fig. 5.12. The fluvial response to the joint action of climatic and tectonic factors. (After Starkel 1987)

It is possible to infer something of the nature of interglacial climates from buried soils. A study of five palaeosols of pre-Wisconsin drift on Mokawan Butte, southwestern Alberta, enabled Eric T. Karlstrom (1987, 1988) to deduce that three of the soils probably formed under interglacial climates that were moister and at least 6 °C warmer than at present, while the two others formed under a semi-arid climate and a modern-type climate respectively. On a broader scale, John A. Catt (1988) looked at interglacial soils developed in periglacial loess at selected sites across Eurasia with a view to reconstructing the type of climate which produced the soils. He found some consistent differences in the degree of soil development between interglacial stages: the soil corresponding to isotope Stage 5e is always more strongly developed than the soil of Stage 1 (the Holocene) in the same area; soils of Stages 7 and 9 are equal or weaker in development to the soil of Stage 5e in most areas; with three exceptions, the soil of Stage 11 is always as strongly developed as, or more strongly developed than, the soils of later interglacials (Stages 1 to 11). Interestingly, the mean values of the strength of soil development for each interglacial in the three chief areas are more variable in central and eastern Europe (2.0 to 6.2) than in either northern Europe (3.9 to 6.0) or Asia (2.0 to 4.3) (Table 5.1), perhaps hinting that central and eastern Europe experienced very variable interglacial climates as well as temperature contrasts between cold and warm stages. It would be risky to draw firm conclusions about climatic differences between interglacials from buried loess soils because, as yet, the resolution of the oxygen isotope record, particularly that part corresponding to the termination of interglacials, is imprecise and so it is tricky to allow for differences in the duration of soil development within individual interglacial stages (Catt 1988). However, this is a promising line of enquiry.

Table 5.1. The relative development of loess soils during successive interglacial stages and a suggested climatic ranking of interglacials. (Catt 1988)

Oceanic warm stages	Mean soil values			Mean soil ranking[a]	Mean climatic ranking[b]
	Central and eastern Europe	Northern Europe	Asia		
1	2.0	3.9	2.0	11	11
5e	3.9	4.3	3.5	8	5
7	4.0	4.2	2.8	9	9
9	3.5	4.0	2.5	10	10
11	3.8	5.0	3.3	6	8
13	6.2	5.0	3.3	1	1
15	5.8	4.0	4.3	3	6
17	5.8	5.3	2.7	4	4
19	5.3	5.5	3.0	4	2
21	4.0	5.0	3.0	6	7
23	6.0	6.0	2.5	1	2

[a]Ranking by mean, worldwide soil values.
[b]Mean, worldwide climatic ranking using Bockheim's (1980) equation.

5.3.2 Soil Landscapes and Climatic Change During the Holocene Epoch

Short-term changes in landscapes and soils are related chiefly to changes in the hydrological regime and concomitant changes in vegetation cover. Changes in the thermal regime are not without significance: during the Little Ice Age, the number of rockfalls and avalanches in western Norway increased (Grove 1972). But the most noticeable relics of Holocene climatic changes in soil landscapes are sequences of river terraces.

Prior to about 1890, alluvial river terraces were ascribed to movements of the Earth's crust. A little later, and terraces in glaciated and unglaciated regions were attributed to climatic change (Davis 1902; Gilbert 1900; W.D. Johnson 1901). William Morris Davis (1902) posited that the slope of the long profile of a river reflects a balance between the erosion and transport of sediments, and believed that the volume and nature of the sediment load are adjusted to climate.

A change from a humid to an arid climate, he surmised, would cause river long profiles to steepen and aggradation to occur in valleys; whereas a change from an arid to a humid climate would cause river long profiles to become less steep and trenches to form in valleys. Later workers were divided as to the relative importance of, on the one hand, flood characteristics, and on the other, sediment supply in explaining the form and sedimentology of alluvial channels and floodplains. Ellsworth Huntington (1914b) opined that valley alluviation in the southwest United States occurred during dry episodes when vegetation was scanty and sediment yields were high; and, conversely, degradation (channel entrenchment) occurred during wet episodes when vegetation was more abundant and the sediment load lower. In contrast, Kirk Bryan (1928) held that channel entrenchment in the Southwest was associated with periods of prolonged drought and occurred because the much-reduced vegetation cover during long dry episodes gave large floods. In turn, the large floods initiated entrenchment, the trenches then expanding upstream. This view was endorsed by Ernst Antevs (1951). Taking yet another tack, C. Warren Thornthwaite and his associates (1942) attributed trenching over the last 2000 years not to major climatic shifts, but to changes in the intensity of storms.

A modern review of the response of river systems to Holocene climates in the United States argues that fluvial episodes in regions of varying vegetation cover occurred roughly at the same times, and that the responsiveness of the rivers to climatic change increased as vegetation cover decreased (Knox 1984). Alluvial episodes occurred between roughly 8000 to 6000, 4500 to 3000, 2000 to 800 years ago. Before 8000 years ago, vegetation change and rapid warming caused widespread alluviation. The magnitude of this alluvial episode generally increased to the west in parallel with increased drying and increased vegetation change. Between 8000 and 7500 years ago, alluviation was broken in upon by erosion. Although of minor proportions in the East and humid Midwest, this erosion was severe in the Southwest. For the next 2000 years, warm and dry conditions in the northern West and the Midwest, and warm and wet conditions in the southern Southwest and parts of the East and Southeast (caused by the persistent zonal circulation of the early Holocene epoch), led to a retardation of alluviation in all places save the Southwest, where major erosion of valley fills occurred. Though being warm and wet at the time, the Southeast did not suffer erosion because forest cover was established. From 6000 to 4500 years ago, all the Holocene valley fills were eroded, except those in the Southwest, where alluviation continued (cf. Holliday 1989b). The reason for this widespread erosive phase lies in the facts that the climate had begun to cool, thus improving the vegetation cover, reducing sediment loads, and promoting trenching; and that the circulation of the atmosphere became more meridional during summer, thus bringing higher rainfall and larger floods. The Southwest was untouched by the erosive phase because the climate there became more arid owing to the northwards displacement of the subtropical high pressure cell. Between about 4500 and 3000 years ago, the rates of erosion and deposition relaxed, but were high again in many regions between 3000 and 1800 years ago. The nature of the intensification of erosion and deposition varied from place to place: for instance, very active lateral channel migration with erosion and deposition of sediments characterized the northern Midwest; alluviation took place in many sites on the western edge of the Great Plains; and erosion and entrenchment occurred in the southern Great Plains of Texas. The intensity of fluvial activity then died down again and stayed at a modest level until 1200 to 800 years ago, when cutting and filling, active lateral channel migration, and so forth, occurred. From 800 years ago to the late nineteenth century, a moderate alluviation took place, after which time trenching started in most regions. A lesson to be learnt from this, and alluvial chronologies in other parts of the world (e.g. Littmann and Schmidt 1989, p. 347), is that the response of the soil landscape to climatic change is generally diachronous, varying from region to region, owing partly to regional variations in the changes of climate and partly to thresholds within the soil-landscape system itself.

5.3.3 Modelling the Response of Hillslope Form to Climatic Change

Most climatic geomorphologists believe that a large enough change of climate will alter the regime of denudational processes in a landscape and so lead to a change of land form. Speculation on this matter can be traced to William Morris Davis's recognition of "climatic accidents" and Albrecht Penck's (1914b) allusion to polygenetic landforms produced by the shifting of climatic zones during ice ages, "pure" climatic forms surviving only at the heart of the world climatic zones where no change of process regime occurred. One way of elucidating the relationship between climate and land form is to use mathematical models. Early endeavours with such models showed that different processes do indeed tend to produce characteristic slope profiles. In particular, slow mass movements and wash processes each tend to produce hillslopes with a characteristic form. Given a steady-state hillslope form under a climatic regime in which wash processes are dominant, a change of climate which favoured the dominance of mass movements could be expected to alter the steady-state form of the hillslope (see Huggett 1985, pp. 162–180).

Little theoretical work has been carried out on the direct link between hillslope form and climatic change, but Frank Ahnert has made progress in this direction. Using his three-dimensional slope development model, Ahnert (1988) found that the duration of the period between climatic changes had a decisive influence on hillslope development. When the duration between climatic shifts was 100 time units or more, then both wash processes and slow mass movement persisted long enough to leave their stamp on hillslope form: wash processes produced a concave profile, while mass movements produced a convex profile (Fig. 5.13a). However, when climate changed every 50 iterations or so, neither of the two sets of processes had time to produce characteristic hillslope forms: the wash slope became concave only in footslope and toeslope positions, while the upper slope retained some of the convexity inherited from the previous phase of mass movement (Fig. 5.13b, c, d). Ahnert also discovered that a change in the characteristic duration of rainfall events in an area can result in a modification of hillsope form. In a series of simulations, rainfall intensity was held constant,

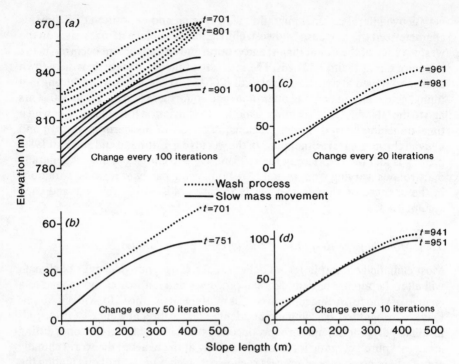

Fig. 5.13. The effect of a change between wash processes and slow mass movement on slope profile development. Change of process every **a** 100 time units **b** 50 time units **c** 20 time units **d** 10 time units. In **b, c,** and **d** only profiles at the end of an interval (just prior to a process change) are shown. (After Ahnert 1988)

while the duration of rainfall events was varied. The longer the duration of rainfall, the longer the slope over which runoff depth increases. In simulation runs which allowed climatic shifts in which "long-duration" rainfall events alternated with "short-duration" rainfall events at intervals of 500, 250, 100, and 50 time units, the following results were obtained: when climatic changed occurred every 250 time steps or longer, then during intervals of short-duration rainfall events, a convex hillslope profile developed; and, during intervals of long-duration rainfall events, a concave hillslope formed (Fig. 5.14a, b); with less protracted periods between climatic changes, the "short-rain" and "long-rain" hillslope forms were far less pronounced (Fig. 5.14c, d). In fact, the more rapid were the climatic swings, the more did the middle part of the hillslope change, first from convex to straight, and eventually to concave. This resulted from the period of "long-rain" development becoming increasingly dominant as the time between climatic switches shortened owing to the fact that the runoff generated by "long-rain" events can denude the slope more than can the runoff generated by "short-rain" events in the preceding interval.

In a different experiment, Ahnert attempted to simulate the natural slope development sequence in the Kall Valley, a tributary of the River Rur (not to be confused with the Ruhr) in the northern Eifel Mountains in Germany. The

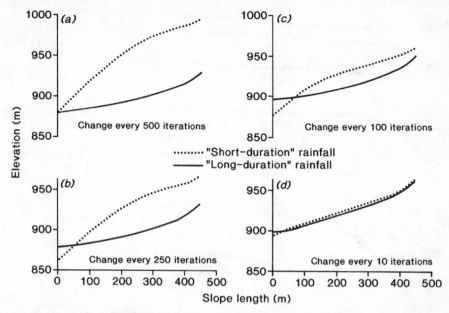

Fig. 5.14. The effect of a change between characteristic "short-duration" rainfall and "long-duration" rainfall on slope profile development. Change of rainfall duration every **a** 500 time units **b** 250 time units **c** 100 time units **d** 10 time units. (After Ahnert 1988)

Kall River is 25 km long and deeply incised into a Tertiary peneplain. On the peneplain, the headwaters of the Kall occupy a very shallow valley, but in moving downstream, the river cuts into the peneplain to form a V-shaped valley. By assuming that before incision during the Pleistogene period the river flowed its entire length on the peneplain, that incision commenced at the river mouth and progressed upstream by headwards erosion, and that headwards erosion is as yet incomplete, Ahnert thought it reasonable to presuppose that, firstly, the valley-side slopes in the uppermost reaches of the river are similar to the valley-side slopes which would have flanked the river along its entire length during Tertiary times; and secondly, as the age of incision is progressively older downstream, the spatial succession of slope profiles from the headwaters to the mouth represent a temporal sequence of hillslope development. The results were rather interesting (Fig. 5.15). It would appear that the changes from periglacial to interglacial climatic conditions during the Pleistocene epoch did not alter the mode of slope development sufficiently to generate different slope forms; rather, climatic swings seem to have caused changes in the intensity of processes without any basic change in the direction of slope development. This would imply that the geomorphological effects of climatic changes on slope development may have been overestimated (Ahnert 1988, p. 399).

Michael J. Kirkby (1989) used a mathematical model to estimate the impact of climatic change on hillslope form and regolith thickness. The main interrelationships included in the model and the interactions between vegetation

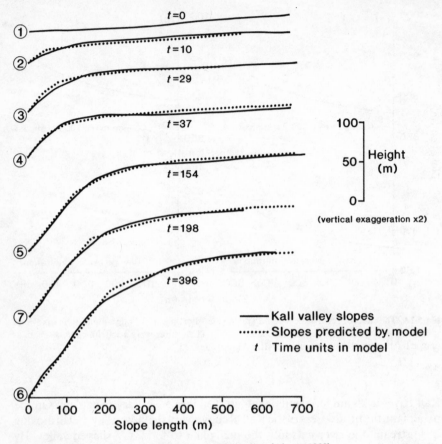

Fig. 5.15. Simulated and observed slopes in the Kall valley. (Ahnert 1988)

components and erosion rates are depicted in Fig. 5.16a and 5.16b. To test the model, Kirkby assumed a set of parameter values applicable to the climate of Luxembourg. He then ran the model to see how slopes and soils would evolve during a period of 200,000 years. In repeated runs, all conditions were held constant bar temperature (acting through potential evapotranspiration), which was varied to span conditions encountered in glacial and interglacial stages, thus providing insight into the rate and type of response to changing climates (Fig. 5.17). It is evident from the results that, as temperature increases from what would be a cool temperate to a semi-arid climate, there is a dramatic increase in the concavity at the foot of the slope and in the overall rates of erosion; in addition, there is a marked decrease in the depth of soil and in the convexity of the slope shoulder and summit. Kirkby concluded that, overall, the effect of such big temperature differences is to produce a range of hillforms which correspond well with accepted views on the qualitative differences between slopes in different climatic zones.

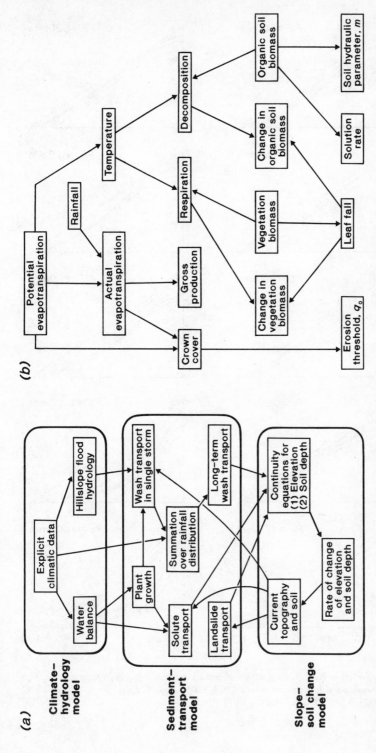

Fig. 5.16. The basic structure of Michael J. Kirkby's model used to simulate the effect of climatic change on slope and regolith form. **a** The chief interactions modelled. **b** The interactions between vegetation and erosion rates. (Kirkby 1989)

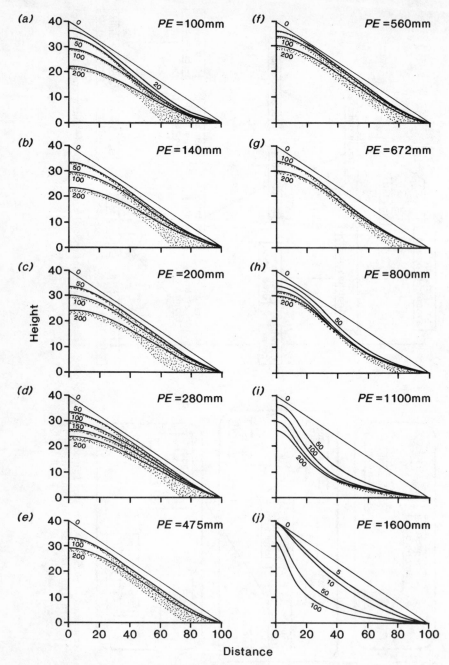

Fig. 5.17. Slope evolution forecast for a range of conditions differing only in temperature. The temperatures associated with each case are **a** 2.4 °C **b** 3.6 °C **c** 5.6 °C **d** 8.4 °C **e** 12.0 °C **f** 13.5 °C **g** 15.8 °C **h** 17.4 °C **i** 21.5 °C **j** 27.2 °C. (Kirkby 1989)

A useful feature of Kirkby's model is the prediction of soil thickness. Insight into soil thickness changes during the late Pleistocene and Holocene epochs has been gained through Donald Lee Johnson's model of soil thickness development (D.L. Johnson 1985; D.L. Johnson and Watson-Stegner 1987; D.L. Johnson et al. 1990). In a nutshell, Johnson saw soil thickness as the balance between processes of removal, upbuilding, and deepening, and thought that pedogenesis (which may be progressive, static, or regressive) should be set in the context of these soil thickness processes. His ideas on how soil has evolved in active geomorphological and tectonic environments over the last 39,000 years are portrayed in Fig. 5.18. With an eye to future developments, I would suggest that Johnson's model would profit from amalgamation with Kirkby's model in that soil thickness processes would be set in the more realistic context of the slope profile, and the effects of vegetation and climate, as well as tectonics, could also be investigated.

5.4 Pre-Pleistogene Soil-Landscape History

William D. Thornbury was of the view that "little of the earth's topography is older than the Tertiary and most of it no older than Pleistocene" (1954, p. 26). In fact, a significant proportion of the land surface is surprisingly ancient. In tectonically stable regions, land surfaces, especially those capped by duricrusts, may persist a hundred million years or more, witness the Gondwanan and post-Gondwanan erosion surfaces in the Southern Hemisphere: Lester Charles King (1983) claims that remnants of erosion surfaces can be identified globally and correspond to pediplanation during the Jurassic period (the Gondwana planation surface), the Early to Mid-Cretaceous period (the Kretacic planation surface), the Miocene epoch (the Rolling land surface), the Pliocene epoch (The Widespread landscape), and the Quaternary sub-era (the Youngest cycle). Remnants of a ferricrete-mantled land surface surviving from the early Mesozoic era are widespread in the Mount Lofty Ranges. Kangaroo Island, and the south Eyre Peninsula of South Australia (Twidale et al. 1974). Indeed, much of southeastern Australia contains many very old topographical features: some upland surfaces originated in the Mesozoic era and others in the early Palaeogene period; and in some areas the last major uplift and onset of canyon cutting occurred prior to the Oligocene epoch (R.W. Young 1983; Bishop et al. 1985).

It is becoming increasingly clear that much of the soil landscape in temperate climates owes many of its features to earlier relief generations (Büdel 1982). In Europe, signs of ancient saprolites and duricrusts, bauxitic and lateritic sedimentation, and the formation and preservation of erosional landforms, including tors, inselbergs and pediments, have been detected (Summerfield and Thomas 1987). The palaeoclimatic significance of these finds has not passed unnoticed: it dawned on David L. Linton (1955), Alain Godard (1965), George H. Dury (1971), Michael F. Thomas (1978), and others that for much of the Cenozoic era, the tropical climatic zone of the Earth extended much further polewards than it

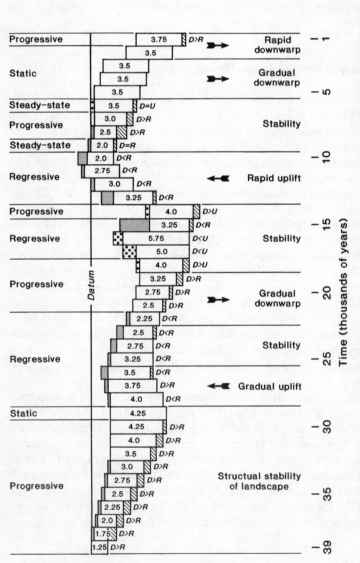

Fig. 5.18. Conceptual diagram showing the relation between the thickness components of an actively evolving hypothetical soil, and how this relates to progressive, regressive, and static pedogenesis. In the diagram, soil thickness, T, reflects the interplay between subsurface deepening, D, surface upbuilding, U, and removals, R, where $T = D + U - R$. Pedons are numbered *1* through *39*, each representing an arbitrarily assigned 1000-year period of soil development. The equalities and inequalities *marked at the pedon bases* show which of the three soil thickness processes (deepening, upbuilding, removals) have been predominant for that pedon and time period. (After D. L. Johnson 1985)

does today. Indeed, evidence from deposits in the soil landscape, like evidence in the palaeobotanical record, indicates that warm and moist conditions extended to high latitudes in the North Atlantic during the late Cretaceous and Palaeogene periods. Julius Büdel (1982) was convinced that Europe suffered extensive etchplanation during Tertiary times. Less controversial are the several vestiges of a tropical weathering regime which have been unearthed. In the British Isles, several Tertiary weathering products and associated landforms and soils have been discovered by Yvonne Battiau-Queney (1984, 1987), G. Frank Mitchell (1980), K.P. Isaac (1981, 1983), A.M. Hall (1985), and John A. Catt (1989). On Anglesey, which has been a tectonically stable area since at least the Triassic period, inselbergs, such as Mynydd Bodafon, have survived several large changes of climatic regime (Battiau-Queney 1987). In Europe, Asia, and North America many karst landscapes are now interpreted as fossil landforms originally produced under a tropical weathering regime during Tertiary times (Büdel 1982; Bosák et al. 1989). The *calas* of the Mallorquin coast, and the western Mediterranean generally, are now thought to be productions of fluviatile dissection during the Neogene and very early Pleistogene periods (Butzer 1962). Tropical landforms north of the Alps underwent a phase of valley formation during earliest Pleistocene times, intense glacial or periglacial activity (or both) and valley incision during the middle and late Pleistocene times, and minor modification during the Holocene epoch; but the large-scale landscape features have been inherited from Tertiary times and are "fossil" forms. Similar relict and fossil soils and landscapes are found in all the chief climatic zones, including arctic and subarctic regions of Canada (Valentine et al. 1987; Tarnocal and Valentine 1989), the subtropical regions of the United States (Nettleton et al. 1989), and the temperate zone of central and northwest Europe (Catt 1989). Remnants of even older soil landscapes also exist. Pre-Cretaceous landforms have survived in some places and have been exhumed by erosion in others (e.g. Battiau-Queney 1987; Lidmar-Bergstrom 1985). Just what proportion of the Earth's land surface pre-dates the Pleistogene period has yet to be ascertained, but it looks to be a not insignificant figure.

An important implication of all this work is that some landforms and their associated soils can survive through various climatic changes when tectonic conditions permit. A problem arises in how to account for the survival of these palaeoforms. Most modern geomorphological theory would dictate that denudational processes should have destroyed them long ago. It is possible that they have survived under the exceptional circumstance of a very-long-lasting arid climate under which the erosional cycle takes a vast stretch of time to run its course (Twidale 1976). A controversial explanation is that much of the Earth's surface is geomorphologically rather inactive. Robert W. Young (1983), for instance, would have us believe that the ancient landscape of southeastern Australia, rather than being an exceptional case, may be typical of a very substantial part of the Earth's surface. If this contention be correct, then widely held views on rates of denudation and on the relation between denudation rates and tectonics would require radical rethinking, and the connections between climate and landforms would be even more difficult to establish.

6 Animals and Plants

It is known . . . that the olive, the vine, the varieties of grain, and the fruit-trees, require entirely different constitutions of the atmosphere.

Alexander von Humboldt (1821, p. 23)

6.1 Climatic Influences on Life

6.1.1 Plants

The notion that climate can influence the constitution of vegetation and therefore control plant distribution dates back to the Ionian philosophers who flourished between the third and fifth centuries B.C. Menestor, in the fifth century B.C., was aware of a correlation between the evergreen and deciduous habits of vegetation and climate (Morton 1981). The first major treatise on plant geography and ecology was dished up by Theophrastus (370–385 B.C.) as an *Enquiry into Plants and Minor Works on Odours and Weather Signs* (1916 edn). In this work, Theophrastus displays a clear understanding of the connection between climate and plant distribution. The next leap forward in understanding plant geography came in the eighteenth century. Carolus Linnaeus (Carl von Linné) set forth elementary interactions of the animal, plant, and mineral kingdoms and their geographical distributions in his essay *Specimen Academicum de Oeconomia Naturae* of 1751. In his elephantine *Histoire Naturelle* (1749–89), Georges Louis Leclerc, Comte de Buffon, indicated that "evolution" and extinction could be caused by a change of climate. Karl Ludwig Willdenow in his *Grundriss der Kräuterkunde* (1792, 1805 edn), and Alexander von Humboldt in collaboration with Aimé Goujaud (called Bonpland) in their *Essai sur la Géographie des Plantes* (1807), all expressed the view that climate and vegetation have changed in the past, basing their beliefs on the evidence of fossil remains. Humboldt's essay was a result of his travels in South America with Aimé Bonpland, and it marked the start of Humboldt's attempts to found a science of plant distribution, the centrepiece of which would be the correlation of geobotanical zones with environmental factors such as temperature, elevation, and barometric pressure. Humboldt and his disciples focussed their attention on temperature zonation, altitudinal and latitudinal, to the exclusion of other environmental factors. Indeed, Humboldt's great achievement was to establish the correspondence between the world's chief thermal belts and its vegetation zones. For most of the nineteenth century, temperature was generally held to exert a decisive influence on the distribution of plants and animals. For example, Alphonse de Candolle, in his *Géographie Botanique Raisonnée* (1855), showed how the present distribution of plants may be explained by present climatic conditions. Like Alexander von Humboldt, he based his physiological classification of vegetation on temperature tolerance, and so it is no surprise that his vegetation types correspond to the major thermal zones of the Earth.

While the thermal ranges of vegetation types were being mapped out, a broader view of the relation of plants to their environment was emerging. The Swiss phytogeographer, Augustin Pyramus de Candolle, in his *Géographie Botanique* (1820), sharpened the meanings of the biogeographical terms "station" (which, at least in the form defined by Charles Lyell (1830–33, vol. ii, p. 130), is tantamount to the modern term "ecological niche"), and "habitation" (which is simply the geographical home of a species). Hewett Cottrell Watson, in a prefatory discussion to his *Remarks on the Geographical Distribution of British Plants; Chiefly in Connection with Latitude, Elevation and Climate* (1835), urged the need to collect information on a range of environmental factors — altitude, climate (temperature and humidity), highest and lowest altitudes of occurrence, and others. Richard Brinsley Hinds, surgeon-naturalist aboard H.M.S. *Sulphur* for 6 years in the Pacific Ocean, published a four-part article entitled *The Physical Agents of Temperature, Humidity, Light, and Soil, Considered as Developing Climate, and in Connexion with Geographic Botany* (1842), and in the next year a book, *The Regions of Vegetation; Being an Analysis of the Distribution of Vegetable Forms over the Surface of the Globe in Connexion with Climate and Physical Agents*. The multiplicity of environmental factors which limit the growth and survival of species was also established, in the context of agricultural crops, by Justus von Liebig in 1862. Here, then, were the seeds of an ecological view of relations between the living and non-living worlds, the germination of which we shall look at in the next chapter.

Late nineteenth-century geographers, choosing not to accept Charles Robert Darwin's discarding of climate as a force capable of shaping the distribution of organisms, took their inspiration from Alexander von Humboldt and adopted his holistic approach to phytogeography. However, many of them, and particularly August Heinrich Rudolf Grisebach, played down Humboldt's aesthetic appreciation of plant societies, and inspected in more detail the link between plant societies and the physical environment. In his *Der Vegetation der Erde nach ihrer klimatischen Anordnung* (1872), Grisebach recognized that the tropical forests, the deciduous forests, and the prairies are all single types of vegetation — distinct "formations", as he styled them, each produced by a vegetational response to a particular climate. To Grisebach, climate, and especially temperature, was the chief determinant of both individual plants and the community as a whole. This seemingly innocuous idea was controversial in its day. In its simplest form it was half mockingly called the theory of "temperature summing". But, in the face of criticism, Grisebach and many of his contemporary phytogeographers upheld Humboldt's isothermal lines wriggling over the globe and kept alive the long-established view that climate fashions the distribution of organisms. Thus they used the term "climate" to mean broad zones of temperature running strictly parallel with latitudinal belts, and were not very interested in other climatic elements. However, in 1884, Wladimir Peter Köppen made a map of climate which took on board seasonality. This spurred the mapping of zonal vegetation on the basis of temperature and other climatic factors. For instance, both Karl Georg Oscar Drude, director of the Royal Gardens at Dresden, in his *Manuel de Géographie Botanique* (1897) and Andreas Franz Wilhelm Schimper, in his *Pflanzengeographie auf physiologischer Grundlage* (1898, 1903 edn) identified formations of vegetation zoned on the basis of moisture and temperature. Indeed, the classification of world climates became an exercise in phytogeography. This is particularly true of Köppen's (1931) long-lasting classification of climates (Fig. 1.1).

Despite Köppen's best endeavours, temperature was still allotted the star role in explaining plant and animals distributions during the closing decades on the nineteenth century. In the United States, Clinton Hart Merriam, on the basis of his explorations in the San Francisco mountains of northern Arizona in the late 1880s, suggested, as Humboldt had done before, that zones of vegetation found on the flanks of mountains correspond to the zones of vegetation found within latitudinal belts. He estimated that each mile of altitude was equivalent to 800 miles of latitude. From here, he derived two primary life areas in the United States and Canada: a southern austral zone and a northern boreal zone. These large bands he subdivided into six more specialized life zones (seven if the tropical zone at the extreme trip of Florida be included), each defined by a certain range of temperature (Merriam and Stejneger 1890; Merriam 1893, 1894). Merriam's life-zone approach proved exceedingly popular, particularly among amateur students of natural history, but many scientists questioned the extent to which vegetation did conform solely to temperature distribution. To be sure, the change of vegetation over the grassland of the United States is more gradual than Merriam would have it, and proceeds, not so much from north to south in response to increasing temperature, as from east to west in response to decreasing rainfall (Shelford 1932, 1945).

In the United States, several important monographs on climate-vegetation connections were published during the first half of the present century. A national survey of the distribution of vegetation in relation to climatic conditions was published by E.B. Livingstone and F. Shreve (1921). This work spotlighted the complexity of the relationships between vegetation and climate and brought out the limited significance of mean annual climatological values, plant species being so sensitive to extremes of weather. Qualitative assessments of plant-climate relations were forthcoming. H.L. Shantz (1923), for example, summarized the relation of plant communities to the depth of penetration of soil moisture in the Great Plains area. In Europe, the Swedish botanist Göte Turesson began looking at the adaptations of plants to local environmental conditions. He felt that hereditary variation within species was probably far greater than previous workers had credited. In a series of experiments he showed variation associated with soil type and climate in a variety of plant species and coined the term "ecotype" to describe genetic varieties within individual species, some of which involve very subtle adaptation to local climatic conditions (Turesson 1922, 1925, 1930). Later studies confirmed Turesson's findings (e.g. Clausen et al. 1948).

6.1.2 Animals

The study of animal distribution advanced less quickly than did the study of plant distribution. By the end of the eighteenth century there were two works of note on the subject: Eberhardt August Wilhelm von Zimmerman's study of man and the quadrupeds entitled *Specimen Zoologicae Geographicae Quadrupedum Domicilia et Migrationes Sistens* (1777) and Johan Christian Fabricius's work on the distribution of insects in Europe called *Philosophia Entomologica* (1778). Early in the nineteenth century, Gottfried Reinhold Treviranus published his *Biologie; oder, Philosophie der Lebenden Natur für Naturforscher und Aerzte* (1802–22), which included a full survey of existing knowledge on the distribution and external conditions of plants and animals and was the forerunner of Humboldt's work (Rehbock 1983, p. 119). The role of environmental factors on the distribution of animals was considered at length by John Fleming, the Scottish zoologist and natural theologian. In his *The Philosophy of Zoology* (1822), Fleming took up the Linnean theme of the interrelatedness of the animal, plant, and mineral kingdoms, and identified temperature, food, and situation (habitat in the general sense) as the principal factors governing the distribution of animals. This eclectic approach was pursued with vim by Edward Forbes in his attempt to explain the distribution of pulmoniferous Mollusca in Britain (e.g. Forbes 1839). Work carried out on the same "ecological" lines during the middle of the nineteenth century was ably summarized by Karl Gottfried Semper in his *Animal Life as Affected by Natural Conditions of Existence* (1881). But it was not really until the early twentieth century that ecological principles were widely applied to the study of animal distributions. The modern era of animal ecology was heralded by a number of essays, including Charles Christopher Adams's report on *Isle Royale as a Biotic Environment* (1909) and his *Guide to the Study of Animal Ecology* (1913), and Victor E. Shelford's paper on *Physiological Animal Geography* (1911) and his *Animal Communities in Temperate America* (1913). The first major work to summarize the application of ecological principles to animal distributions was Richard Hesse's *Tiergeographie auf ökologischer Grundlage* (1924), which appeared in English as *Ecological Animal Geography* (Hesse et al. 1937).

6.2 Biogeographical Regularities

The work briefly discussed above (and much other work which for lack of space must pass unmentioned) has evolved into modern ideas about the way in which organisms interact with their environment. Emphasis has moved to the notion of "stress", a disputatious term which may be taken as "external constraints limiting the rates of resource acquisition, growth or reproduction of organisms" (Grime 1989). A wealth of modern ecological studies look at species populations under marginal conditions where climatic stress occurs (e.g. Root et al. 1986; Hill et al. 1988; Read and Hill 1988, 1989; Read and Hope 1989). The literature on the response of animals and plants to climatic stress is huge. Given that this book is chiefly concerned with systems of intermediate and large size, it would be inappropriate here to summarize the material which has been gathered on individual species. Rather, the approach adopted will be to dive in at the regional level and deal with broad biotic patterns which appear to relate to climate, namely, biogeographical regularities and, in the next section, patterns of species richness.

6.2.1 Ecogeographical Rules

During the nineteenth century it was noticed that the form of many warm-blooded animal species varies in a regular way with climate. These regularities led to the proposing of a number of ecogeographical "rules". Much later, Julian Huxley (1942) coined the term "cline" to describe a gradual change in measurable characteristics (size, colour, and so forth) of a species in response to a gradual change in climate, altitude, and other environmental factors (see also Rensch 1937–38). In effect, ecogeographical rules describe common patterns of clinal variation in animal populations produced by regular gradients of climatic factors, notably temperature and moisture.

Gloger's Rule. The first ecogeographical rule was established by Constantin Wilhelm Lambert Gloger in 1833. It states that races of warm-blooded animals in warmer regions are more darkly coloured than races in colder or drier regions. The explanation of this regularity is that animals in warmer regions require more pigmentation to protect them from the light. Gloger's rule was first observed in mammals such as wolves, foxes, and hares, but the same phenomenon has been observed in insects. In butterflies, for example, it is commonly found that the cold-season forms of populations which produce many generations a year (multivoltine populations) resemble forms produced by populations living at high altitude or latitude which produce one generation per year (univoltine populations). This indicates that the temperature effect is mainly a physiological response through melanin growth and deposition, and is not genetic (Lane and Marshall 1981, p. 11). The ladybird, *Harmonia axyridis*, found in southeast Asia, has a polymorphic elytral patterning based on black and orange pigmentation. In Japan, the extreme dark and light forms display a clinal variation running from north to south, the dark form (*H. a. conspicua*) being most frequent in the south, the light form (*H. a. succinea*) in the north. This cline can be explained by the effect of humidity and, possibly, temperature (Lane and Marshall 1981, p. 11).

Bergmann's Rule. Also known as the size rule, Bergmann's rule states that races of a warm-blooded species of animal are larger in cold climates than in warm climates. It was established by Carl Bergmann in 1847 and has been found to apply to a wide range of birds and to mammals (Rosenzweig 1968a; J.H. Brown and Lee 1969; Kendeigh 1969; McNab 1971; Murphy 1985; Koch 1986). Examples abound. Here are a few noted by Richard Hesse and his co-authors (1937, p. 387): puffins vary in size from a giant form, *Fratercular arctica naumanni*, which lives in Spitzbergen and northern Greenland (wing length 175–195 mm); through smaller forms living on Bear Island, the Norwegian coast, Iceland, and southern Greenland (wing length 158–177 mm), and a still smaller form living on the Channel Islands and Heligoland (wing length 155–166 mm); to a dwarf form (wing length 135–145 mm) that winters in Mallorca. In central Europe, the larger mammals, including the red deer, roe deer, bear, fox, wolf, and wild boar, increase in size towards the northeast and decrease in size towards the southwest. The skull length in the wild boar ranges from 560 mm in Siberia to 324 mm in southern Spain. Foxes introduced into Australia are only half the size of their English ancestors. In North America, the pocket gopher, *Geomys bursarius*, is 296 mm long north of latitude 46° N, 284 mm long between latitudes 40° and 46° N, and just 256 mm long south of latitude 40° N. Human populations also comply to Bergmann's rule, each subspecies having its own clinal system (Coon 1953, 1966; D.F. Roberts 1973). Of course, not all species obey Bergmann's rule: the capercaillie, *Tetrao urogallus*, for instance, is smaller in Siberia than in Germany.

Allen's Rule. Allen's rule, or the proportional rule, extends Bergmann's rule to include protruding parts of the body, such as necks, legs, tails, ears, and bills. Joel A. Allen (1877), after whom the rule is named, found that protruding parts in wolves, foxes, hares, and wild cats are shorter in cooler regions. To take an example, the relative size of ears increases in the desert fox, *Canis zerda*, the European fox, *Canis vulpes*, and the polar fox, *Canis lagopus* (Fig. 6.1). The jackrabbit, subgenus *Macrotolagus*, which resides in the southwestern United States, has ears one third its body length; in the common jackrabbit, *Lagus campestris*, which ranges from Kansas to Canada, the ears are the same length as the head. Another observation conforming to Allen's rule is that mammals with great surface areas relative to body mass, such as bats, are found chiefly in the tropics.

Hesse's Rule. Also known as the heart-weight rule, Richard Hesse's rule is basically an extension of Bergmann's rule. It states that extra metabolic work done to maintain heat in a cold environment causes a greater volume and mass of heart in animals living there as compared with their counterparts in warmer

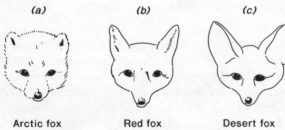

(a)   *(b)*   *(c)*

Arctic fox   Red fox   Desert fox
(Canis lagopus)   *(Canis vulpes)* *(Canis zerda)*

Fig. 6.1. Heads of **a** Arctic fox, **b** red fox, **c** desert fox. (After Hesse et al. 1937)

regions. Thus, the relative heart weight of the sparrow, *Passer montanus*, is 15.74 per thousand in Leningrad, 14.00 per thousand in northern Germany, and 13.1 per thousand in southern Germany (Hesse et al. 1937, p. 392).

Mayr's Rules. Ernst Mayr (1942) added a few extra rules which apply only to birds: in colder climátes, the number of eggs in a clutch are larger, the digestive and absorptive parts of the gut are larger, the wings longer, and migratory behaviour more developed.

6.2.2 Climatic Interpretations of Ecogeographical Rules

Classically, Bergmann's and Allen's rules are interpreted in terms of relations between body surface area and environmental temperature, and expressed in simple mathematical terms. As an object increases in size, its surface volume becomes relatively smaller (increasing by the square) than the volume (increasing by the cube). When the protuberances are relatively shorter, then the surface area is reduced even more. The traditional physiological explanation is that this mechanism reduces heat radiation so that animals in colder climates have a relatively smaller surface area from which they may lose heat. However, there has been some doubt cast as to whether size change with ambient temperature truly be a consequence of heat conservation. The opposing view is that relative reduction in surface area is hopelessly small for an effective reduction in heat loss, and holds that the principle mechanisms for regulating body temperature are insulation and vascular control. Andrew R. Cossins and K. Bowler (1987, p. 122) contest that Bergmann's rule cannot be explained as an adaptive strategy for energy conservation because, notwithstanding differences in weight-specific metabolism, a larger animal will have a greater overall energy requirement than a smaller animal. They allege that large size may relate more to the greater ability to store energy, a factor which would seem to be rather crucial in a harsh environment. There have been no critical experiments to test the rival ideas, but Carleton Ray (1960) investigated whether the rules apply to poikilotherms, excluding fish, by rearing both vertebrates and invertebrates at different temperatures. His experiments showed that body length increased between 10 and 50 percent, and body weight increased by well over 100 percent, for a 10 °F (5.5 °C) decrease in temperature. Protruding body parts differed less dramatically with temperature, showing an increase 2 to 9 percent less than body length. Nonetheless, the correlation between laboratory experimental results and observations made in Nature were good, and affirmed that ecogeographical rules apply to poikilotherms as well as to homeotherms, as suggested by Bernhard Rensch (1932) in the case of the size and relative thickness of the shell of land snails and by C.C. Lindsey (1966) in the case of vertebrates. Other researchers have educed factors, such as the presence or absence of potential competitors, food size, habitat productivity, species diversity, and equilibrial niche size, to explain Bergmann's and Allen's rules (e.g. Rosenzweig 1968a; McNab 1971; Boyce 1978; Koch 1986), but climate surely is an important factor, as the work discussed below strongly suggests.

6.2.3 Modern Studies of Bergmann's Rule

A careful study of the relationship between climate and geographical size variation in birds in the eastern and central United States was made by Frances C. James (1970). Taking wing length as an indicator of body size, James produced a series of isophenetic maps which showed increasing size northwards and westwards from Florida in the following species: the hairy woodpecker (*Dendrocopos villosus*), downy woodpecker (*Dendrocopos pubescens*), blue jay (*Cyanocitta cristata*), Carolina chickadee (*Parus carolinensis*), white-breasted nuthatch (*Sitta carolinensis*), and eastern meadowlark (*Sturnella magna*). She found that in all cases there is a tendency for larger (or longer-winged) birds to extend southwards in the Appalachian Mountains, and for smaller (or shorter-winged) birds to extend northwards in the Mississippi River valley. In the cases of the downy woodpecker, female white-breasted nuthatches, and female blue jays, she noticed a tendency for relatively longer-winged birds to extend southwards into the interior highlands of Arkansas, and relatively shorter-winged birds to extend northwards into other river valleys. These subtle relations between intraspecific size variation and topographical features indicate that the link between the two phenomena may involve precise adaptations to very minor climatic gradients, and have also been detected in a study of the red-cockaded woodpecker (Mengel and Jackson 1977). Frances C. James also correlated the wing length in the downy woodpecker and seven other bird species with a variety of climatic variables: monthly, seasonal, and annual dry-bulb and wet-bulb temperatures, vapour pressure, vapour pressure deficit, and absolute humidity. The outcome of this analysis was that the variation in wing length correlated most highly with those variables, such as wet-bulb temperature, which register the combined effects of temperature and humidity. This suggested that size variation is not dependent solely upon dry-bulb temperature, that moisture levels are significant, too.

Clinal patterns akin to those found by Frances C. James in birds emerged from a study of three mammal species carried out by James R. Purdue (1980). The species investigated were the eastern cottontail (*Sylvilagus floridanus*), fox squirrel (*Sciurus niger*), and grey squirrel (*Sciurus carolinensis*). Body size in all three species, as measured in several skeletal elements, displayed strong east-west patterns along the western edge of the eastern deciduous forest with the eastern cottontail and grey squirrel increasing in size and the fox squirrel decreasing in size. The two squirrel species also evinced a north-south trend in size: the grey squirrel is larger in the north, the fox squirrel smaller. The fact that these clinal patterns are similar to those found by James points to some common, underlying cause. James confessed her ignorance as to the causes of clinal variation in the birds she studied, but managed to ascertain that winter climatic conditions appear to be important. This finding was borne out by Paul L. Koch's (1986) investigations of clinal variation in the opossum (*Didelphis virginiana*), striped skunk (*Mephitis mephitis*), white-tailed deer (*Odocoileus virginianus*), eastern mole (*Scalopus aquaticus*), and eastern grey squirrel (*Sciurus carolinensis*), the size of all of which species, bar the eastern mole, correlated most highly with the average temperature of the coldest month. Whatever cause them, it is certain that morphological clines may evolve very swiftly. Clines in house sparrows, *Passer domesticus*, have evolved in North America within 100 years and resemble the clines found in Europe (Johnston and Selander 1971), while the European wild rabbit (*Oryctolagus cuniculus*), introduced in eastern Australia a little over a century ago, already displays clinal variation in skeletal morphology. The rapidity of clinal evolution revealed by these empirical studies has been reproduced using genetic models of populations which show that, even in the presence of gene flow, clines can develop within a few generations (Endler 1977).

While the size, colour, and other characteristics of a species can be looked at individually in relation to climate, in reality they will generally respond to climate in concert. A particularly good example of the overall adaptive response of a species to climate is furnished by mole rat populations (Nevo 1986). Subterranean mole rats of the *Spalax ehrenbergi* complex living in Israel comprise four morphologically indistinguishable incipient chromosomal species (with diploid chromosome numbers 2n = 52, 54, 58, and 60). These four chromosomal species appear to be evolving and undergoing ecological separation in different climatic regions: the cool and humid Galilee Mountains (2n = 52), the cool and drier Golan Heights (2n = 54), the warm and humid central Mediterranean part of Israel (2n = 58), and the warm and dry area of Samaria, Judea, and the northern Negev (2n = 60). All the species are adapted to a subterranean ecotype: they are basically cylindrical with short limbs and no external tail, ears, or eyes. Their size varies according to heat load: large individuals in the Golan Heights to smaller ones in the northern Negev. These size differences presumably minimize the risk of overheating in different climatic conditions. The colour of the pelage ranges from dark on the heavier black and red soils in the north to light on the lighter soils in the south. The smaller body size and paler pelage colour mainly associated with 2n = 60 helps to mitigate the heavy heat load in the hot steppe regions approaching the Negev desert (Nevo 1986). The mole rats show several adaptations at the physiological level — basal metabolic rates decrease progressively towards the desert, thus minimizing water expenditure and overheating; more generally, the combined physiological variation in basal metabolic rates, non-shivering thermogenesis, thermoregulation, and heart and respiratory rates, appears to be adaptive at both the macroclimatic and microclimatic levels, and both between and within species, thus contributing to energy optimization. Ecologically, territory size correlates negatively and population numbers correlate positively with productivity and resource availability. Behaviourally, activity patterns and habitat selection appear to optimize energy balance, and differential swimming ability appears to overcome winter flooding, all paralleling the climatic origins of the different species. In summary, the incipient species are reproductively isolated to varying degrees representing different adaptive systems which can be viewed genetically, physiologically, ecologically, and behaviourally. All are adapted to climate, defined in terms of humidity and temperature regimes, and ecological speciation is correlated with the southwards increase in aridity stress. Whether climate play such a dominant role in the evolution of the majority of species has yet to be discovered.

6.3 Patterns of Species Richness and Climate

6.3.1 The Species-Energy Theory

The striking geographical variability of species number (diversity or "richness") has been known at least since Alfred Russel Wallace wrote his *Tropical Nature and Other Essays* (1878). Studies on several groups of animals have unmasked latitudinal, altitudinal, and (to a lesser extent) longitudinal gradients of species richness, as well as a tendency for fewer animal species to occur on peninsulas. A primary thrust of modern ecology has been to track down the factors which are responsible for producing this pattern. Historical processes (speciation and dispersal), climate, climatic variability, topography, biotic processes (primary productivity, competition, and so forth), disturbance, and the richness of other groups of organisms have all been suggested as important factors in explaining species richness. There is little doubt that all these factors can operate on a local scale. But a number of recent studies have shown clearly that, for both animals and plants, present-day species richness for largish regions can be explained very well in terms of climatic factors, especially available energy.

The idea that regional species richness may be determined by energy was proposed by G. Evelyn Hutchinson in 1959, and developed by Joseph H. Connell and E. Orias (1964) and James H. Brown (1981). The first person to set down a "species-energy" hypothesis to explain species richness was David Hamilton Wright (1983). Wright's hypothesis states that, subject to water supply and other factors being not limiting, diversity within terrestrial habitats is to a great extent controlled by the amount of solar energy available, declining with latitude in accordance with the polewards decrease of solar radiation receipt. In support of his hypothesis, Wright produced highly significant regression equations relating plant and bird diversity (taken from a worldwide sample of 36 islands) to solar energy.

Against Wright's hypothesis, it might be objected that solar energy and diversity are both known to decline with latitude and one would expect them to be correlated. However, John R.G. Turner, Catherine M. Gatehouse, and Charlotte A. Corey (1987), looking at the diversity of butterflies and moths in Great Britain, showed that the species-energy relationship holds only during that part of the year when the insects are absorbing energy, and cannot be detected during diapause; and that diversity is highly correlated with sunshine and temperature data, variables which are to some extent independent of latitude. In a later study, Turner and two other colleagues, Jack J. Lennon and Jane A. Lawrenson, found that the varying distribution of small insectivorous birds in Britain in summer and winter corresponds well with the prediction of the species-energy theory (Turner et al. 1988), though their thesis has met with stiff opposition from some quarters (see Cousins 1989; Elkins 1989; Turner and

Lennon 1989). They avoided the problem of picking up the latitudinal decrease in species richness by hypothesizing that a correlation between energy and species diversity would only apply during winter months when organisms are actively absorbing energy, and would not apply in the case of summer visitors, or might become negative in the case of hibernating insects. Their findings confirmed the predictions of the species-area theory, the diversity of the British birds studied being correlated with various seasonal climatic variables. To explain the correlations between species richness and energy, they adopted a simple stochastic theory: given a constant turnover (colonization and extinction) of species in any area, regions in which populations are bigger will have more species in the steady-state because smaller populations become extinct more frequently. Bird populations might be larger where more energy is available, either because of the indirect effect of productivity, or through the direct effect of the lower metabolic energy needed to maintain body functions when the air is warmer — cold weather kills birds. The results obtained by Turner and his colleagues provide some support for this view, for if birds merely sought immediately favourable weather, resident species should correlate only with the current season, whereas their distributions both in winter and in summer correlate with winter and summer weather, suggesting that climate influences their ability to support viable populations round the year. Furthermore, if species richness in some way be determined by the thermal budget of organisms, then homeotherms will show only weak correlations with sunshine since birds do not use sunlight for temperature control, but the species richness of ectotherms, which do regulate their temperature by basking in sunlight or by living in the boundary layer of air heated by the Sun, should show correlations with both temperature and sunshine. And indeed, the bird species studied displayed weak correlations with sunshine hours, whereas the butterflies and moths were strongly correlated with with both temperature and sunshine.

6.3.2 Species Richness in North America

David J. Currie and Viviane Paquin (1987) tested the species-energy theory by inspecting the distributions of North American trees. They divided the continental United States and Canada into 336 quadrats, each $2\frac{1}{2}° \times 2\frac{1}{2}°$ in size south of latitude 50° N, and $2\frac{1}{2}°$ (latitude) × 5° (longitude) in size north of latitude 50° N. Taking range maps of the 620 indigenous tree species (a tree species being defined as any ligneous plant growing to 3 m or more anywhere within its range), they counted the number of species within each quadrat. Next, they obtained physical and climatic attributes of each quadrat using published maps. The spatial pattern of tree species richness in North America is shown in Fig. 6.2. A pronounced latitudinal gradient occurs only in the east. Maximum tree species richness occurs on the high plateau of the Appalachian Mountains, whereas minimum tree species richness occurs in areas immediately to the east of the Rocky and Sierra Nevada Mountains. Peninsulas do not contain notably fewer species. This pattern contrasts with species richness patterns for birds and mammals where maxima occur in the Rockies and Appalachians, local minima

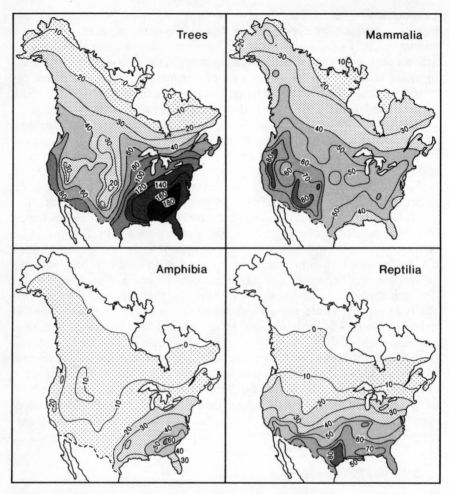

Fig. 6.2. The species richness of North American trees, mammals, amphibians, and reptiles. *Contours* join points with the same approximate number of species per quadrat. (After D. J. Currie and Paquin 1987 and D. J. Currie 1991)

occur on peninsulas, and fewer species occur in the southeast (Fig. 6.2). In an attempt to account for the richness pattern of trees, Currie and Paquin correlated species richness with physical and environmental factors. The strongest correlation was with annual evapotranspiration. The relation between tree species richness and annual evapotranspiration was best described by the following non-linear regression (Fig. 6.3):

$$y = \frac{185.8}{\left\{1.0 + e^{(3.09 - 0.00432\,x)}\right\}} \quad (r^2 = 0.762),$$

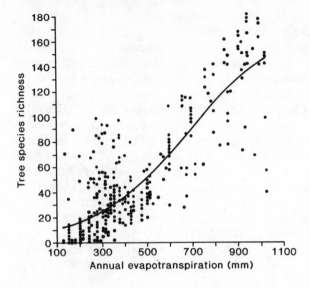

Fig. 6.3. North American tree species richness as a function of annual actual evapotranspiration. The *solid line* is a fitted logistic model. (D. J. Currie 1987)

where y is tree species richness and x is annual realized evapotranspiration (mm). Most of the residual variation, explored by multiple regression, is explained by the range of elevation within a quadrat and the distance from the coast, especially in low latitudes. No significant variation in the data could be attributed to quadrat area, seasonality of climate, whether the quadrat had been glaciated, or whether the quadrat were on a peninsula. A crucial question D.J. Currie and V. Paquin asked is why evaporation should be so important in explaining tree species richness. A likely answer is that evapotranspiration is highly related with terrestrial primary productivity and is thus a measure of energy consumption by a community. This idea is consistent with the hypothesis that energy is partitioned among species and the total available energy thus limits the number of species in an area. The limits to species richness may be varied by other factors. Mountains, for example, being physically complex, allow even more partitioning of energy than the available energy level would suggest and so can support a greater number of species.

A surprising result of D.J. Currie and V. Paquin's work, and a result found as well in the studies carried out by John Turner and his associates, is the apparent insignificance of historical processes such as glaciation and dispersal. To explore this finding more fully, D.J. Currie and V. Paquin examined the tree species richness in Great Britain and Ireland in $2\frac{1}{2}° \times 5°$ quadrats. The accepted view is that the British and European flora are impoverished relative to the North American flora owing to barriers to post-glacial recolonization in Europe. However, the observed tree species richness in the British Isles is very close to the number predicted by regression equations established for the North American data. Although it would be rather unwise to discard historical factors out of hand, it would seem that the most parsimonious interpretation of the results is that tree species richness in North America and the British Isles varies among regions primarily as a function of varying amounts of energy available to the community (D.J. Currie and Paquin 1987).

   In a further study of the energy theory of species richness, D.J. Currie (1991) mapped the number of species of mammals, reptiles, amphibians, and birds in North America (Fig. 6.2; bird species richness not shown) and correlated them with 21 descriptors of the environment. The environmental variables were chosen to test a variety of hypotheses concerning species richness (Table 6.1). The maps reveal the well-known decline of species richness with latitude, but the relationship is monotonic only in the cases of the reptilia and trees: birds and mammals showed pronounced richness peaks at latitudes 44° N and 39° N respectively, while amphibian richness peaked more gently at latitude 34° N. The pattern of tree richness has already been discussed. It would seem reasonable to suppose that the richness of animals would vary as a function of the richness of vascular plants as a whole which, though not mapped by D.J. Currie, should covary with tree species richness. However, it is evident from Fig. 6.2 that only amphibian richness shows a clear monotonic relation with tree species richness.

Table 6.1. Factors hypothesized to influence species richness. (After D.J. Currie 1991)

Factor	Rationale	Variables used
Climate	Benign conditions permit more species	Mean annual: temperature, precipitation, potential and actual evapotranspiration, total surface solar radiation receipt; and elevation
Climatic variability	Stability allows specialization	Difference between mean January and mean July temperature and precipitation
Habitat heterogeneity	Physically or biologically complex habitats furnish more niches	The difference between the maximum and minimum values of all the above-listed variables
History	More time allows more complete colonization and the evolution of new species	Quadrat glaciated of inundated during the Wisconsin — yes or no?
Energy	Richness is limited by the partitioning of energy among species	Mean annual solar radiation
Competition	(i) Competition favours reduced niche breadth; (ii) competitive exclusion eliminates species	Difficult to assess on a large scale and therefore neglected
Predation	Predation retards competitive exclusion	Difficult to assess on a large scale and therefore neglected
Disturbance	Moderate disturbance retards competitive exclusion	Difficult to assess on a large scale and therefore neglected Addititonal variables: whether or not quadrat is on a sea coast or on a peninsula (Nova Scotia and Michigan were treated as peninsulas)

Another hypothesis would be that animal species richness is related to primary productivity. In fact, except in the case of the amphibians, no tight relationship between between species richness and productivity was apparent in the study (but see J.G. Owen 1988). By far the strongest relations, as measured by non-parametric correlation coefficients, were between vertebrate species richness and virtually all the climatic variables, with the three strongest correlates being annual potential evapotranspiration, solar radiation, and mean annual temperature, all of which are aspects of the regional energy balance. Annual potential evapotranspiration, which is a measure of crude, integrated, ambient energy, alone accounted for about 79 percent of variability in species richness. Even at the level of family and order, richness was highly correlated with annual potential evapotranspiration (Figs. 6.4 and 6.5). Thus, the outcome of this study is a strong vindication of the species-energy hypothesis: of all the factors which could affect species richness, energy (or, at least, surrogate climatic measures of it) stands out as the chief factor. This is fairly easy to appreciate in the case of plants: according to the species-energy hypothesis, in regions of the same area, species richness is determined by the energy flux; primary productivity in plants is a direct measure of energy capture and has been found on a large scale to correlate most strongly with annual potential evapotranspiration. The case of animal species richness is less straightforward because several hypotheses should be considered (Table 6.1). However, on the basis of D.J. Currie's researches, the only hypothesis which seems tenable is that animal species richness is determined largely by regional ambient energy levels. But why should vertebrate richness be influenced more by crude atmospheric energy than by energy available in the form of food? A possible answer is that the thermal budget of individuals, which depends on atmospheric energy, is more important in deter-

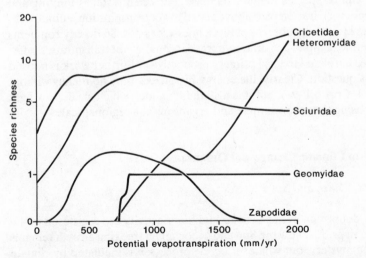

Fig. 6.4. The species richness of five North American mammalian families as a function of potential evapotranspiration. (After D. J. Currie 1991)

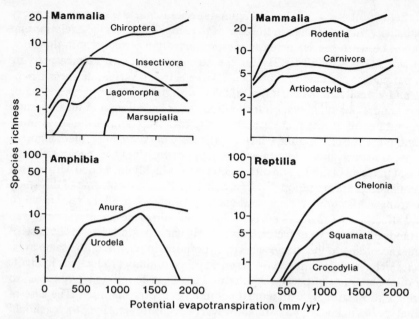

Fig. 6.5. The species richness of orders of amphibia, reptiles, and mammals living in North America as a function of potential evapotranspiration. (After D. J. Currie 1991)

mining species richness than is food energy. Another speculative answer was proffered by D.J. Currie: so long as water is not limiting, both plant species richness and animal species richness increase, but when water is limiting and potential evapotranspiration rises above actual evapotranspiration, animal richness continues increasing whereas plant richness does not. So do environments in which potential evapotranspiration exceeds actual evapotranspiration offer energy sources unrelated to local primary productivity? Further work is needed to answer this question. Clearly, the energy hypothesis is by no means without problems (D.J. Currie 1991), but it is a stimulating idea which sheds new light on the relationships between climate and organisms at a regional scale.

6.4 Short-Term Climatic Change and Organisms

6.4.1 Animals, Plants, and Solar Cycles

There is little doubt that cycles of solar and lunar activity are registered in the growth of tree rings. Just how the Sun and Moon affect tree ring growth remains something of a mystery, but whatever be the process, it is mediated by climate (Schove 1961; Fritts 1976). Regularities in the thickness of tree rings were first noticed by the astronomer Andrew Ellicott Douglass at the end of the nineteenth century. He realized that each ring resulted from the annual growth of the tree, mostly during springtime, and that the thickness of the ring depended on climatic factors; he looked for, and found, an 11-year period in the trees he studied (Douglass 1909, 1918, 1919). Edouard Brückner's (1890) 35-year climatic cycle was discovered in the annual growth rings of a yew tree, 200 years old, growing on a southwest slope overlooking Lower Sondley in the Forest of Dean (Anonymous 1928). As revealed by a Mr. E.G. Burtt, the rings had apparently grown more rapidly in dry intervals than in wet ones, and clear growth maxima occurred in the years 1790, 1830, 1860–1870, and 1900. In the United States, the growth patterns of a huge number of trees were investigated, largely under the guidance of Ellsworth Huntington. These studies vindicated Brückner's cycle, an approximate 35-year cycle being evident in the growth rings of the giant *Sequoias* of California. Curiously, the Brückner cycle signal is, in the main, actually stronger in tree-ring sequences than in rainfall data (Brooks and Glasspoole 1928, pp. 181–185). More recent work has established the presence of the signals of all the solar cycles (and other signals as well) in tree ring sequences. The 22-year Hale cycle in solar magnetic polarity and its 11-year sunspot hemicycle are paralleled in rhythmical tree growth (Zeuner 1952; Schove 1961, 1983), as well as in isotopic variations within trees' growth bands (Libby 1983), although no evidence for the Hale cycle in tree rings has been found prior to 1600 (Schove 1987, p. 359). The 178.73-year cycle of solar inertial motion has been detected in the 10,000-year record of radiocarbon preserved in tree rings (Fairbridge and Shirley 1987). Dendrochronological evidence for a 200-year weather cycle exists, though it is rather inconclusive in the A.D. period (Schove 1983, pp. 318, 327).

The effect of solar cycles on plant distribution is unclear, but a few pieces of evidence suggest a link of some kind. Pollen production by alder (*Alnus*), oak (*Quercus*), and hazel (*Corylus*) seem to exhibit cycles of the right length, though whether the distribution of these species alter as well is unknown (Wijmstra et al. 1984). Severe droughts in the American Great Plains, correlated with sunspot cycles of 46 and 91 years, led to changes in plant occurrence (Abbot 1963). And during the Little Ice Age, when sunspots were few in number, species extinctions occurred in Europe (Lamb 1982). A correlation between sunspots and the quality of wine vintages has also been established (Stetson 1937). Other studies have forged a link between sunspot activity and cycles of change in animal populations. Using power spectrum analysis, A.J. Southward and his colleagues (1975) found a 10- to 11-year cycle in the surface temperatures of the English Channel which influenced the numbers of warm-water fish species, such as hake and red mullet, and cold-water fish species, such as cod and haddock. Paul R. Hurt and his colleagues (1979) looked for periodicities in the records of annual populations of blue crab (*Callinectes sapidus*) in Chesapeake Bay for the years 1922 to 1976. They discovered variations with periods of 18.0, 10.7, and 8.6 years. Within the limits of experimental error, the period of 18.0 years accords with the 18.6-year lunar nodal cycle, the 10.7-year period corresponds (probably) to the sunspot cycle, and the 8.6-year period relates to the 8.8-year period of the Earth-Moon-Sun tidal force. The 18.0- and 8.6-year signals are almost identical to the signal for annual rainfall at Philadelphia. This accordance led Hurt and his colleagues to suggest that phases of minimum rainfall allow high tides to wash nutrients into the surface waters of the Bay, and allow the waters to become more saline, thus promoting growth of the crab population.

A controversial link between solar activity and the outbreak of viral diseases has been suggested by those masters of the outrageous hypothesis, Fred Hoyle and Chandra Wickramasinghe (1990). These astronomers claim that the 11-year pulse in the solar wind drives viruses, arrived from deep space in the tail of a comet and trapped in the upper atmosphere, down to the surface, where epidemics break out. In support of this notion, they point to the remarkable coincidence of sunspot numbers and the occurrence of influenza pandemics since 1761.

6.4.2 Animals, Plants, Volcanoes, and Meteorites

We saw in Chapter 2 that the dust and gases produced by volcanoes and by bolide bombardment are likely to perturb the world climate system, so causing it to change. Little is known about the indirect influence of such climatic perturbations on animals and plants. Climatic changes induced by volcanic eruptions may have an effect on beings living far from the zone of direct damage, but this is by no means certain. A coincidence has been noted between Egyptian and Biblical reports of darkness and rains of ash at the time of the Exodus and the explosive eruption of Santorini (Thera) in the Aegean Sea in the second millenium B.C. (e.g. D.J. Stanley and Sheng 1986). This large explosion may have contributed also to the downfall of Minoan Crete (Marianatos 1939), and have had environmental repercussions as far away as China (Pang and Chou 1985). Less dramatic changes may occur as well. The occurrence of frost damage in the annual rings of species such as *Pinus longaeva* and *P. aristata* exhibits a good relationship with the timing of known volcanic events (LaMarche and Hirschboeck 1984). And studies of tree rings in Irish bog oaks seem to reveal volcanic events at 4375 B.C., 3195 B.C., 1628 B.C., 107 B.C., and 540 A.D. (Baillie and Munro 1988), which dates may correspond to peaks of activity in Greenland ice cores (Hammer et al. 1980, 1987).

The impact of a bolide may have both immediate and delayed effects on climate and thus organisms. Within seconds of piercing the atmosphere, a bolide would create an extraordinarily strong wind. A bolide with a diameter of about 14 km would generate a superwind capable of flattening forests within a range of 500 to 700 km (Emiliani et al. 1981). The impact itself would create a blast wave producing overpressures capable at their peak of destroying forests and killing animals (Napier and Clube 1979). Particularly vulnerable would be large land vertebrates with a small ratio of strength to weight, a fact which has been used to explain the selective extinction of large dinosaurs at the close of the Cretaceous period (Russell 1979). A wave of intense heat would also radiate from the site of impact, killing all exposed life forms within the lethal radius. For impacts of bodies 10 km in diameter the lethal radius could include areas of continental size. The intense heat would trigger wildfires which would release soot into the atmosphere (Sect. 2.5.2). The darkness caused by the thick dusting of the atmosphere would lead to reduction or a collapse of photosynthesis (Alvarez et al. 1980; Emiliani et al. 1981) and a breakdown of food chains (Russell 1979). Michael R. Rampino and Tyler Volk (1988) have calculated that a rapid global warming of 6 °C would have followed a large-body impact at the Cretaceous-Tertiary boundary because most of the marine calcareous phytoplankton would have been killed, causing a severe drop in the production of dimethyl sulphide, and so a reduction in the generation of condensation nuclei in the marine atmosphere and a concomitant fall in the marine cloud albedo. The dramatic change in climate may be the cause of the relatively sudden disappearance of large parts of the fauna and flora at the close of the Cretaceous period, and radical climatic changes caused by the same mechanism might have occurred at the Precambrian-Cambrian, Devonian-Carboniferous, and Permo-Triassic boundaries as well. Looking back to the remotest past, it seems likely that during the earliest period of the Earth's history, very heavy bombardment by enormous asteroids probably produced lethal conditions by evaporating large volumes of ocean water, thus preventing the evolution of ecosystems (Sleep et al. 1989).

6.5 Medium-Term Climatic Change and Organisms

6.5.1 Bergmann's Rule and Holocene Faunas

Geographical variations in morphological features in mammals are known from fossil populations. Nine species from the New Paris No. 4 fauna in Pennsylvania, for instance, tend to increase in size towards the north, although four species tend to get larger in the opposite direction (Lundelius et al. 1983). Bergmann's rule can also be seen in operation in the changing size of members of a population during time. In this case, it is assumed that the size adjustments track climatic change. In Missouri, archaeological specimens of the grey squirrel (*Sciurus carolinensis*) increase in size from the early to middle Holocene epoch; specimens of the eastern cottontail (*Sylvilagus floridans*) decrease in size during the same time interval and then increase in size to modern proportions during the late Holocene epoch (Purdue 1980). James R. Purdue (1986, 1989) believes that changes in the size of the white-tailed deer (*Odocoileus virginianus*) in central Illinois during Holocene times were strongly influenced by insolation-driven summer climate acting through food resources, in particular summer forage. Other examples are discussed by Holmes A. Semken (1984, pp. 190–192). Longer-term changes of body size have been charted by Richard G. Klein (1986) in fossil populations of carnivores from South Africa. Klein found that in 14 of the 17 modern carnivore populations, carnassial tooth length (which should directly reflect body size) tends to increase with latitude south. The mean carnassial length in fossil samples of the same species tend to be greater in samples which accumulated under relatively cool conditions. In the case of the black-backed jackal, *Canis mesomelas* (Fig. 6.6), carnassial teeth show a tenden-

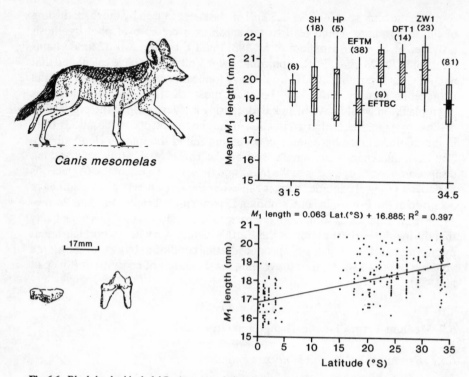

Canis mesomelas

17mm

Fig. 6.6. Black-backed jackal (*Canis mesomelas*): length of carnassial teeth, M_1, versus latitude. The *bottom graph* shows modern samples with a regression equation fitted. The *top graph* shows fossil samples. The sites are: *SH* Sea Harvest; *HP* Hoedjies Punt; *EFTM* Elandsfontein Main; *EFTBC* Elandsfontein Bone Circle; *DFT1* Duinefontein 1; *ZW1* Swartklip 1. (After Klein 1986)

cy towards large size in "cool" samples from Equus Cave layers 1B and 2B (not shown in the diagram), Elandsfontein Bone Circle, Sea Harvest, Duinefontein 1, and Swartlip 1; and show a tendency towards small size, comparable with the modern size range, in "warm" samples from Equus Cave layer 1A and Elandsfontein Main. Brown hyena (*Hyaena brunnea*) samples from the Equus Cave, Sea Harvest, and Elandsfontein sites display the same pattern of size as the jackals, though the relationship is less trustworthy owing to the smaller size of the samples (Fig. 6.7).

6.5.2 Pleistocene Extinctions and Extirpations

The extinctions at the end of the Pleistocene epoch affected much of the world's terrestrial fauna, and particularly the megafauna. In North America, 43 genera died, including 73 percent of the megafauna; in South America, at least 46 genera became extinct, including 80 percent of the megafauna; and in Australia, 26 genera disappeared, including 86 percent of the megafauna. There were few true

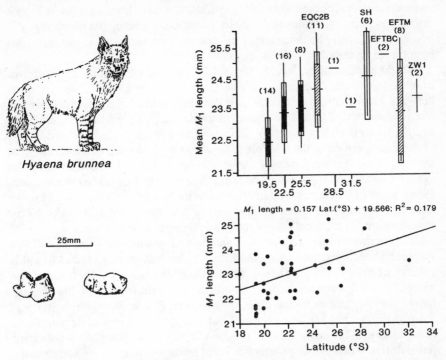

Fig. 6.7. Brown hyena (*Hyaena brunnea*): length of carnassial teeth, M_1, versus latitude. The *bottom graph* shows modern samples with a regression equation fitted. The *top graph* shows fossil samples. The sites are as for Fig 6.7 with the following addition: *EQC2B* Equus Cave Level 2B. (After Klein 1986)

extinctions in Europe, Africa, and Asia, though there were many extirpations. Many taxa survived the extinction event but did not come through unscathed: surviving animals and plants changed their geographical ranges and abundances and reassembled into new communities.

The cause of the late Pleistocene extinctions has been debated for nearly 150 years. The argument has always been divided between climatic change or else human impact as the causative agent (R. Owen 1846; Lyell 1863), though middle-of-the-road positions have also been taken wherein environmental change weakens populations rendering them vulnerable to Man the hunter, who delivered the coup de grâce (e.g. Kurtén and Anderson 1980, p. 363). In North America, for instance, the rapid disappearance of the megafauna around 10,000 years ago has been explained by the overkill hypothesis, which suggests that the Clovis hunters were responsible (Martin 1984a,b), and by the climatic hypothesis, which sees a sudden (square-wave) environmental change about 10,000 years ago as the culprit (Wendland 1978). Late Cenozoic extinction episodes do correlate with bouts of rapid climatic change, particularly those associated with rapid glacial terminations (Vrba 1984; S.D. Webb 1984). But in some cases climatic change coincides with the arrival of humans, thus making it difficult to point the finger at the guilty party. Only where climatic change did not coincide with the appearance of humans, or vice versa, can the rival hypotheses be tested.

In many parts of the world, terminal Pleistocene extinctions can be shown to relate to climatic change, on both a local and a regional scale. The correlation between local changes of climate and the extinction of individual species has been demonstrated for the American mastodon (J.E. King and Saunders 1984) and the Irish elk (Barnosky 1986). In Ireland, that magnificent beast and largest of the European deer, the Irish elk (*Megaloceras giganteus*), became extinct at the same time that the Younger *Dryas* cold spell took hold some 10,600 years ago, but 1000 years before the first humans arrived in Ireland (Barnosky 1985, 1986). Signs of butchering and hunting were looked for on hundreds of known Irish specimens of the giant elk but nary a one was found (Barnosky 1985). Investigation of pollen data support climatic change as the root cause of the Irish elk's extinction. It suggests that the quantity and quality of forage, as well as a shortened feeding season, decreased during the Nahanagan Stadial (roughly equivalent to the Younger *Dryas*). The teeth of the giant elk suggest that it was an opportunistic browser, supplementing its diet with large amounts of grass. Browse plants, such as *Juniperus, Empetrum,* and *Betula*, were common before the Nahanagan Stadial, when they all but disappeared. Grasses declined, too. Compared with the environment immediately before the Nahanagan cold spell, fewer nutrient-rich plants were available for fewer weeks in spring and summer, and fewer browse plants were available in winter. Studies of living artiodactlys indicate that these changes would have put the elk populations under stress because the energy intake required to sustain the large bodies and build up fat reserves for the next winter would have been increasingly difficult to maintain. Eventually, deaths, caused chiefly by winter-kill, would have outnumbered births, and extinction would have ensued. Taphonomic data support this conclusion: attritional age-frequency distributions, the presence of antlers, and bone weathering in most of the Irish specimens, and the disproportionately high number of young adults found in a site at Balybetagh, attest to death during the winter, probably owing to malnutrition. Thus climatic deterioration, acting through changes in vegetation, seems to have led to the expulsion of the Irish elk from Ireland. Stephen Jay Gould's (1974, p. 216) suggestion that the growth of woodland impaired the free travel of the large-antlered animals seems less believable, especially as the pollen record indicates a decrease, rather than an increase, of tree cover at the time that the Irish elk waned (Barnosky 1986, p. 133). If a shorter feeding season did cause the demise of the Irish elk, then it may also explain why the elk never returned to Ireland once food plants were again plentiful. The argument that the elk could not have returned because the Irish Sea stood in its path is not entirely satisfactory because *Megaloceras giganteus* was unable to survive anywhere by early Holocene times. It seems more plausible that by 10,000 years ago, summer insolation at latitude 50° N was beginning to decrease towards its present values, and the length of the spring "green-up", the time when plants contained maximum levels of nutrients so vital to the fitness of large artiodactlys with large antlers, became shorter than it had been in the late Pleistocene epoch. It is probably no coincidence that the Irish elk's presence in Ireland, and its maximum abundance in Britain, was associated with a time of maximum summer and minimum winter insolation from about 12,000 to 10,000 years ago (Barnosky 1986, p. 133).

Climatic change on a regional scale has been shown to have been responsible for late Pleistocene extinctions in situations where human activity is unlikely to have been involved. In Australia, major extinctions of the megafauna occurred around 20,000 years ago when climates became less equatable, some 20,000 years after the first human occupance of the area. Perhaps the best-documented regional relationship between climatic change and mass extinctions comes from the mid-Appalachian Mountain area (Guilday 1984) and the north-central Great Plains (Wendland et al. 1987) of the United States. The vegetation of the mid-Appalachian region, comprising much of Pennsylvania, West Virginia, western Virginia, western Maryland, and parts of Ohio, Kentucky, and Tennessee, changed through the late Pleistocene epoch into the Holocene epoch in response to the amelioration of climate. From 18,000 to 10,500 years ago, the northernmost parts of the region resembled a periglacial tundra, while the rest was a parkland with spruce (*Picea*), jack pine (*Pinus banksiana*), fir (*Abies*), birch (*Betula*), and an understorey of woody shrubs, grasses, sedges, and herbs. The composition of the vegetation has no modern analogues, but is indicative of boreal conditions (cf. Sect. 7.2.2). By 10,000 years ago, the vegetation began to shift from open coniferous forest to the present-day closed-canopy deciduous forest in response to post-glacial warming. This climatically driven change in the vegetation precipitiated a radical change in the fauna. Eighteen large and one small mammal species became extinct in the region, while three large and ten small species were extirpated. Some mammals survived the change: four species are now rare and local boreal relicts; nine species have become less common or have undergone ecological readjustment expressed as reductions of ranges (Guilday 1984, p. 254). The same pattern of faunal change occurred in the northcentral Great Plains region (Graham and Mead 1987; H.E. Wright Jr. 1987). From about 18,000 to 10,500 years ago, the fauna of Iowa consisted of 70 percent boreal species and 20 percent steppe or deciduous species. Around 10,500 years ago, the figures were 30 percent boreal species and 50 percent steppe or deciduous species, and have stayed more or less at those values ever since (Wendland et al. 1987). In Illinois and Missouri, the change in faunal composition was less dramatic but still evident. Climate model simulations lend support to the view that these faunal changes were a result of climatic change (Barnosky 1989, p. 245). The extinctions and range disruptions can be explained by the changes of climate predicted for North America (Kutzbach 1987; COHMAP Members 1988). From about 18,000 to 15,000 years ago, adiabatic warming of air flowing off the Laurentide ice sheet, coupled with the particular combination of the Croll-Milankovitch variables, gave rise to a climate cooler than the present on average, but with less extreme differences between summer and winter: July temperatures were 7 to 10 °C cooler than at present, January temperatures 5 to 10 °C cooler. By 9000 years ago, the ice sheet had retreated and, owing to changes in the Croll-Milankovitch variables, the climate had become more seasonal, with warmer summers and cooler winters than those experienced today. The faunal changes in the northcentral Great Plains can be accounted for by these climatic shifts (Barnosky 1989, pp. 245–246). As the climate became more seasonal, communities should have reassembled as animals not adapted to cold winters moved south and those not adapted to hot summers moved north. This is what did happen as the late Pleistocene disharmonious floras and faunas broke up. In general, the large herbivores which survived were ruminants such as bison, deer, moose, and sheep which can live on vegetation of low diversity. They could follow their preferred food plants despite the restructuring of plant communities. On the other hand, herbivores requiring a broad range of food plants within their normal grazing range — mammoths, mastodonts, horses, camels, sloths, and peccaries — became extinct or were extirpated over wide areas (Guthrie 1984). Herbivores, such as cervids with enormous antlers, whose physiology was locked into the "old" pattern of seasonality also became extinct. In turn, the demise of the large herbivores led to the extinction of the large carnivores and scavengers which preyed on them. To be sure, the disappearance of the proboscideans is likely to have disrupted intricate grazing food webs, thus adding to the change of community composition and thence extirpation and extinction (Graham and Lundelius 1984; Owen-Smith 1987).

Patrick J. Bartlein and I. Colin Prentice (1989) made the very interesting suggestion that species evolve to withstand environmental perturbations caused by frequently occurring combinations of the Croll-Milankovitch pulses, such as those which forced glacial and interglacial cycles during the late Pleistocene epoch, but not to infrequent combinations of them. The rare combinations of forcings, perhaps coupled with other causes of climatic change such as epierogeny, may therefore lead to widespread extinctions. This might have been what happened at the close of the Pleistocene epoch when an infrequent combination of Croll-Milankovitch forcings induced a rather extreme climatic change, involving the maximum change that would be expected in insolation and ice volume, and far greater than typical transitions from glacial to interglacial regimes.

To pursue the climatic hypothesis of late Pleistocene extinctions, detailed predictions of species' distributions in space and time are needed, as are fine-grained records of late Pleistocene faunas with which to test the predictions (S.D. Webb and Barnosky 1989, p. 428). It would seem that megafauna and microfauna have responded differently to climatic change in different regions, and only by building up detailed maps of distributions at different times will a full picture of changing biotas emerge. Examples abound of climatic change having different effects on faunas in different regions; for instance, capybaras, spectacled bears, tapirs, llamas, peccaries, and such cats as jaguars, ocelots, and margays disappeared from northern temperate latitudes but survive in the neotropical realm (S.D. Webb 1985). It would also seem likely that a climatic change, should it adversely affect a species, will not affect the entire species range at the same time, but will affect one part and then another. To establish this effect requires very detailed, sequential stratigraphical records from widely dispersed sites, but it has been ascertained in some species. Large tortoises (*Geochelone*) moved incrementally southwards: they lived on the Great Plains during the last interglacial, and survived in Florida and Central America into the last glacial, and they survive today on the Galápagos Islands (Hibbard 1960). As the details of faunal changes are filled in, and late Pleistocene and Holocene climatic simulations are improved, so will hypotheses of biotic change since the Last Glacial Maximum be easier to test.

7 Biomes and Zonobiomes

The flora and fauna of a district are determined mainly by the character of the climate, and not by the nature of the soil, or the conformation of the ground. It is from difference of climate that tropical life differs so much from arctic, and both of these from the life of temperate regions. It is climate, and climate alone, that causes the orange and the vine to blossom, and the olive to flourish, in the south, but denies them to the north, of Europe. It is climate, and climate alone, that enables the forest tree to grow on the plain, but not on the mountain top; that causes wheat and barley to flourish on the mainland of Scotland, but not on the steppes of Siberia.

James Croll (1875, p. 2)

7.1 The Coming of Ecology

The word "ecology" was coined by Ernst Haeckel, Darwin's disciple, in 1866. But, as Haeckel conceded, the concept of ecology is at least a century older. Notions of the "plenitude of Nature", food chains, and equilibrium of numbers were first grappled with by some of the most celebrated natural historians of the eighteenth century, such as Carl Linneaus, as well as by men no less great but not so well known in their day, such as Gilbert White of Selbourne (Worster 1985). During the nineteenth century, ecological ideas were developed by, among others, Alexander von Humboldt, Edward Forbes, Henry David Thoreau, Herman Melville, and Charles Robert Darwin. It was not until the twentieth century that ecological thought had a profound impact on science thanks to the pioneering endeavours of men such as Eugenius Warming, a Dutch plant geographer, Frederic Clements, a Nebraskan plant ecologist, and Charles Christopher Adams, whose work was mentioned in the previous chapter. Warming's treatise *Plantesamfund: Grundtraek af den Økologiske Plantegeografi* (1895), which was rendered into English as *Oecology of Plants: an Introduction to the Study of Plant Communities* (1909), made three chief points: firstly, that plants make structural and physiological adjustments to the state of habitat in which they live — to light, heat, humidity, soil, terrain, and animals; secondly, that animals and plants are part of a community, an assemblage of species sharing similar environmental tolerances; and thirdly, that plant communities depend more on the water content of the soil than on temperature. The first and third points imply an important role for climate in explaining many aspects of plant geography. Warming perceived the tendency of unrelated plants to undergo "epharmonic convergence" (that is, to evolve the same life-forms in similar environments) largely as an adaptive response to climate. For instance, the American cactus and the South African euphorbia, both living in arid regions, have adapted by evolving fleshy, succulent stems and by evolving spines instead of leaves so as to conserve precious moisture. Warming's focus on water as a determinant of plant distribution is mirrored in his classification of the major plant communities of the world and in his designation of hydrophytes, xerophytes, mesophytes, and other such terms which are still in common use.

Although Warming's work had a distinctly ecological flavour, it did not stress the mutual relations between animals and plants. "Biome", a blanket term to describe the biotic community as a whole (the combined communities of plants and animals), was coined as late as 1939 by Frederic E. Clements and Victor E. Shelford in their book *Bio-Ecology*. Clements was a plant ecologist, Shelford an animal ecologist. Their collaboration achieved a giant step forward in the breaking down of artificial barriers between subject areas within ecology and treating plants and animals together. Nonetheless, Clements stood by his botanical guns, insisting that in any biome it be the plants that determine which animals are included, and not the animals that determine the plants. To him, plants are the "immediate and most direct translation of climate into food, as well as an essential buffer against environmental extremes" (Worster 1985, p. 215).

The development which seems to have brought the study of entire biotic communities to the fore was the study of energy flow in ecosystems. The energetic basis of ecology had slowly grown during the late-eighteenth and nineteenth centuries after it had been realized that plants exchange matter with the atmosphere, lithosphere, and hydrosphere, Nicolas Théodore de Saussure (1804) had given the correct equation for photosynthesis and, eventually, the equation of primary production had been worked out (Rabinovitch 1971; Lieth 1975a,b). Once the factors of primary production were comprehended, serious investigation of plant production could get underway. Primary production was first measured by the Frenchman Jean Baptiste Joseph Dieudonné Boussingault (1844). A few decades later, in a study very advanced for its time, Ernst Wilhelm Ferdinand Ebermayer (1876, 1882) measured the primary productivity and nutrient uptake of Bavarian forests. Edgar N. Transeau (1926) appears to have been the first person to make measurements of energy in plants. The terms "producer", "consumer", and "reducer" or "decomposer" were introduced by August Thienemann in his *Limnologie* (1926; see also Thienemann 1939), while Charles S. Elton, in his seminal *Animal Ecology* (1927), showed how ecological roles were integrated in a food chain. But the person responsible for setting ecologists thinking about energy flow as a unifying concept in understanding the structure and function of communities was Arthur George Tansley. In 1935, Tansley coined the term "ecosystem" to encompass the relations between all organisms living in a prescribed area *and* their physical environment, to present the "biological and non-biological aspects of the environment as one entity, with strong emphasis on measuring the cycling of nutrients and the flow of energy in the system — whether it be a pond, a forest, or the earth as a whole" (Worster 1985, p. 378). Then, with the appearance of Raymond Lindeman's famous paper on the trophic-dynamic aspects of ecosystems in 1942, the immense value of an energetic approach became patently clear. As we shall see, later work has spotlighted the strong relationship between the rates biospherical processes, especially photosynthesis and net primary productivity, and climatic variables.

7.2 Communities and Climate

7.2.1 Life Zones

Some work on the relations between climate and communities has continued the tradition established by Clinton Hart Merriam in using climatic parameters to define life zones. Merriam's life-zone scheme has indeed been much embroidered (e.g. Holdridge 1947, 1967; Box 1981). Leslie Rensselaer Holdridge's system defines numerous life zones using several climatic measures of the radiation and water balances of the Earth's surface (Fig. 7.1). Other ecologists prefer to define fewer community units. Heinrich Walter (1985) divides the Earth's vegetation into nine distinct zonobiomes each occurring in a climatic zone (Table 7.1). In identifying his zonobiomes, Walter took the seven chief climatic belts: the equatorial rain zone, the summer-rain zone on the margins of the tropics, the subtropical dry regions, the subtropical winter-rain region, the temperature zone with year-round precipitation, the subpolar zone, and the polar zone. He subdivided the very large temperate zone into three parts and combined the subpolar and polar zones to give a single Arctic zone, thus recognizing nine climatic zones in all, ecologically designated as zonobiomes. The relation between zonobiomes and climatic variables has been elucidated by Mikhail I. Budyko (1974, p. 364). In his analysis, Budyko used two climatic parameters: the radiative index of dryness, which characterizes the relative values of the components of the heat and water balances; and the net annual radiation balance of the Earth's surface, which characterizes the amount of energy available to power surface processes (Sect. 3.1.2). The general zonal disposition of world's principal soil and vegetation types are defined well by isolines on the world map of the radiative index of dryness R/LP (Fig. 3.1): index values of 0.33 and less correspond tundra; values from 0.33 to 1.0 correspond to forest; values from 1.0 to 2.0 correspond to steppe; values from 2.0 to 3.0 correspond to semi-desert; and values in excess of 3.0 correspond to desert. However, to obtain a picture of the absolute intensity of natural processes, it is necessary to include the net radiation, R, a separate factor. This is done in Fig. 7.2, which shows the geobotanical conditions associated with the parameters R/LP and R. The vertical lines, corresponding to different radiative index of dryness values, distinguish the principal zonobiomes: tundra, forest, steppe, desert, semi-desert, and desert. In the forest and steppe zones, the differences in the radiation balance, R, produce marked changes of vegetation, though the general character of the vegetation remains the same. Interestingly, the hydrological regime in the zones is in theory related directly to the parameter R/LP, and the absolute runoff values can be determined from the parameters R and R/LP (Fig. 7.2).

7.2.2 Vegetation and Temperature

Recent research has vindicated Alexander von Humboldt's conclusion that the distribution of vegetation types is constrained by temperature parameters. In a

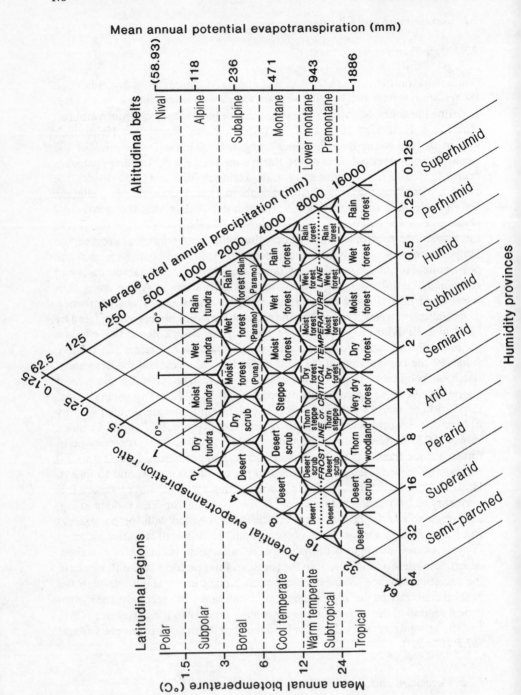

Table 7.1. Heinrich Walter's zonobiomes. (After Walter 1985)

Zonobiome		Zonal vegetation	Zonal soil type
I	Equatorial with diurnal climate, humid[a]	Evergreen tropical rain forest	Equatorial brown soils (ferrallitic soils, latosols)
II	Tropical with summer rains, humido-arid	Tropical deciduous forests or savannas	Red clays or red earths (savanna soils)
III	Subtropical-arid (desert climate), arid	Subtropical desert vegetation	Sierozems
IV	Winter rain and summer drought, arido-humid	Sclerophyllous woody plants	Mediterranean brown earths
V	Warm-temperate (maritime), humid	Temperate evergreen forests	Yellow or red podzolic soils
VI	Typical temperate with short period of frost (nemoral)	Nemoral broad-leaved deciduous forests	Forest brown earths and grey forest soils
VII	Arid-temperate with a cold winter (continental)	Steppe to desert with cold winters	Chernozems to sierozems
VIII	Cold-temperate (boreal)	Boreal coniferous forests (taiga)	Podzols (raw humus-bleached earths)
IX	Arctic (including Antarctic), polar	Tundra vegetation (treeless)	Tundra humus soils with solifluxion

[a]The term "humid" is used to describe a climate with much rainfall, and "arid" to describe a dry climate with little rainfall. Where both terms are used, the first refers to summer conditions and the second to winter conditions.

study of the distribution of humid to mesic forests in eastern Asia, Jack A. Wolfe (1979) found that mean annual temperature, mean annual range of temperature, and the mean temperatures of the warm months and the cold months are important determinants of extant vegetation types (Fig. 7.3). For example, broad-leaved deciduous forest develops in humid and mesic regions in eastern Asia if the cold-month mean be less than 1 °C and the warm-month mean be more than 20 °C. Where the cold-month mean is more than 1 °C, broad-leaved deciduous forest gives way to broad-leaved evergreen forest; where the warm-month mean is less than 20 °C, broad-leaved deciduous forest gives way to coniferous evergreen forest. However, broad-leaved deciduous plants may occur as subordinate elements in broad-leaved evergreen and coniferous evergreen forests. The relations between forest types and temperature parameters differ in other regions. In subhumid areas of Europe and western North America, where temperature parameters are similar to those in eastern Asia, broad-leaved evergreen forest is the climax vegetation. In southeastern North America,

◀───

Fig. 7.1. Leslie R. Holdridge's correlation of life-zones with climate. Mean annual biotemperature is defined as the mean of unit-period temperatures with the substitution of zero for all temperatures below 0 °C and above 30 °C. Puna and paramo life zones occur only in tropical subalpine areas. (After Holdridge 1947, 1967)

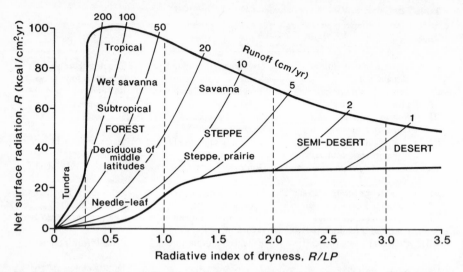

Fig. 7.2. Geobotanical zones defined in relation to net surface radiation, R, and the radiative index of dryness, R/LP. Runoff in the different zones is also shown. (After Budyko 1974)

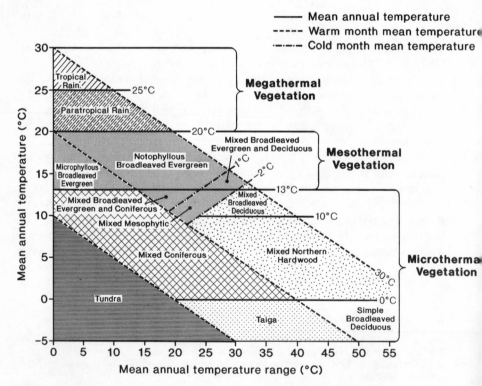

Fig. 7.3. Temperature parameters of humid to mesic forests of eastern Asia. (After Wolfe 1979, 1985)

broad-leaved deciduous forest (or pine forest or a mixture of both) occurs in most areas where the cold-month mean is more than 1 °C, and broad-leaved evergreen forests occur only in the southern part, usually as shrubs or small trees. The northwards extension of broad-leaved evergreen forests in southeastern North America is probably prevented by intense Arctic cold fronts. Nonetheless, the mesothermal forest vegetation of southeastern North America has a physiognomy comparable to the east Asian mixed broad-leaved evergreen and deciduous forest. And in the northern range of the southeastern North American mesothermal forests, broad-leaved evergreens are almost absent, the dominant tree being oak (*Quercus* sp.) with pinnate leaf-lobes which form forests physiologically analogous to Asian mixed broad-leaved deciduous forest. The intense cold fronts which affect southeastern North America also lead to areas which in eastern Asia would support mixed mesophytic forest being physionomically akin to Asian mixed hardwood forests and lacking broad-leaved evergreen small trees and shrubs as a subsidiary component. In western and central Europe, microthermal vegetation should occur where the cold-month mean is less than 20 °C, but the climax vegetation is actually broad-leaved deciduous forest of low diversity. In eastern Asia, areas with the same temperature parameters would be dominated by mixed coniferous forests. Historical factors are normally called on to explain the presence of deciduous forest in western and central Europe: the region was covered by coniferous forest before the Pleistogene period, but since the Ice Age it has been recolonized by deciduous species. Given the temperate requirements of broad-leaved deciduous forests in humid and mesic regions, it is no surprise to find that deciduous forests do not occur in the Southern Hemisphere because there are no regions in which the cold-month mean is less than 1 °C and the warm-month mean is greater than 20 °C.

The Australian flora, too, is adapted to thermal regimes. This was shown by H.A. Nix (1981) by taking the major climatic elements of solar radiation, temperature, and precipitation and identifying the parameters of them most relevant to the biology and ecology of heat adaptation, and then establishing three broad organismic groups: megatherms with thermal optimum temperatures of around 28 °C and a lower threshold temperature of 10 °C; mesotherms with a thermal optimum in the 18 to 22 °C range and a lower threshold temperature of around 5 °C; and microtherms with a thermal optimum in the 10 to 12 °C range and lower thresholds at or below 0 °C. While mean temperatures provided useful insights into the adaptation of elements of the Australian flora, Nix found that extremes of temperatures were of greater biogeographical significance. For example, the mean maximum temperature of the hottest week and the mean minimum temperature of the coldest week correlated best with floral type.

7.2.3 Communities as Dynamic Continua

Large-scale communities are undoubtedly influenced very strongly by climate, as Walter, Budyko, Wolfe, and many others have demonstrated. It is remarkable that pollen data for 500 years ago in the eastern United States parallels the major vegetational regions recognized today, and even more remarkable that steep gradients in pollen abundances mark the ecotones between the regions (see T. Webb 1987, p. 179). However, an important issue which we cannot pass over without mention is whether communities actually exist or whether they be merely abstractions discerned by ecologists within a continuum of animal and plant species? Views on this matter fall into two schools: the organismic school, which holds that communities are integrated units with discrete boundaries; and the individualistic school, which submits that communities are collections of populations with the same environmental needs. Space prohibits a full discussion of this vexed issue. Suffice it to note two lines of evidence suggesting that the individualistic school may be nearer the truth. Firstly, studies of living species' abundances and ranges along environmental gradients have revealed that, in reality, species are individuals, each distributed in its own way according to its own genetic, physiological, and life-cycle characteristics, and to its own relations with the biotic and abiotic milieu in which it lives (e.g. Whittaker 1962, 1970). Along an environmental gradient, individual species appear and disappear according to their own requirements, and the distributions of different species overlap: there are no sharp divides where one set of species ceases and another set commences, though there may be ecotones. As a result, communities intergrade continuously rather than forming distinct zones with well-marked boundaries. That does not mean that demarking communities within the vegetational continuum is impossible:

"It is useful to recognize life-zones, as Merriam did, as major kinds of communities in relation to temperature. But the zones are continuous with one another: distributions of the major plant species by which we recognize the zones overlap broadly, and other plants and animals do not form groups with distributions closely similar to those of the dominants. The zones are kinds of communities man recognizes, mainly by their dominant plants, within the continuous change of plant populations and communities along the elevation gradient. The zones may be compared to the colors man recognizes, and accepts as useful concepts, within the spectrum of wave lengths of light, which are known to be continuous". (Whittaker 1970, pp. 37–38)

The "colours" of the species' distribution spectrum are the commonly recognized community units — the world's biomes and zonobiomes.

A second line of evidence which tends to vindicate the individualistic school of thought on community structure comes from the analysis of faunal and floral changes during the last 18,000 years in North America. Reconstructing the changing distributions of animals and plants since the height of the last glaciation has brought out the transitory nature of communities. Species abundances and distributions are constantly changing, partly in response to climatic change. Biomes are temporary communities of animals and plants living under a particular climate. They will come and go in answer to environmental, and especially climatic, changes. There is no reason to suppose that there be anything special about the biomes we see today. They are simply the result of species' adapting to the prevailing climatic regions and gradients which developed after the last deglaciation. As an example it is instructive to consider a late Pleistocene biome which existed in northcentral United States between about 18,000 to 12,000 years ago. The vegetation of the area was rich in spruce and sedges (Watts 1983; T. Webb 1987; T. Webb et al. 1987). It formed a spruce parkland or "boreal grassland" (Rhodes 1984), perhaps similar to the vegetation found in the southern part of the Ungava Peninsula in northern Quebec today. The fauna of this biome, as well as the flora, was disharmonious: species which now inhabit grassland or deciduous woodland, including prairie voles, sagebrush voles, and the eastern chipmunk, lived cheek-by-jowl with species which now occur in boreal forests and arctic tundra, including arctic shrews, lemmings, voles, and ground squirrels (Lundelius et al. 1983, 1987). The "boreal grassland" biome has no modern analogue. This does not mean that the animals and plant species which comprised it were not in harmony with the prevailing Pleistocene environments; it simply means that they were maintained by climates which have no modern counterparts (Graham and Mead 1987, p. 371). During the late Pleistocene epoch, disharmonious faunas were found over all the United States, save the far west, where vertebrate faunas bore a strong resemblance to their modern-day equivalents, and date from at least 400,000 years ago to the Holocene epoch. Several different disharmonious faunas have been mapped, each with a characteristic combination of species and each associated with a particular climatic regime (Graham et al. 1987). The disharmonious biotas were supported by equable climates of a kind which are found nowhere today (Hibbard 1960; Dalquest 1965; Guilday et al. 1964; Lundelius 1976; Lundelius et al. 1983). Russell W. Graham (1976, 1979) attributes the disharmonious communities to the response of individual species to changing environmental conditions during the late Pleistocene epoch. In turn, the environmental changes which came about at the end of the Pleistocene epoch were probably responsible for the disintegration of the disharmonious communities: the climate became less equable (there were stronger seasonal contrasts) and individual species readjusted their distributions. In other words, the increased seasonality of climate in the United States which arose during the early Holocene epoch sorted and sifted the members of the disharmonious communities into something approaching their modern ranges. The break-up of the late Pleistocene communities was a time-transgressive process, possibly because of the latitudinal lapse rates in climatic adjustments to the ablation of the Laurentide and Cordilleran ice sheets: it occurred first in the southern Plains and southern Rocky Mountains, possibly starting before 11,000 years ago, but did not shift northwards until 10,000 years ago or less (Lundelius et al. 1983).

7.2.4 Plant Physiognomy and Climate: a Predictive Model

Viewing biomes as dynamic entities calls into question the value of delimiting large-scale communities of animals and plants using "master" climatic variables. It is undoubtedly true that each climatic life-zone is associated with a distinct type of community. But can a climatic classification of communities be used as a predictive tool? The answer is yes, in that given a particular combination of climatic factors, we might say that steppe vegetation would be expected to flourish. This kind of information can be serviceable, especially when trying to reconstruct past temperature regimes, but it does not get at the root of the matter, that is, the mechanisms by which climate wields a degree of control over plant distributions: it tells us what sort of vegetation to expect under a given climatic regime, but it does not say anything about the processes by which plants adapt to climatic conditions; it establishes that the life-forms (physiognomy) of plants are strongly correlated with temperature and effective rainfall, but it does not disclose the processes which lead to a particular life-form's being suited to a particular climate, and so is uninformative as to the mechanisms by which climate influences plant distribution (cf. Woodward and McKee 1991). For these reasons, and presuming that the palaeoecological evidence be correctly interpreted, then all that can be said on the basis of climatic correlations is that a substantial change of climate will create new biomes, but just what those novel biomes would look like is impossible to say. To predict life-form from climate, a model of a different stamp is required.

It was realized at least as early as the late nineteenth century that climate must exert control over plant distributions through its effect on basic physiological processes (e.g. Schimper 1898), and it has been known since early in the present century that physiognomic features of vegetation, notably leaf-form of climax vegetation not limited by water stress, are related to mean annual temperature (Bailey and Sinnott 1915; Wolfe 1979). But not until recently was a model built to predict the type of plant which would be expected to occur under given climatic conditions. The builder of the model was F. Ian Woodward. He took as his starting point the critical and climatically controlled limitations to the completion and perpetuation of a successful life-cycle (Woodward 1987, p. 63). The model begins with the well-verified hypothesis that the mass accumulated by a plant, m, depends on the incoming solar radiation, S:

$$m = E\int_0^t Si \, dt \, ,$$

where E is an coefficient representing efficiency of energy conversion (usually about 2.5 percent), and i is the fraction of incoming radiation intercepted by the canopy. Defining i as

$$i = 1 - e^{-kL},$$

where k is the extinction coefficient for solar radiation and L is the leaf area index, the accumulated plant mass may be described by

$$w = E\int_0^t S(1 - e^{-kL}) \, dt.$$

The leaf area index will be strongly influenced by the amount of water available to the plant and thus to the local water budget. For this reason, effective precipitation is, arguably, the critical climatic variable in predicting canopy development measured as the leaf area index. Applying his model, Woodward predicted the leaf area index for meteorological stations dotted over the Earth's surface on the basis of local balances of precipitation and evapotranspiration. Although quite promising, the predictions were adrift in the tundra zone, where actual leaf area indices are lower than those predicted, and in areas of seasonal precipitation such as southeast Asia, where actual leaf area indices are higher than those predicted. Taking these errors on board and using minimum temperatures to discriminate between physiognomic types of vegetation (Table 7.2), Woodward further developed his model and came up with a new set of predictions (Fig. 7.4). The revised leaf area index predictions were much closer to actual leaf area indices: the model brought out the drought deciduous regions of Africa, India, and America, the winter deciduous forests of Europe, western Russia, and eastern North America, and the deciduous-coniferous forests of north Russia; but it erred in predicting a winter deciduous forest along the Pacific northwest

Table 7.2. Cardinal minimum temperatures and expected dominant physiognomy. (Woodward 1987)

Temperature range (°C)	Phenomenon	Expected physiognomy
>15	Temperature not limiting	Broad-leaved evergreen when rainfall adequate
–1 to 15	Chilling	Broad-leaved evergreen when rainfall adequate
–15 to 0	Freezing and supercooling	Broad-leaved evergreen
–40 to –15	Freezing and supercooling	Broad-leaved deciduous
<–40	Freezing and supercooling	Evergreen and deciduous needle-leaved (coniferous)

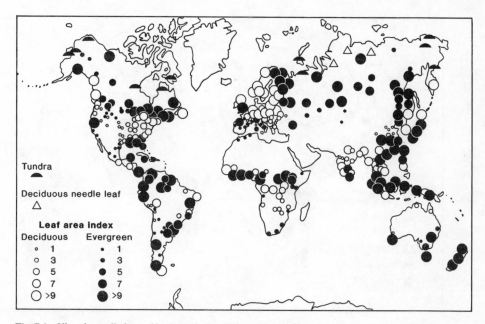

Fig. 7.4. Climatic predictions of leaf area index. (Woodward 1987)

of America where an evergreen coniferous forest actually grows, and also in giving maritime areas of Scandinavia and the British Isles a more southerly type of vegetation. The broad, world-scale predictions of major physiognomic types are presented in Fig. 7.5 in which herbaceous vegetation is arbitrarily assumed to have a leaf area index of 1, shrubs a leaf area index of 3, and forests a leaf area index above 3. Tundra embraces a wide range of physiognomies but is portrayed as herbaceous on the map, that being the "average" physiognomy. With the few exceptions mentioned, the predicted vegetation maps are remarkably accurate and seem to vindicate the model. Further work is needed to elucidate the controls on the distribution incorporated in the model (see Woodward 1989, 1991; Woodward and McKee 1991). The problem is that studies should be conducted on vegetation, rather than individual plants, and, as Woodward (1987, p. 107) said, vegetation is an unwieldy subject for experimentation.

7.2.5 Plant Productivity and Climate

Attempts to understand primary productivity as a function of plant growth and environmental properties commenced at the start of the twentieth century. F.F. Blackman and V.H. Blackman were leading modellers in the field, the latter developing a model of plant growth equivalent ot the law of compound interest in economics (V.H. Blackman 1919; Leith 1975a). Later, a new approach emerged with the advent of biophysics and allometric growth models (e.g. von

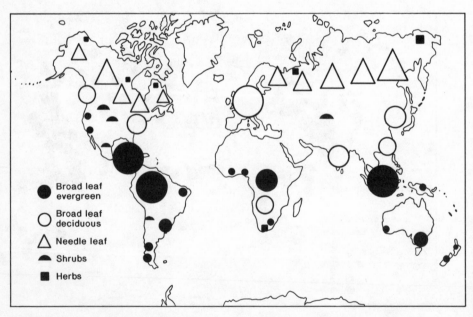

Fig. 7.5. Climatic predictions of the physiognomy of vegetation types. (After Woodward 1987)

Bertalanffy 1951). These developments are of interest here only in that they led to the first attempts to derive global forest yields from environmental data. A prime example of this is the work of Sten Sture Paterson (1956, 1975), who devised a CVP (Climate-Vegetation-Productivity) index, I, defined by the equation

$$I = \frac{Tv \cdot P \cdot G}{Ta \cdot 12},$$

where Tv is the temperature of the warmest month, Ta is the annual range of temperature, P is the mean annual precipitation, and G is the length of the growing season computed using a version of Emmanuel de Martonne's indice d'aridité adapted for separate months of the year.

Subsequent models refined the prediction of productivity. For instance, John Lennox Monteith (1965) showed that daily photosynthesis can be estimated from daylength and insolation. And, using data sets taken from the United States International Biological Programme Eastern Deciduous Forest Biome memo reports for productivity profiles of Wisconsin, New York, Massachusetts, Tennessee, and North Carolina, J.R. Reader (1973) found a significant relationships between net primary productivity, y, and length of growing period, x (Fig. 7.6). The usefulness of evapotranspiration in predicting primary production was explored by Helmut Lieth (1961, 1972; Lieth and Box 1972) and Michael L. Rosenzweig (1968b), who undertook correlation studies of net primary produc-

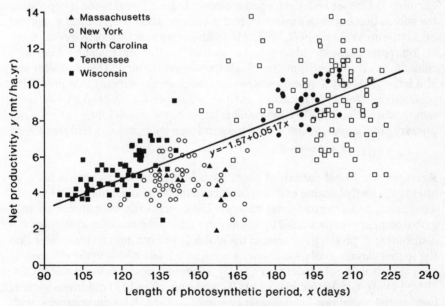

Fig. 7.6. Net primary productivity versus photosynthetic period in North American deciduous forests. (After Reader 1973)

tivity and actual evapotranspiration. Rosenzweig, impressed by Leslie R. Holdridge's life-zone classification which underlines the degree to which the abiotic environment influences in large measure the characteristics of mature vegetation in terrestrial communities, sought an environmental variable which would correlate well enough with primary terrestrial production to be used as predictor. He noted Jack Major's (1963) comment that, qualitatively, the amount of vascular plant activity is related to actual evapotranspiration, and the fact that actual evapotranspiration depends on water supply and available energy. Holdridge (1959) had earlier demonstrated that "biotemperature" and potential evapotranspiration are linearly related. As potential evaporation is used in the calculation of actual evapotranspiration, Rosenzweig deemed it valid to consider actual evapotranspiration, Ea, as an environmental variable for predicting production. Using C. Warren Thornthwaite and John R. Mather's (1957) method for calculating actual evapotranspiration by means of latitude, mean monthly temperatures, and mean monthly precipitation, he set about estimating Ea for a number of sites in the United States for which suitable production figures were available. The result was a productivity prediction equation of the form:

$$\log_{10}y = (1.66 \pm 0.27)\log_{10}x - (1.66 \pm 0.07),$$

where y is the net above-ground primary productivity (g/m^2) and x is actual evapotranspiration in (mm) (Fig. 7.7).

The establishment of relations between primary productivity and environmental conditions was given a spur by the International Biological Programme (IBP). As part of this programme, Helmut Lieth, then at the University of North Carolina at Chapel Hill, developed a computer-based world model expressing the correlation between environmental parameters and biospherical processes over large areas (Lieth 1972, 1973). He and his team took productivity data from various parts of the world collected as part of the IBP and matched them with climatic data from nearby meteorological stations as compiled by H. Walter and H. Lieth (1961–66). Two curves were derived from the data set, one predicting productivity from precipitation and the other predicting productivity from temperature (Fig. 7.8). The following relationship was also found between net primary productivity, y, and annual evapotranspiration, x (Leith and Box 1972):

$$y = 3000 \{1 - e^{-0.0009695(x-20)}\}.$$

Recently, the global pattern of production in vegetation has been a focus of attention, chiefly because of its importance to the human population in terms of agriculture and environmental change. These concerns were discussed in a global context as early as 1862 by Justus von Liebig, who assumed a roughly even distribution of productivity around the globe. It was not until 60 years later that the spatial variations of productivity were appreciated, owing to the research of people such as H. Schroeder (1919, 1975), Vladimir Ivanovich Vernadsky (1924, 1944, 1945b), and W. Noddack (1937). And only after 1950 did investigations into spatial variations in productivity surge forward, Russian scientists, with Leonid Efimovich Rodin, Natalia Ivanova Basilevich, and their comrades in the vanguard, making significant progress during the 1960s (see Bazilevich and

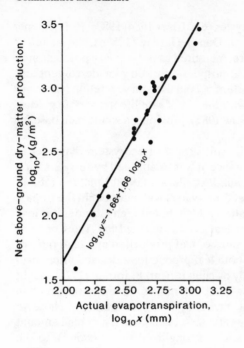

Fig. 7.7. Net above-ground productivity versus actual evapotranspiration for a range of environments in the United States. (After Rosenzweig 1968b)

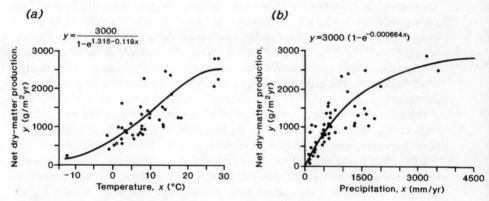

Fig. 7.8. Net dry-matter production as a function of **a** mean annual temperature and **b** mean annual precipitation. Data sets for sites scattered around the world. (After Lieth 1972)

Rodin 1967; Rodin and Basilevich 1968; Bazilevich et al. 1971; Rodin et al. 1975; Bazilevich and Titlyanova 1980), and Helmut Lieth and his colleagues producing a series of net productivity maps using the empirical relationships between productivity and climatic variables mentioned above during the early 1970s (e.g. Lieth 1975c). Lieth's "Miami model" of global net primary productivity, which uses the temperature and precipitation equations, and his "Thornthwaite Memorial model", which uses the actual evapotranspiration equation, are still used in models of global biogeochemical cycles (e.g. Esser 1984, 1987). Refinements have been made, for example by L.D. Danny Harvey (1989d), the better to simulate the response of the biosphere, measured as net primary production in its component parts, to increases in atmospheric carbon dioxide content and temperature. Estimates of the distribution of land use and vegetation types, and their biomass, have been improved with the aid of satellite imagery (e.g. Matthews 1983, 1984; Matthews and Rossow 1987), providing a sound data base for use in climate studies (Matthews 1986).

Of course, productivity is not the sole aspect of ecosystem dynamics to depend very much on climate. For instance, it was established by Jerry S. Olson (1963) and Leonid E. Rodin and Natalia I. Basilevich (1968b) that the ratio of leaf fall to litter accumulation on forest floors varies with latitude. In the tropics, the ratio of leaf fall to litter accumulation is high, but at high latitudes it is low. This trend results from high rates of leaf production and low rates of litter accumulation in the tropics, and low rates of leaf production and high rates of litter accumulation at high latitudes, both leaf production rates and litter accumulation rates being affected primarily by climate. Carl F. Jordan (1971) sought other worldwide patterns in terrestrial plant energetics. To avoid the problem of irregularities resulting form local environmental conditions, he chose to employ ratios of plant energetics. The ratio he selected was the annual amount of energy bound in trunks and branches as long-lived tissue (wood), to the amount bound in leaves, fruits, flowers, bark, and twigs as short-lived tissue. The ratio was calculated for selected communities and the results plotted against two environmental factors — solar radiation available during the growing season and precipitation (Fig. 7.9). The chief findings of the study were that the ratio of wood production to litter production in forests generally increases as the solar energy available during the growing season decreases, but decreases with decreasing precipitation. In the light of his investigation, Jordan was able to show that absolute production of different parts of plants follows this trend: as available solar energy decreases, so wood production remains the same but litter production decreases; and as precipitation decreases, so litter production remains constant but wood production decreases. An interesting aspect of this work is that by considering ratios it is possible to offer an explanation of the distribution of gymnosperms and angiosperms: in general, gymnosperms have a higher wood-to-litter production ratio than angiosperms, and so may have an adaptive advantage in regions with limited radiation during the growing season, a fact which would explain why gymnosperms tend to replace angiosperms at higher latitudes and at higher altitudes on mountains lying in mesic regions.

7.2.6 Plants and the Biosphere: Model Interactions

It was mentioned earlier that, at the current state of the art, modelling the general circulation of the atmosphere is handicapped by the lack of realistic interactions between vegetation and the air. An effort to rectify this problem was made by Robert E. Dickinson (1984), who constructed a model which emphasizes the

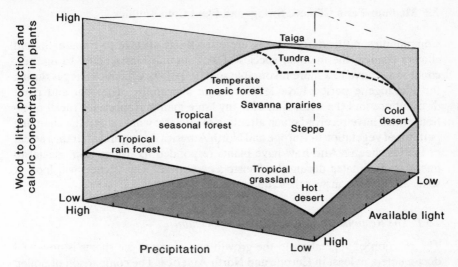

Fig. 7.9. Production ratios and caloric concentrations plotted against the amount of light available during the growing season and the amount of precipitation. *Any point of the surface* represents a combination of light and precipitation on the *horizontal axes*, and the ratio and caloric concentration on the *vertical axis*. Types of communities commonly associated with particular combinations of light and precipitation are indicated. (After Jordan 1971)

role of stomatal resistance in the process of evapotranspiration and recognizes the pivotal role of interception loss in the land phase of the water cycle. Later, building on the foundation laid by Dickinson, a "simple biosphere model" was fabricated. Designed and tested by Piers J. Sellers and his colleagues (P.J. Sellers et al. 1986; P.J. Sellers and Dorman 1987), this model simulates, in a biophysically realistic way, the transfer of energy, mass, and momentum between the atmosphere and the vegetated surface of the Earth. The model contains seven prognostic state variables: the temperature of the vegetation canopy, the temperature of the ground cover and soil surface, an interception water store for the canopy, an interception water store for the ground cover, and three stores for soil moisture. The simple biosphere model has been coupled to a general circulation model and produced significantly different results from the same general circulation model run with a conventional hydrological model (Sato et al. 1989). It was found that, when coupled to a simple biosphere model, the general circulation model created a more realistic partitioning of energy at the land surface, producing more sensible heat flux and less latent heat flux over vegetated land, so giving a much deeper day-time planetary boundary layer and reduced precipitation rates over continents (see also P.J. Sellers et al. 1989; Dorman and P.J. Sellers 1989). Thus incorporating realistic models of plant processes in atmospheric general circulation models is a valuable exercise and will doubtless lead to improved models of the climate system within the next few years.

7.3 Medium-Term Climatic Swings and Plant Communities

Communities, particularly biomes, are often finely attuned to climate. Should climate change, the member species of a community must adapt to the new conditions, move to a more favourable spot, or perish. Climatic changes during the Pleistogene period have led to drastic alterations of animal and plant distributions and the extinction of many large mammal species. Crucial questions which have received much attention of late are: what caused the changing patterns of vegetation in Europe and North America during and after the retreat of the ice sheets? And how have plants responded to swings from glacial to interglacial climates during the Pleistogene period? This section will look at some attempts to answer these questions.

7.3.1 The Response of Vegetation to Deglaciation

The reaction of vegetation to the growth and decay of ice sheets is now well documented, at least in Europe and North America. The comparison of pollen diagrams with oxygen isotope records from deep-sea cores has revealed some interesting and somewhat unexpected patterns of response. In Europe, vegetation reacted, perhaps in less than 100 years, to the deterioration of climate at the onset of the last ice age corresponding to the transition between isotope Stages 5e and 5d (Duplessy et al. 1986). The timing of climatic deterioration was different in North America and Europe. The western margins of the Atlantic Ocean cooled first, leading to the establishment of a cold and humid climate over North America and favouring the growth of ice sheets. The eastern margins of the Atlantic Ocean remained warm for several thousand years after the western parts had chilled, thus preserving an interglacial climate and biota in Europe. Indeed, glacial conditions were well established in North America before cold water reached the European seaboard and cooling occurred. Vegetation changed rapidly in response to the climatic fluctuations associated with the last deglaciation. However, during times of climatic improvement, when forest growth was possible, vegetation seems to have reacted more slowly, just a few species, such as birch and pine which had survived in "refugia" within their main range, apparently taking advantage of the more clement conditions. In both continents, deglaciation involved wild fluctuations of temperature in excess of 10 °C within a few centuries, but the exact chronology of these short-lived climatic excursions has not yet been fully charted.

It is to be expected that the migration of vegetation in response to changes of climate associated with deglaciation would be time-transgressive: one would not expect the timing of events to be the same at all places. Nor would one expect the sequence of vegetational change to be exactly the same in all parts of a continent, let alone through a hemisphere or the entire world. That is not to say that the pattern of changes from 18,000 years to 8000 years ago were not broadly synchronous. Reliable radiometric timescales have revealed the close timing of many major changes in far-flung regions. But within the main sequence of events accompanying deglaciation, the detailed changes of flora between, and even within, continents are distinctly time-transgressive, varying from place to place. One might be justified in presuming, for instance, that in southern Europe the effect of the warming of the Atlantic waters which led to deglaciation would be felt first in Spain, then in southern France, and finally in Italy. Evidence for a "wave" of warming moving into southern Europe has been gathered by William A. Watts (1986). Deglaciation appears to have commenced relatively early in the Pyrenees: local ice caps and cirque glaciers had retreated or had disappeared before 14,000 to 13,000 years ago, and there is no sign of their subsequent rejuvenation. The first vegetation to appear was dominated by grass, wormwood (*Artemisia*), and herbs; there were no trees. Birch, pine, and oak started to invade the region after about 12,000 years ago, and continued to do so until about 9500 years ago when oaks came to dominate, in some places in association with hazel. The Spanish vegetation betrays no suggestion of the cooling associated with the Younger or Older *Dryas* periods, whose effects seem to have been far more strongly felt in northwest Europe, although the ocean record clearly displays the Younger *Dryas* event in the Bay of Biscay (Duplessy et al. 1981). In Italy, the change from a "glacial" to an "interglacial" environment was rapid with no severe fluctuations of climate. At Lake Monticchio in southern Italy, full glacial vegetation was steppe-like and dominated by herbs which appear to have grown under a very dry climate (at least during the summer growing season) and possibly, but not necessarily, colder than today. The herbs give way to birch, then oak, and then in the Holocene epoch, as more water became available during the summer, to more mesic deciduous trees such as beech, lime, and ash. This early mesic phase may correlate with the northwards expansion of the African monsoon, indicated by sapropel horizons in the eastern Mediterranean marine cores and palynology, which was possibly caused by orbital forcing (Rossignol-Strick 1983) (Sect. 3.3.1). In mid-Holocene, oak became dominant and the more mesic trees retreated, probably because the climate became drier and warmer during summer. In Italy, pine is not an important component in the succession of plants induced by deglaciation, and, as was the case in Spain, there is little hint in the pollen records of the Younger *Dryas* event. As Watts (1986, p. 111) asserted, events in southern Europe have been seen through the distorting vision of northwest European perceptions.

A superbly well-documented sequence of floral change is that complied by Thompson Webb III (1985, 1986b, 1987) for eastern North America from 18,000 to 500 years ago (Fig. 7.10) (see also Delcourt and Delcourt 1987; Huntley and Webb 1988). Taking pollen data measured in radiometrically dated sediment cores extracted from lakes and bogs, Webb developed a series of isopoll maps (showing contours of equal pollen percentages). For each site, the ages of various pollen samples were used to interpolate the pollen percentages for each 500-year interval (500, 1000, 1500, and so on). The resulting maps have several interesting features. Selected pollen types from 500 years ago reflect the present major patterns of vegetation. Indeed, the correspondence is so good as to justify using maps of fossil pollen to interpret the broad-scale patterns and composition of vegetation at earlier times (see Sect. 7.2.2). The data for sagebrush (*Artemisia*), prairie forbs, sedge family (Cyperaceae), spruce (*Picea*), pine (*Pinus*), birch

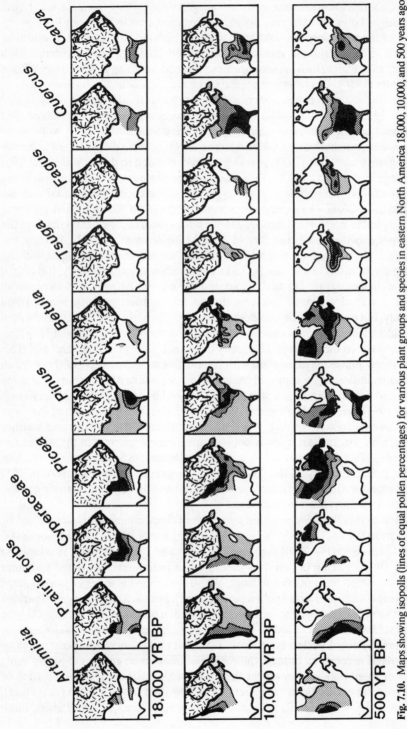

Fig. 7.10. Maps showing isopolls (lines of equal pollen percentages) for various plant groups and species in eastern North America 18,000, 10,000, and 500 years ago. In all cases, *dark areas* correspond to places of highest pollen abundance. The groups shown are sagebrush (*Artemisia*), prairie forbs (the sum of *Ambrosia*, *Artemisia*, other Compositae, Chenopodiaceae, and Amaranthaceae pollen), sedge (Cyperaceae pollen), spruce (*Picea*), pine (*Pinus*), birch (*Betula*), hemlock (*Tsuga*), beech (*Fagus*), oak (*Quercus*), and hickory (*Carya*). (After T. Webb 1986b, 1987)

(*Betula*), hemlock (*Tsuga*), beech (*Fagus*), oak (*Quercus*), and hickory (*Carya*) pollen illustrate how the changing distribution and abundance of plant taxa and populations altered the spatial pattern, local composition, and the overall structure of vegetation so leading to the disappearance and appearance of major biomes and ecotones (Fig. 7.10). For instance, the *Picea* parkland biome, which existed from 18,000 to 12,000 years ago in a broad swathe south of the ice sheet, disappeared within two millenia after 12,000 years ago; and the modern boreal forest biome began to develop across a large tract of central Canada only after 6000 years ago. The pattern of vegetation in eastern North America took on a distinctly modern look between 12,000 and 10,000 years ago when a major northeast-to-southwest vegetational gradient formed between central Canada and Nebraska, later to extend to northeastern Canada. This gradient reflects the pattern of summer isotherms and could develop only when the Laurentide ice sheet had shrunk to a small enough size. Over the past 10,000 years, gradual but significant changes have occurred, including a shift of the border between forest and prairie, first eastwards and then westwards (T. Webb et al. 1983); the emergence of northern hardwood forests from Nova Scotia to Minnesota as recorded in pollen percentages for birch, beech, and hemlock (Gaudrea and Webb 1985); and the emergence of the southern conifer biome in the southeast states, shown by increased values of pine pollen, which led to the development of a marked northwest-to-southeast vegetation gradient from Nebraska to Florida, in part reflecting the increase in winter temperatures in the southeast during the last 9000 years caused by an 8 percent increase in solar radiation receipt during Northern Hemisphere winters (T. Webb et al. 1987). Thompson Webb's work reveals that over thousands of years, vegetational change involves a continual overlapping and separation of the abundance distributions of different taxa, although fairly steep ecotones always seem to divide major biomes. It seems reasonable to suppose that the same dynamics will have operated over millions of years as well. In other words, the insight gained from detailed mapping during the past 18,000 years may provide a window through which to look at comparable intervals in the geological past (T. Webb 1987, p. 178).

Naturally, it was not just in high and middle latitudes that the impact of glacial climates was felt: the climate of the tropics changed, too. Generally, plants growing in the tropics responded to climatic change by tracking the shifting zones of rainfall: extinctions were less common than in areas lying nearer the ice sheets. In Africa, the history of climatic and vegetational change, pieced together from palaeohydrological and palynological data by Anne-Marie Lézine (1988a,b, 1989), involved swift shifts of the belts of vegetation in the wake of wandering climatic zones. Palaeohydrological data indicate several periods with rainfall higher than at present (Lézine and Casanova 1989). Two of these wetter phases are associated with abrupt changes in vegetation: 9000 years ago the vegetation responded to an increase in rainfall accompanying an increase in the Atlantic monsoon flux — the Sahelian wooded grassland (which had been confined to latitude 10° N under arid conditions of the Last Glacial Maximum) rapidly migrated northwards, Guinean elements reaching latitude 16° N and Sahelo-Sudanian elements attaining latitude 21° N, the margin of the modern Sahara;

and 2000 years ago the vegetation quickly shifted to its modern distribution in response to generally drier conditions.

7.3.2 The Response of Vegetation to Glacial-Interglacial Cycles

A number of studies indicate that during glacial and interglacial cycles vegetation tracks orbitally forced changes of climate. Barbara Molfino and her colleagues (1984) looked at pollen from three cores from the Grand Pile bog in the Vosges Mountains of northeastern France. Three time series of pollen were analysed: herbs (an amalgam of *Artemisia*, *Rumex acetosella*, and Chenopodiaceae), pine (*Pinus*), and birch (*Betula*). The power spectra of herbs and pine showed peaks significant at the 95 percent level. The peaks corresponded to the following approximate periodicities: for herbs, 23,400 and 9200 years; for pine, 18,600, 9300, and 6400 years. The 9300-year and 6400-year periodicities may be second and third harmonics of the 18,600-year component, and are taken by Molfino and associates as a sign that continental vegetation responds non-linearly to precessional orbital forcing. In another study, Martine Rossignol-Strick and Nadine Planchard (1989) established a link between global ice volume (as measured in oxygen isotope ratios in marine cores) and continental climate by studying oxygen isotope ratios and the abundance of deciduous oak (*Quercus*) pollen (a good indicator of terrestrial climate) in a well-dated core spanning 55,000 to 9000 years ago taken from the Tyrrhenian Sea. They found that pollen abundance peaks of grass (Poaceae), wormwood (*Artemisia*), fir (*Abies*), and oak (*Quercus*) are locked in phase with oxygen isotope depletion events, suggesting that vegetational cycles closely accompany deglacial pulses of global extent.

Theoretical insight into the tie between vegetation changes and climatic forcing in the Croll-Milankovitch frequency band has come from Thompson Webb III's (1986a) formulation of the problem. Making the not unreasonable assumption that the rate of vegetational response will vary in proportion to the degree of climatic change, Webb equates the rate of vegetational response, dV/dt, with the difference in vegetational composition at time t, V, and the composition of the vegetation when, after a sudden change of climate at $t = 0$, equilibration with the new climatic regime is attained, V_1:

$$\frac{dV}{dt} = -(1/\lambda)(V - V_1).$$

Response functions exist that can estimate V_1 for a given climatic state. Integration of the equation between $t = 0$ (when $V = V_0$) and t yields:

$$V - V_1 = (V_0 - V_1)\, e^{-t/\lambda}.$$

The coefficient λ defines the response time of the vegetation. The basic model can be modified to include periodic climatic changes such as those produced by orbital forcing. Webb writes

$$V_1 = f(\text{climate}) = a_0 + a_1 \sin(2\pi t/S)$$

and

$$dV/dt = -(1/\lambda)[V - \{a_0 + a_1 \sin (2 \pi t/S)\}],$$

where S is the period of sinusoidal forcing and a its amplitude. Integration gives

$$V - V_1 = C_1 e^{-\lambda t} + (a_1/C_2) \sin\{(2 \pi t/S) - \gamma\}, \tag{7.1}$$

where C_1 is defined by the initial conditions (but can be ignored if t be very much larger than γ), and

$$C_2 = \{1 + (2 \pi\lambda/S^2\}.$$

The tangent of the term $2 \pi\lambda/S$ is the phase angle, γ. So in Eq. (7.1), the vegetation responds sinusoidally with an amplitude reduction, a_1/C_2, and a phase lag, γ, which depend on λ/S, the ratio of vegetational response time to the period of forcing (Fig. 7.11). For orbital forcing in the Croll-Milankovitch frequencies, S lies in the range 20,000 to 100,000 years, and values of vegetational response, λ, in the order of 400 to 1000 years can be tolerated because they effect only small reductions in the amplitude of the response (T. Webb 1986a, p. 79). Equation (7.1) models the continuous nature of climatically forced changes in vegetation during the Pleistogene period and earlier times. No taxa can survive for long unless their response times, λ, be short enough to enable them to track the continuously changing location of habitats favourable to their growth. Given this fact, the question of disequilibrium conditions seems to hinge, not upon the long-term tracking of suitable habitats, but upon selected intervals of time when regional changes of climate might be so fast that the lags in the response of vegetation influence significantly the composition of vegetation for 1000 years or more. Sinusoidal forcing does incorporate times of relatively rapid climatic and vegetational change. During spells of quick change, the vegetational lag time will be very much in evidence. This, it might be argued, is one reason why

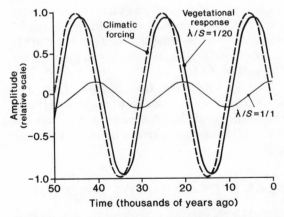

Fig. 7.11. Vegetational response to sinusoidal climatic forcing as predicted by Eq. (7.1) in the text. The amplitude is scaled from –1.0 to 1.0. (After T. Webb 1986a)

vegetational change trailed so far behind climate in the period of rapid climatic change lasting from about 14,000 to 9000 years ago (T. Webb 1986a, p. 80).

7.4 Long-Term Climatic Change and Biotic Communities

As we have seen, most animals and plants are sensitive to climatic conditions. The more sensitive ones are admirably suited for enlistment as indicators of past climates. Indeed, where the fossil record permits, analyses of plant community structure, vegetational and leaf physiognomy, and growth rings and vascular systems in wood make it possible to define terrestrial palaeoclimatic parameters, such as mean annual temperature and mean annual range of temperature, with better resolution than most sedimentological techniques (cf. Spicer 1989; Chaloner and Creber 1990). However, the using of fossil organisms as a guide to past climates is not a practice to be pursued with abandon, for it is plagued by pitfalls. For instance, it is not uncommon for fossil forms to have conferred on them the climatic ranges of their "nearest living relatives" (Chaloner and Creber 1990). Clearly, the older the fossils, the less justifiable the technique because the more likely is ecological tolerance to have altered owing to evolutionary changes. Modern, frost-sensitive, evergreen, relict cycads, for example, are a poor model of the more diverse Mesozoic cycads, some of which may have been deciduous and resistant to frost (Kimura and Sekido 1975). Only where several unrelated lineages of conservative groups, such as the gymnosperms, behave in concert can inferences about climatic trends safely be made (Spicer and Parrish 1990a, p. 333). Care must also be taken not to fall into the trap of circular argument by explaining away faunal and floral changes as a consequence of the very climatic changes which themselves were inferred from the fossil animals and plants. Independent predictions of climate are required to sidestep this problem. For times before the Pleistogene period, simulations using atmospheric general circulation or energy balance models run with appropriate boundary conditions provide a yardstick against which to test palaeoclimates as reconstructed from palaeobiological evidence. Equally, the success of a general circulation model can be assessed by gauging the similitude between the climate it predicts and the climatic pattern pieced together from palaeoclimatic indictators. Thus theory and observation can be made to work in tandem.

Many climatic changes have been established from the abundance and distribution fossil forms. Of the multeity of interrelationships between long-term changes in climate and organisms which could be examined here, just two will be pursued: biotic changes in late Mesozoic and Tertiary times; and climatic hypotheses of mass extinction.

7.4.1 Biotic and Climatic Changes in the Cretaceous and Palaeogene Periods

A striking feature of the biosphere in late Mesozoic and early Cenozoic times is the presence of warmth-loving animals and plants at high latitudes, both in the Northern and Southern Hemispheres. The presence of warm, moist climates reaching northwards into areas which are now barren, gelid, and ice-capped has been a perennial problem for palaeoclimatologists at least since Georg Adolf Erman (1833–48) found evidence of formerly milder climates at high latitudes in Asia and Oswald Heer (1868–83) reported his finds of Tertiary plant fossils in the Arctic region. Later palaeobotanical work has confirmed the warmth of polar regions during the Cretaceous and Palaeogene periods (e.g. Wolfe 1977; Wolfe and Upchurch 1986, 1987a,b; Upchurch and Wolfe 1987; Spicer and Parrish 1986, 1990a,b), though colder conditions (with cool summers, variable growing conditions, and a mean annual temperature in the range 2 to 6 °C), appear to have prevailed in northern Alaska during the late Cretaceous (Spicer and Parrish 1990b).

Several explanations of the relatively warm high-latitude climate have been offered. They include a transport of large volumes of warm ocean waters from the Pacific Ocean to the Arctic Ocean during the Mesozoic and early Cenozoic eras due to the greater separation of Alaska and Siberia (see Donn 1987), a decrease of axial tilt (Wolfe 1978, 1980), long-term changes in solar luminosity (Newkirk 1980), changes in the position of the poles (Donn 1982, 1987), and an increased carbon dioxide content of the atmosphere. The failure of general circulation models to reproduce warm polar climates with changed geography, obliquity, or a realistic increase in carbon dioxide content has already been mentioned (Sect. 3.4.2). This would suggest that the palaeobiological indicators are being wrongly read or that the theoretical models are in need of an overhaul. Each of the hypotheses of warm polar climates seems plausible but so far vindication has proved elusive: in all cases, conflicting evidence can be presented. For instance, the suggestion has been moved that palaeobotanical evidence is compatible with a low obliquity (Allard 1948; Wolfe 1978, 1980; J.G. Douglas and Williams 1982; Petersen 1984). To Jack A. Wolfe (1978), the presence of broad-leaved evergreen forests at high latitudes during the early Cenozoic era is compatible with an obliquity gradually changing from about 10° in the Palaeocene epoch to about 5° in the middle Eocene epoch. He reasoned that at present broad-leaved evergreen forests do not occur north of latitude 50° N and most occur equatorwards of latitudes 40° to 45° N because they cannot tolerate lengthy dark winters: decreased obliquity would produce more light at high latitudes and so enable notophyllous broad-leaved deciduous forests to flourish at latitude 60° N during the early Cenozoic era. Several lines of evidence appear to undermine this hypothesis. On the modelling front, experiments with atmospheric general circulation models suggest that the poles would actually be colder if obliquity were reduced (Sect. 2.5.8). In any case, fresh palaeobotanical evidence dispenses with the need for a reduced obliquity: if the obliquity were low then the annual range of annual temperature would be smaller and tree growth-rings might be expected be less prominent, but this appears not to be the case (Creber and Chaloner 1984). Moreover, studies of middle Cretaceous fossil plants on North Slope, Alaska, have shown that all the taxa there, with the exception of a microphyllous conifer probably capable of entering dormancy, appear to have been deciduous, dying back to underground perennating organs or surviving as seeds over the winter (Spicer and Parrish 1986, 1990a; Spicer

1987). This wholly deciduous flora lying at a palaeolatitude between 74° and 85° N precludes the necessity of educing a reduction of obliquity to provide a more even distribution of light during the year, though it does not disprove that obliquity was reduced at that time.

Temperature changes during the Cenozoic era have been mapped out using proxy data from deep-sea marine organisms and from land floras. Agreement between the two groups of organisms is reassuringly close (Wolfe and Poore 1981). A thermal maximum in the early Eocene epoch, and a major vacillation in mean annual temperature during the Oligocene epoch, are evidenced in both marine organisms and land flora. Additionally, land flora indicate a low mean annual range of temperature during the Palaeocene and Eocene epochs, with an increase through to Oligocene times when the mean annual range was greater than today. After a warm period in the middle of the Miocene epoch, overall mean annual temperatures declined in extratropical areas, though some short-lived warmer spells did occur. The land flora also suggest that the seasonal contrast of temperatures decreased during the Neogene period. Within this broad pattern of climatic change during the Cenozoic era, two times of major and relatively rapid upheavals of climate occurred, one at the Eocene-Oligocene boundary and the other in the late Miocene epoch. The late Eocene event is clearly shown by the global changes of vegetation mapped by Jack A. Wolfe (1985). In the Palaeocene epoch, diverse notophyllous broad-leaved forests were found at latitude 57° N; by the middle Eocene epoch they had extended to between latitudes 70° and 75° N, and multi-stratal vegetation extended to between latitudes 55° and 60° N (Fig. 7.12a,b). During cool intervals of the Eocene epoch, broad-leaved evergreen forests retreated to equatorwards of about latitude 50° N and were replaced polewards of that latitude by densely stocked coniferous forests (Fig. 7.12c). Semi-deciduous tropical to paratropical forests occupied southeastern Asia and southeastern North America in the middle and late Eocene epoch where they replaced broad-leaved evergreen forests. After the phase of cooling at the end of the Eocene epoch, vegetational zones in the Northern Hemisphere were as follows: dense coniferous forests occupied areas to the north of latitudes 50° to 60°; microthermal broad-leaved deciduous forests (unknown in Eocene times) grew south of the coniferous forests to latitude 35°; broad-leaved evergreen forests grew south to latitude 20°; and multi-stratal forests grew in the belt between latitude 20° and the equator. A similar change of vegetational zones occurred in the Southern Hemisphere (Fig. 7.12a to d).

The cause of the late Eocene event is unclear. Several views have been tabled. Nils-Axel Mörner (1984c), noting that the late Eocene and late Miocene were times of major marine regressions, proposed that the regressions led to the lowering of the geoid under the continents, a drop in the level of the ground water table, and so to drought. Jack A. Wolfe (1978) favoured the notion that the Eocene event was caused by a sharp increase of obliquity to between 25° and 30° at that time. Such a change of obliquity would have given rise to the following modifications of climate, all of which in Wolfe's eyes are attested to in the palaeobotanical record: an increase in winter temperature at low latitudes, thus raising the mean annual temperature; an increase in winter temperatures roughly equal to a decrease in summer temperatures at latitude 43°, thus leaving mean annual temperature unchanged; a decrease in summer temperatures at high latitudes, thus depressing mean annual temperatures; and a decrease in the mean annual range of temperature proportional to latitude. However, there is a demonstrable connection between major episodes of volcanic activity during the Cenozoic era and changes in fossil floras (Axelrod 1981), and the late Eocene event might have resulted from this cause. Another possibility is that the late Eocene event was induced by meteorite impacts. Gerta Keller and her colleagues (1987) examined microtektites and related crystal-bearing microspherule layers in deep-sea sediments of the west equatorial Pacific Ocean, offshore New Jersey, and southeastern Spain. Their investigations led to the confirmation of the presence of at least three microspherule layers in late Eocene sediments. These layers do not coincide precisely with planktonic foraminiferal species extinctions, but a major change of faunal assemblage is associated with the oldest of the three. Abundant pyrite in one of the sites of the youngest North American tektite layer suggests that reducing conditions prevailed, perhaps owing to a sudden influx of organic matter (dead plankton) to the ocean floor, while the middle ("crystal-bearing") layer coincides with five radiolarian extinctions. All three microspherule layers are associated with a decrease in carbonate content, possibly resulting from sudden changes of plankton productivity, dissolution as a result of sea-level and climatic fluctuations, or excess dust and water vapour in the atmosphere owing to impacts. Evidence of the effect of the impacts upon late Eocene climates is equivocal, the signal being largely obscured by vital effects and diagenesis. However, the marine plankton indicate climatic cooling: ecologically sensitive and general warmer-surface-water species decline and eventually die out. Another line on the cause of the late Eocene event was taken by Steven M. Stanley (1984, 1986). He proffered the hypothesis that oceanic cooling might have resulted from the separation of Antarctica and Australia (Kennett and von der Borch 1986). Once the continents had separated, the oceanic gyre which runs round Antarctica would have trapped water from the great oceanic gyres in the Southern Hemisphere and caused them to cool while circling Antarctica and sink into the deep ocean. It is unclear how such a deep-sea refrigeration mechanism would have been tied to contemporaneous cooling in the Northern Hemisphere. Nonetheless, it is evident that a major change in the thermal structure of the oceans and global climates occurred at the close of the Eocene epoch.

Floral and faunal changes in many parts of the world point to a cooling and, in some regions, drying of climate during the Miocene epoch. In the broad expanses of the North American midcontinent, S. David Webb (1983) discovered that the shift to cooler and drier climatic conditions in Miocene times (Wolfe 1978) drove a change of predominant vegetation type from forest to savanna to steppe, and parallel changes in the diversity of browsing and grazing ungulate taxa (Fig. 7.13). In northern Pakistan too, the fauna changed significantly about 7.4 million years ago when larger browsing animals (tragulids, suids, Okapi-like giraffes, low-crowned bovids, and others) were replaced by grazing

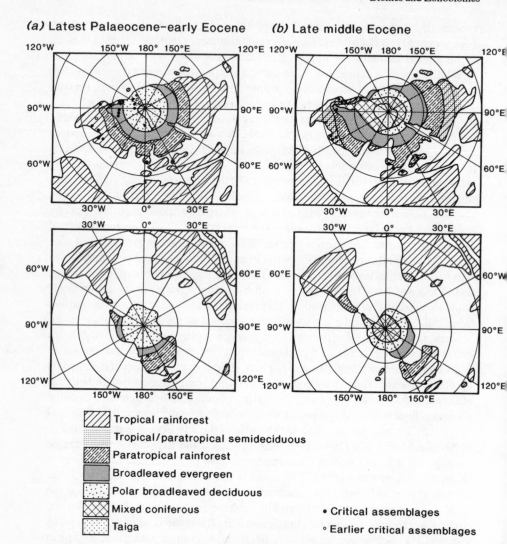

Fig. 7.12. Suggested vegetation maps for four stages of the Tertiary sub-era. **a** Latest Palaeocene to early Eocene epochs (55 to 50 million years ago). **b** Late middle Eocene epoch (46 to 41 million years ago). **c** Eocene cool intervals (50 to 46 million and 41 to 37 million years ago). **d** Early Miocene epoch (22 to 18 million years ago). (After Wolfe 1985)

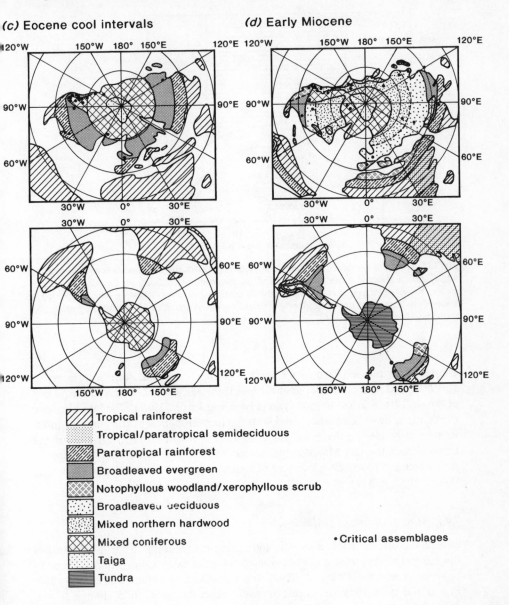

(c) Eocene cool intervals   **(d)** Early Miocene

Tropical rainforest
Tropical/paratropical semideciduous
Paratropical rainforest
Broadleaved evergreen
Notophyllous woodland/xerophyllous scrub
Broadleaved deciduous
Mixed northern hardwood
Mixed coniferous
Taiga
Tundra

• Critical assemblages

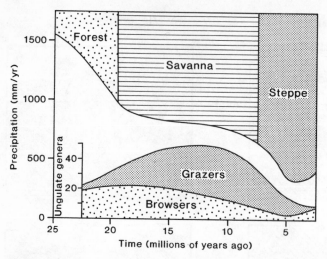

Fig. 7.13. Changing ungulate communities in North America during the Miocene epoch. Increasing aridity during Miocene times produced a shift in dominant vegetation from forest to savanna to steppe. Fossils of large ungulate genera show an increasing richness of grazing taxa. When steppe became established the browsing taxa were severely reduced and even the number of grazing taxa was cut. (After S. D. Webb 1983)

animals, as witnessed by a new tragulid with high-crowned dentition (Barry et al. 1985), and rodents underwent a rapid turnover, probably as a consequence of the shift from forest to grassland (Flynn and Jacobs 1985). The late Miocene event might have been associated with the intensification of the summer monsoon, and orogeny may have played a role in some areas: the rising of the Himalayas in the late Miocene times was probably responsible for the climatic shift which led to the development of steppe vegetation in northwestern China (Wolfe 1985, p. 371).

7.4.2 Mass Extinctions and Climate

Several terrestrial processes are thought capable of leading to a level of stress in the biosphere severe enough to bring about mass extinctions. There appear to be three main possibilities, all of which act by effecting a change of climate. The first is the possibility that plate tectonic processes lead to a change in the geographical placement of land masses, the building of mountain ranges, and the changing of sea level, thus triggering drastic changes of climate which place the biosphere under considerable environmental stress. The second possibility is that a change of climate following an outburst of protracted volcanism will be so radical as severely to stress the biosphere. The third possibility is that dramatic climatic and other changes are induced by cosmic processes, again with dire consequences for the biosphere (cf. Huggett 1989a, 1990).

Climatic change was for a long time deemed the principle cause of extinctions in the history of life. An influential advocate of this notion was George Gaylord Simpson (1953). While the link between climate and extinction has been watered down in recent decades, there is still much evidence which points to climatic change as potential disruptor of ecological stability. The idea that the redeployment of land masses causes changes of climate large enough to precipitate bouts of rapid extinction is currently regaining ground. If a continent drift from one climatic zone to another, extinctions are likely: the northwards drift of India probably led to the demise of several groups of land plants (Knoll 1984). Plate motions affecting the Gondwanan land mass and southwestern Laurasia at the close of the Triassic period brought South America and southern Africa into low latitudes and produced increasing aridity on those continents. The drier climate in turn brought about floral changes: new plants evolved, better fitted to the arid conditions. This floral evolution had repercussions higher up the food chain: the mammal-like reptiles and rhynchosaurs became extinct because they were unable to feed on the lowland bush vegetation which had previously supported them (Tucker and Benton 1982). Steven M. Stanley declared that periods of global cooling, caused by the encroachment of continents on one or other of the poles (Sect. 3.4.2), have been a prominent cause of biotic crises in the marine realm and have had a far greater effect on the marine biosphere than have reductions in the area of sea floor associated with global regressions of the sea. The oldest biotic crises identified by Stanley occurred during the Palaeozoic era in late Ordovician, late Devonian, and late Permian times (Stanley 1988a,b). Each of these biotic crises was a protracted affair in which tropical marine biotas, including stenothermal calcareous algae, declined greatly, and reef communities were decimated. As the Ordovician and Permian crises wore on, so warm-adapted taxa were displaced towards the equator, and as the Devonian crisis got under way, so tropical taxa died out in New York State while cold-adapted hyalosponges expanded. In the aftermath of each marine crisis, biotas became cosmopolitan, little or no reef growth took place, and the deposition of carbonates was reduced. Stanley (1988a, 1988b) makes much of the coincidence between these crises and the occurrence of glacial episodes, apparently triggered in each case by the proximity of a major continent to one of the Earth's rotatory poles. To him, this coincidence and the similar pattern of taxial change in each crisis (preferential loss of tropical taxa, replacement at low latitudes of warm-adapted by cold-adapted forms, and an aftermath in which cosmopolitan faunas prevailed, limestone production was diminished, and reef production was for a long while suppressed) implicates climatic deterioration, probably resulting from plate movement, as the primary agent involved in these Palaeozoic extinction events. As discussed in the previous section, a similar sequence of events may account for extinctions in late Eocene times. Biotic crises at the boundary of the Pliocene and Pleistocene epochs, too, have been linked with the expansion of glaciers and ice sheets (Stanley 1986; Raffi et al. 1985).

Changes in climate produced by protracted periods of volcanism may have resulted in environmental stress severe enough to cause mass extinctions. Certainly, large-scale volcanism and the changes of climate it would bring about has been educed to explain the extinction event at the close of the Cretaceous period. The detailed pattern of extinctions at the time suggests a more or less gradual increase in extinction rate for many groups of organisms, followed by a catastrophe lasting a few tens of thousands of years. In the marine realm, the extinction of planktonic foraminiferal species spanned 300,000 years below, and some 200,000 to 300,000 years above, the Cretaceous-Tertiary boundary (Keller 1989). To account for the pattern of extinctions, a scenario of environmental deterioration caused by increased volcanism over an extended period has been painted (McLean 1981, 1985; Geldhill 1985; Officer et al. 1987). Flood basalt fissure eruptions can produce individual lava flows with volumes greater than 100 km^3 and are capable of lofting large masses of aerosols, particularly sulphates, into the lower stratosphere. The consequences of large injections of sulphates high into the atmosphere would be potentially disastrous (Stothers et al. 1986): large amounts of acid rain would fall, the alkalinity of the surface ocean would drop, the atmosphere would cool (its cooling being enhanced by injections of ash from contemporary volcanoes), and the ozone layer would be depleted. Towards the end of the Cretaceous period, flood basalts poured out over large parts of India and the North American Tertiary Igneous Province (Courtillot and Cisowski 1987). The greatest outpourings coincide with the iridium peaks found around the Cretaceous-Tertiary boundary. The protracted period of volcanism might have led to the extinction of plankton and the ecological catastrophe among terrestrial plants, a gradual deterioration of the environment putting many species under stress and the short sharp burst of volcanism at the very end of the Cretaceous period delivering the coup de grâce (Officer et al. 1987). However, a detailed study of planktonic foraminiferal species' extinctions across the Cretaceous-Tertiary boundary in continental shelf sections at El Kef, Tunisia, and Brazos River, Texas, made by Gerta Keller (1989) reveals patterns of extinction more consistent with increased ecological stress resulting from the late Maastrichtian marine regression and global cooling, than with either volcanism or impacts. Keller concludes that the Cretaceous mass extinctions might in fact have resulted from multiple unrelated causes including sea level regression, global cooling, an impact of limited extent, and extensive volcanism! Such a multi-cause theory may well prove correct, but subjecting it to empirical test will be tricky.

8 Synthesis

Nature considered *rationally*, that is to say, submitted to the process of thought, is a unity in diversity of phenomena; a harmony, blending together all created things, however dissimilar in form and attributes; one great whole (τὸ πᾶν) animated by the breath of life.

Alexander von Humboldt (1849, vol. i, pp. 2–3)

Several important issues arise from the discourse of the foregoing chapters. The chief among them are the dynamicity and interrelatedness of systems comprising the biosphere, the problem of scale, and the origin of cyclicity. To conclude this survey of climate, Earth processes, and Earth history we shall discuss these pivotal issues with a view to possible future lines of research by developing from existing mathematical frameworks a tentative, general model of the biosphere.

8.1 Relations in the Biosphere: a General Model

It may be advantageous to put the relations between the world climate system and other geosystems on a formal footing. The biosphere may be thought of as a system consisting of a huge number of elements standing in relation to one another. The system elements may be measured, each measure being a state variable of the system. In the most general case, the time rate of change in any state variable, dx_i/dt, is a function of all other state variables, x_i. It follows that, if the biosphere be fully interconnected, a change in any one state variable will effect a change in all other state variables comprising the system. Such a system may be represented by a set of simultaneous differential equations of the style:

$$\frac{dx_1}{dt} = f_1(x_1, x_2, \ldots x_n)$$

$$\frac{dx_2}{dt} = f_2(x_1, x_2, \ldots x_n)$$

$$\cdot \quad \cdot \quad \cdot \quad \cdot \quad \cdot$$
$$\cdot \quad \cdot \quad \cdot \quad \cdot \quad \cdot$$
$$\cdot \quad \cdot \quad \cdot \quad \cdot \quad \cdot$$

$$\frac{dx_n}{dt} = f_n(x_1, x_2, \ldots x_n).$$

Or in matrix form

$$dx/dt = f(x).$$

As so clearly shown by Ludwig von Bertalanffy (1951, 1973), systems of equations of this kind are found in many fields, the measures used to define the state factors varying from one application to another. They were developed in ecology by Robert M. May (1973) and have proved a most valuable tool of investigation of large-scale dynamics (e.g. Aurada 1982, 1988; Moore et al. 1989). Accordingly, we shall develop a system of such equations to provide the architecture for a general model of dynamics in the biosphere.

At the most general level of formulation, the state of the biosphere at time $t+1$, b_{t+1}, can be said to depend on some function of its previous state, b_t, and any driving or forcing variables, z, acting on the biosphere from outside its boundaries, z:

$$b_{t+1} = b_t + z.$$

Expressing this equation as a rate of change we have

$$\frac{db}{dt} = b + z.$$

Driving variables include heat emanating from the Earth's interior, volcanic dust, and radiation and particles from extraterrestrial sources. While there appears to be much merit in regarding climate as a state of the biosphere, a view which would be fully endorsed by the proponents of the Gaia hypothesis, it is helpful when formulating relationships between climate and other systems to separate life-supporting systems, which includes the world climatic system, from life systems. Doing this, the time rate of change of the system variables in the life systems, o (for organisms), may be expressed as a function of the state of life systems themselves as well as the state of the life-supporting systems and driving variables. There are many ways of subdividing the life-supporting components of the biosphere. For the purposes of illustration, we shall follow the divisions of the life-supporting systems in the terrestrial biosphere adopted and adapted by Hans Jenny (1941; 1980, p. 202) from the writings of Vasilii Vasielevich Dokuchaev, Chas F. Shaw (1930), and Reinhold Tüxen (1931–32): climate, cl, soil, s, relief, r, and parent material, p. We may write:

$$\frac{do}{dt} = f(cl, s, o, r, p) + z.$$

This formulation is really a development of Jenny's acclaimed state factor equation of soil formation, but differs in using the time rate of change rather than including time as a "state factor". The components ("state factors") climate, soils, organisms, relief, and parent material will themselves vary with time. The time rate of change for the climatic component may be written:

$$\frac{dcl}{dt} = f(cl, o, s, r, p) + z.$$

Similar expressions could be written for the other components. The full set of equations looks like this:

$$\frac{dcl}{dt} = f(cl, o, s, r, p) + z$$

$$\frac{do}{dt} = f(cl, o, s, r, p) + z$$

$$\frac{ds}{dt} = f(cl, o, s, r, p) + z \qquad\qquad (8.1)$$

$$\frac{dr}{dt} = f(cl, o, s, r, p) + z$$

$$\frac{dp}{dt} = f(cl, o, s, r, p) + z.$$

These equations express the fact that all components in terrestrial life and life-supporting systems interact to some extent; they are all interrelated and they are all influenced by various driving variables external to the biosphere. Of course, to apply the model to an actual situation, each component must be defined by a suitable set of state variables — mean annual precipitation, soil acidity, biomass, or whatever. As formulated in Eq. (8.1), the system constitutes a general state factor model because any component is expressed as a function of all other system components and of driving variables. The model is very broad but does serve as a formal starting point for developing far more elaborate models with which to explore the interactions between climate and other Earth systems. It may also be used to probe the problem of scale and the question of cyclicity.

To show how the model helps in understanding the dynamics of the biosphere, we shall discuss a similar model developed independently by Jonathan D. Phillips (1989a), who borrowed his formulation from R. Slingerland's (1981) investigation of the stability of river hydraulic geometry. Phillips modelled what he termed a "soil-ecosystem", but his model could equally apply to the biosphere represented by Eq. (8.1). His model is useful, not the least because it demonstrates the possibility of solving a general state factor equation for a particular scale of operation, as will be seen in the next section. Starting from the premise that soil, climate, organisms, topography, and parent material are all interdependent to varying degrees, Phillips postulated that a soil-ecosystem could be represented as a series of (probably non-linear) partial differential equations where the n individual state variables are expressed as functions of each other:

$$dx/dt = f(x), \qquad\qquad (8.2)$$

where x is a vector of state variables, dt refers to temporal changes over a given timescale, and f is a matrix specifying the interaction between the state variables. If analysis be carried out in a qualitative manner, the form of f need not be known providing it is possible to say whether individual state variables in the vector x increase, decrease, or remain constant when another variable changes. Now if C be a steady state, then a deviation, x', from this state (a change in the soil-ecosystem) will be given by

$x' = x - C.$

Putting this equation into Eq. (8.2), rearranging, and expanding as a Taylor series yields

$$dx'/dt = f(x' + C)$$
$$= f(C) + Ax' + g(x'),$$
(8.3)

where $g(x')$ is a polynomial vector with terms of second order or higher, each small compared with x' and vanishing at $x' = 0$; and A is an $n \times n$ interaction matrix whose elements have the form

$\delta f_i (C)/\delta x'_i$

and define the interaction between system variables. Now, when the system is in, or very near, a steady state, x' and $g(x')$ are very small compared with Ax' and it is permissible to consider the simpler, linearized system

$$dx'/dt = Ax',$$
(8.4)

solutions of which take the form

$$x'(t) = vC \exp (\lambda t),$$
(8.5)

where v are eigenvectors, and λ are the eigenvalues of the matrix A. The response of a steady-state soil-ecosystem to small perturbations is determined by the eigenvalues in Eq. (8.5): should any of the eigenvalues be negative, the deviation will increase at an exponential rate through time; should all the eigenvalues be negative, then the system will remain stable, and perturbations be countered by negative feedback mechanisms; should any eigenvalue have a zero real part and all other eigenvalues be negative, the system may be unstable or metastable. Although this analysis applies to small deviations from equilibrium, it nonetheless enables the determination of the conditions under which one state variable will be stable in response to changes in one or more of the others (cf. Phillips 1989a, p. 172).

To illustrate the use of the model in a simple, qualitative way, Phillips considers the interaction of soil and soil-forming factors (climate, organisms, relief, and parent material), the very variables defined in Eq. (8.1). First, he presents the system as a signed, directed digraph (Fig. 8.1). Negative links mean that, for example, relief adjusts to parent material (via the repose angle); a positive link means that, for example, climate causes a deviation or increase in soil. Notice that soil and vegetation are mutually adjusting. Phillips then draws up an interaction matrix based on Fig. 8.1 in which entries, a_{ij}, are positive, negative, or zero (Table 8.1). He determines the characteristic equation for the system for several different cases: without any of the scale-limited links, with only the broad-scale links, and with only the short-term or local links, the links operating over a range of scales being included in all cases. In every case the determinant of the interaction matrix was zero, which means that the system is under-determined, that is, it has more variables than it has equations to deter-

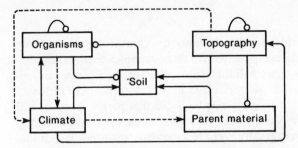

Fig. 8.1. A signed and directed digraph showing interactions between soil-ecosystem components. *Arrows* signify positive links; *closed circles* stand for negative links. *Solid lines* are relationships which act over a range of spatial and temporal scales; *dashed lines* are connections which act only at restricted scales. (Phillips 1989a)

Table 8.1. The interaction matrix derived from Fig. 8.1 (Phillips 1989a)

	Soil	Organisms	Climate	Relief	Parent material
Soil	0	$-a_{12}$	0	0	0
Organisms	$-a_{21}$	$-a_{22}$	$(a_{23})^a$	0	0
Climate	a_{31}	a_{32}	0	a_{34}	(a_{35})
Relief	a_{41}	0	(a_{43})	$-a_{44}$	$-a_{45}$
Parent material	a_{51}	0	0	0	0

[a]Brackets indicate links which act over restricted scales.

mine them. This under-determination arises because soil, biota, and topography can be described in terms of themselves and climate and parent material, but climate and parent material cannot be described in that way. Only by assuming a constant climate and parent material, and so a relatively short timescale, can an interaction matrix be derived which is non-trivial, the characteristic equation of which implies that a soil system is likely to be unstable under constant climate and parent material: should any state factor change, then the soil will seek a new steady state. This finding supports the view that Hans Jenny's state factor equation is insoluble insofar as it is impossible to let all factors vary. However, more limited models, based on up to three factors, may yield solutions. In short, Phillip's application of systems theory and qualitative modelling shows that barriers to solving the general equation are the range of timescales over which the state variables interact, and the fact that the interior, exterior, and temporal components of the state factors are not easily combined in a single model. This is very useful to know and sets a challenge for future modellers of the biosphere. It also establishes a starting point for the investigation of scale effects, as we shall now see.

8.2 The Problem of Scale

The biosphere and the systems of which it is made are characterized by a richly complex nexus of interacting components. All state variables used to describe the components of the biosphere will therefore influence, to a greater or less extent, all others. Now, many of the variables will vary over distinctly different spatial scales. To illustrate the point, it may be noted that parent material and temperature vary considerably only over many kilometres, whereas soil acidity and moisture content may vary greatly over a few metres or even less. In defining the biosphere, or a subsystem of it, only a selection of all possible components in the whole system is included. This is because it is impossible to identify, let alone consider, every component of which the biosphere is composed. In theory, we may say that the biosphere may be described by n state variables whose interaction is represented in a $n \times n$ interaction matrix, A; but, in practice, when prescribing components of the biosphere to include in an actual investigation, some components of the total system must of necessity be omitted. It is to be hoped that the chief components will have been singled out, but there will remain many more components which may have some influence. So, to write a set of equations describing a system which is actually going to be investigated, some variables must be omitted (see Phillips 1986a, 1988a). This abstracted system contains m ($m < n$) state variables and will be described by m equations or by an $m \times m$ interaction matrix $A(a)$. The interaction matrix for the whole system (which exists only in theory) may thus be broken down into four submatrices: $A(a)$ represents the interactions among the abstracted variables; $A(d)$ the interaction among the omitted variables; $A(2)$ the influence of the omitted variables on the abstracted variables; and $A(3)$ the influence of the abstracted variables on the omitted variables. Now, if the abstracted and omitted variables run on distinctly different scales, then $A(a)$ and $A(2)$ will be of a very different order than $A(d)$ and $A(3)$. Under these conditions, it follows (Schaffer 1981, pp. 389–393; Phillips 1986a, p. 165) that

$$\Delta(\lambda) = [\det\{A(a) - \Delta/I\}][\det\{A(d) - \lambda/I\}],$$

where $\Delta(\lambda)$ is a characteristic equation of a matrix similar to A, and I is an identity matrix. Similar matrices have identical eigenvalues, and the product of the eigenvalues gives the determinant. This demonstrates that eigenvalues of A comprise separate, scale-related components. The important implications of this proof are that factors controlling the systems of the biosphere operate or vary at spatial and temporal scales which differ substantially and are, in effect, independent of one another, and that this independence precludes a general solution to complete system models. This piece of theory justifies, at least in principle, the division of the systems of the biosphere into hierarchical levels, as was tentatively done in Fig. 1.3.

Whole system models of the biosphere cannot be solved unless two conditions be met: firstly, the different scales of operation, or tiers in the systems holarchy, must be ascertained; and secondly, the links between the components operating at the different scales must be established. Identifying the spatial scales of systems comprising the biosphere is not easy, but several techniques are now available for doing this and are proving successful (e.g. Oliver and Webster 1986; Phillips 1987; O'Neill et al. 1989). More difficult is the task of linking the components at different scales of operation. At least three ways of achieving scale-linkage have been tried: spatial models (Phillips 1988b), spatial sampling (Oliver and Webster 1986; Phillips 1986b; O'Neill et al. 1989), and maximum expected variation (Phillips 1990). Discussion of these is beyond the scope of the book. The point is that the holarchical nature of systems in the biosphere is now accepted and methods for analysing the holarchy are being invented and tested. Furthermore, the value of taking an holarchical approach to systems, which embodies scale differences, is becoming clear in a range of environmental disciplines: ecology (T.F.H. Allen et al. 1984; O'Neill 1988; O'Neill et al. 1986; D.J. Currie 1990), biology (Vrba and Gould 1986; Eldredge 1985), geomorphology (Schumm and Lichty 1965; Haigh 1987; Schumm 1988; Phillips 1986a,b, 1988b, 1989b; de Boer and Campbell 1989; Thornes 1990), hydrology (V.K. Gupta et al. 1986; Beven 1987), and pedology (Haigh 1987; Kachanoski 1988).

8.3 The Origin of Cyclicity

Many of the variables which drive the biosphere — the zs in Eq. (8.1) — are strongly periodic in nature. Their period lengths span the full gamut of timescales — from days to millions or billions of years. These pulses of forcing are echoed to varying degrees in the changing state of the biosphere. Indeed, the biosphere and its component systems display a vast range of cyclical changes, many of which are attributable to external forces. Writes Ferenc Benkö:

"According to both theoretical considerations and concrete investigations, many of the Earth's physical features are subject to cyclic, periodically recurrent changes. The cyclic change of some of them is beyond doubt; if that of others can be debated, this requires considerations of more particular investigations. It is evident that the short-period and recently measurable cycles can be proved directly and easily, while the long-period ones can be identified less unambiguously. The widening knowledge, however, proves even more convincingly the cyclic character of the changes". (Benkö 1985, p. 62).

Cyclicity in the biosphere results from a combination of periodic external forcing, both extraterrestrial and terrestrial in origin, and possibly complex interactions within the biosphere itself (cf. Pantič and Stefanović 1984). In general, cyclical behaviour in the biosphere is traditionally attributed to periodic external forcing, though some kind of internally created pulsebeat cannot be discounted, especially for shorter timescales. The biosphere is forced from the heavens by variations in solar and gravitational energy and by injections of cosmic material; and it is forced from deep within the bowels of the Earth by those geological processes which lead to changes in the crust (Fig. 8.2).

A major pulse in the biosphere is the long-term switching from icehouse to greenhouse states. The root cause of the periodicity of glaciations is normally sought in geological or galactic processes (cf. Sect. 3.4.2). Alfred G. Fischer (1981, 1984a) postulates that during the last 700 million years, climatic conditions

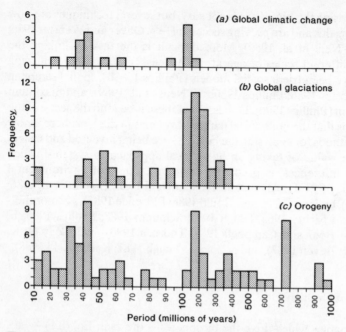

Fig. 8.2. Periodicity in selected terrestrial phenomena. **a** Global climatic changes: dominant periods within the ranges 35 to 45 million and 100 to 150 million years. **b** Global glaciations: dominant periods in the range 100 to 250 million years. **c** Orogeny: dominant periods in the ranges 30 to 50 million and 150 to 200 million years. Notice that the scale changes on the *horizontal axis*. (Data from Benkö 1985)

resembling those of today (an icehouse state) have alternated with climatic conditions like those during the Cretaceous period (a greenhouse state). These megaswings of climate have taken place roughly every 300 million years, giving two great climatic cycles during the Phanerozoic aeon (Fig. 8.3). As to the cause of this great cycle, Fischer looks to the interplay of the rates carbon dioxide addition and withdrawal from the ocean-atmosphere system. Carbon dioxide is added to the system by volcanism, and withdrawn from the system by weathering. These two processes are governed by very different factors, but their action will always strive towards an equilibrium level of atmospheric carbon dioxide. Volcanism increases during bouts of accelerated mantle convection when plate fragmentation and plate movement occur. At these times, the supply of carbon dioxide into the climate system (atmosphere-ocean) increases. Associated with plate activation is an increase in the volume of mid-oceanic ridges leading to marine transgressions. Less land area then being available, the atmosphere cannot lose carbon dioxide by weathering as fast as previously. The net effect of increased mantle convection is thus a rise in the amount of carbon dioxide in the atmosphere-hydrosphere system, until a new balance be reached wherein the greater intensity of weathering offsets the smaller area being weathered to counterbalance the volcanic additions. The carbon dioxide level of the atmos-

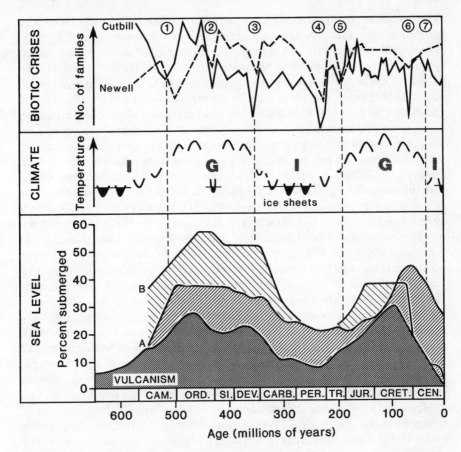

Fig. 8.3. Two supercycles of Phanerozoic history. Biotic crises (*numbered*) are shown as falls in the number of marine animal families. Climate is inferred to alternate between greenhouse (*G*) and icehouse (*I*) states, in each of which occur minor fluctuations. Times of ice sheet formation are shown by blackened troughs. Sea levels according to (*A*) Vail et al. (1977) and (*B*) Hallam (1977). Vulcanism as judged from granite emplacement in North America (Engel and Engel 1964). (After Fischer 1981, 1984)

phere may rise to three or four times its present level by this process, thus creating a supergreenhouse effect and a much warmer climate. During times of sluggish mantle convection, the number of plates becomes smaller, the volume of mid-oceanic ridges diminishes, and the continents become aggregated. Volcanism becomes subdued and consequently carbon dioxide emissions decline. Sea level drops, thus exposing more land to the atmosphere and increasing the withdrawal of carbon dioxide from the air. When the carbon dioxide content of the atmosphere has fallen low enough, the greenhouse state is broken, and the climate system assumes an icehouse state with ice sheets and glaciers.

Fischer's supercycles appear ultimately to have been driven by processes within the Earth's core and mantle, possibly the same processes which drive the supercontinent cycle posited by Damian Nance and his colleagues (1988). The two Phanerozoic supercycles have lesser cycles superimposed: one with a period of around 30 million years, and several others in the Croll-Milankovitch frequency band with periods apparently related to the Earth's orbital perturbations. The 30-million cycle was detected in pelagic environments but can be expected to occur in other environments as well. In the organic world it is expressed in the global diversity of planktonic and nektonic taxa, including globigerinacean foraminifers and ammonites, and in the episodic development of superpredators 10 to 18 m long. In the inorganic world it is expressed in temperature fluctuations, variations in carbon isotope ratios, and carbonate compensation depths. A.G. Fischer and Michael A. Arthur (1977) concluded that the instability associated with this periodicity pervades the geological record, but declared that the cause of the instability remains unclear. Given that in many ways the 30-million-year cycle mimics the supercycle of carbon dioxide content, Fischer suggested that it may involve carbon dioxide fluctuations, too. A link with periodic convection in the mantle seems possible, and a model based on fluctuating temperatures at the core-mantle boundary furnishes an explanation of the correlation in 30-million-year periodicities of magnetic-field reversals, climate, and mass extinctions (Loper et al. 1988). Moreover, episodes of major flood-basalt outpourings over the last 250 million years, when subjected to time series analysis, display a possible periodicity of roughly 30 million years (Rampino and Stothers 1986). It may be no coincidence that this is roughly in accord with the periodicity of impact cratering (Urey 1973; Rampino 1987): extraterrestrial and geological forcing of the biosphere may be interconnected. Indeed, there is now mounting evidence that all terrestrial processes are locked into planetary and galactic rhythms. This is especially true of the 250-million-year and the 30-million-year cycles of terrestrial evolution (McCrea 1981; Clube and Napier 1986), the orbitally driven cycles in the Croll-Milankovitch frequency band, and the many cycles with periods of days to centuries which, by processes at present guessed at, appear tied into the cyclical activity of the Sun, Moon, and planets. The origin of other cycles remains obscure. What, for example, causes the longer Klüpfel cycles (Klüpfel 1917; Bayer and Seilacher 1985) with periods of about a million years which seem to record increasing storm agitation of ocean sediments? Longer cycles of solar activity may drive such cycles, but more research is required to settle the matter.

    The pulse of glaciations has also been attributed to the passage of the Solar System about the centre of the Galaxy during the course of a galactic year. As long ago as 1909, Friedrich Nölke postulated that ice ages may be caused by galactic dust. Later, Harlow Shapley expressed the view that cosmic processes might influence the Earth's climate, and William Trowbridge Merrifield Forbes (1931) drew attention to a possible connection between the revolution of the Solar System about the galactic centre, then estimated to take 230 million years, and an apparent 210-million-year pulse of major glaciations. This link was explored further by Johannes Herman Frederik Umbgrove (1939, 1940, 1942), Harlow Shapley (1949), G.F. Lungershausen (1957), G.P. Tamrazyan (1967), Johann Steiner (1967, 1973, 1978, 1979; Steiner and Grillmair 1973), and George

E. Williams (1975a, 1981a). The more recent work has the advantage of a more accurate chronology of events and gives the mean period of mean ages of glaciations as about 155 million years. With a galactic year of 303 million years, that means that two glaciations occur during every revolution of the Solar System about the centre of the Galaxy (Williams 1975a, 1981a) (Sect. 3.4.2).

Exceedingly long cycles of terrestrial processes, admittedly of a rather speculative nature, may be locked into intergalactic beats: George E. Williams (1975a) thinks the properties of the interstellar medium, or the energy output of the Sun (or both), may be sufficiently influenced by the tidal action of the Large and Small Megallanic Clouds, companion galaxies to the Milky Way, to have far-reaching climatic consequences for the Earth and other terrestrial planets. Although the gravitational torque involve is minuscule, it is just possible that the pole of the ecliptic and the plane of the Solar System may track the Large Megallanic Cloud in its orbit around the Galaxy thereby causing secular changes in the obliquity of the ecliptic and consequent very long-term changes in the Earth's palaeoclimates and tectonism. It is intriguing that a terrestrial geotec-tonic megarhythm of roughly 600 to 800 million years, a putative 1300-million-year pulse of very long-term climatic change, and a 2500-million-year period of postulated secular change in the Earth's obliquity, are, respectively, about one quarter, one half, and equal to the estimated orbital period of the Large Megallanic Cloud (Williams 1981a, p. 13).

It seems reasonable to suppose that galactic and geological processes would drive the biosphere through periodic changes. However, it has recently become apparent that the biosphere may display long-term periodic behaviour owing to the nature of its inner workings. Many of the interactions within the biosphere are non-linear and involve bifurcations and multiple steady states. One stable state may persist for millions of years until a minor chance perturbation happen to tip the balance of forces and nudge the system into a new stable state. This kind of behaviour in systems held away from equilibrium by the constant import of energy (dissipative systems) has been well known in climatology since Edward Lorenz first chanced upon the "Butterfly Effect" in his early simulations of weather patterns (see Gleick 1988) and has been demonstrated in computer simulations of the "biosphere" by Antonio C. Lasaga (1983; see also Kump 1989).

Galaxy, core and mantle, inside the biosphere — from which of these sources emanate the forces of change in the climate system? It is possible that all three sources are a seat of change. As we have seen, plausible links can be established at many timescales between galactic processes and geological processes on the one hand, and change in the biosphere on the other. Less clear at present are the internal dynamics of the biosphere itself and the periodic changes which they engender, but it seems that immanent processes can drive the climate system over days and over aeons. If galaxy, core and mantle, and the biosphere do play a role in climatic change, then is it possible to discover the part each plays, both singly and in combination with the others? Experiments with climate models, especially those which link life systems and life-supporting systems, enable the role of forcing factors to be tested. We have seen the value of this approach in the case of obliquity changes, the effect of mountain ranges, and several other factors. It seems not unreasonable, therefore, to answer the question in the affirmative and suggest that a much deeper understanding of atmospheric change and its connection with change in other systems in the biosphere should not be too long in arriving. But is it the right question to ask? A radical aspect of the new and revolutionary branch of physics known as non-linear dynamics is the replacement of cause-and-effect relationships by sets of embedded systems (a systems holarchy) which resound with resonances (Huggett 1990, p. 198). The implications of this idea are momentous. From the viewpoint of non-linear dynamics, the Solar System is embedded within the galactic system and exists because of its resonant interactions with the entire system. If the Solar System were independent there would be no resonances, but this is undeniably not the case because terrestrial phenomena — from tides to the solar wind — beat in sympathy with stellar influences. Non-linear dynamics does away with the view that there is some kind of galactic clock which, by the regular motion of planets and stars, acts as a pacemaker of Earth history. There is no need, arguing from the basis of non-linear dynamics, for events in the biosphere and galactic or geological events to be precisely correlated. Rather, coincidences amongst their chronological permutations are strongly suggestive of dynamical correlations among them all, and provide "direct evidence of dissipative processes that have woven together the record of terrestrial evolution into a coordinated theme" (H.R. Shaw 1987). The involvement of galactic and geological processes in the biosphere is relative to a subsuming pattern within which the criteria of chaotic dynamics operate in a systematic and ongoing cyclical state. The inclusion of non-linearities in a general model of the biosphere, such as that represented by Eq. (8.1), will create multiple steady states, bifurcations, and chaotic behaviour. Experiments with relatively simple models of the separate parts of the biosphere — atmospheric systems (e.g. Saltzman and Maasch 1990), global biogeochemical systems (e.g. Lasaga 1983; Kump 1989), ecological systems (e.g. R.M. May 1981), and geomorphological systems (e.g. Scheidegger 1990) — have shown this to be the case. A challenge for the future is to develop dynamical models for the world climate system as whole, to consider the conjoint action of all parts of the biosphere. The Geosphere-Biosphere Programme has taken a big step in this direction; prosecutors of the Gaia hypothesis have taken an even bigger step. The conceptual frame, the methodological machinery, the will, and the money to study the biosphere as a whole are all there. Whether an integrated approach will prove more fruitful than the traditional, analytical approaches has yet to be seen, but preliminary results are most encouraging. Perhaps Alexander von Humboldt's vision of Nature as "one great whole (τὸ πᾶν) animated by the breath of life" was far nearer the mark than many scientists have, until recently, been willing to grant.

References

Abbot CG (1963) Solar variation and weather. Smithsonian Institution, Washington DC, Smithsonian Miscellaneous Collections 146, no 99

Abbot CG, Fowle FE (1913) Volcanoes and climate. Smithsonian Institution, Washington DC, Smithsonian Miscellaneous Collections 60, no 29

Adams CC (1909) An ecological survey of Isle Royale, Lake Superior. Wynkoop, Hallenbeck, & Crawford, Lansing, Michigan

Adams CC (1913) Guide to the study of animal ecology. Macmillan, New York

Adhémar JA (1842) Révolutions de la mer. Carilan-Goeury et V. Dalmont, Paris

Aharon P (1984) Implications of the coral-reef theory record from New Guinea concerning the astronomical theory of ice ages. In: Berger A, Imbrie J, Hays J, Kukla G, Saltzman B (eds) Milankovitch and climate: understanding the response of astronomical forcing, Part 1. D. Reidel, Dordrecht, pp 379–389 (Proceedings of the NATO Advanced Research Workshop on Milankovitch and Climate, Palisades, New York, 1982. NATO ASI Series C, Mathematical and Physical Sciences, vol 126)

Ahnert F (1970) Functional relationships between denudation, relief, and uplift in mid-latitude drainage basins. Am J Sci 268:243–263

Ahnert F (1987) An approach to the identification of morphoclimates. In: Gardiner V (ed) International geomorphology 1986, Part II. John Wiley, Chichester, pp 159–188 (Proceedings of the First International Conference on Geomorphology)

Ahnert F (1988) Modelling landform change. In: Anderson MG (ed) Modelling geomorphological systems. John Wiley, Chichester, pp 375–400

Aksu AE, Mudie PJ, Macko SA, de Vernal A (1988) Upper Cenozoic history of the Labrador Sea, Baffin Bay, and the Arctic Ocean: a paleoclimatic and paleoceanographic summary. Paleoceanography 3:519–538

Albrecht F (1947) Die Aktionsgebiete des Wasser- und Wärmehaushaltes der Erdoberfläche. Z Meteorol 1:97–109

Algeo TJ, Wilkinson BH (1988) Periodicity of mesoscale Phanerozoic sedimentary cycles and the role of Milankovitch orbital modulation. J Geol 96:313–322

Allard HA (1948) Length of day in climates of past geological eras and its possible effects upon the changes in plant life. In: Murneek AE, Whyte RO (eds) Vernalization and photoperiodism. Chronica Botanica, Waltham, Massachusetts, pp 101–119

Allen JA (1877) The influence of physical conditions in the genesis of species. Radical Rev 1:108–140

Allen TFH, O'Neill RV, Hoekstra TW (1984) Interlevel relations in ecological research and management: some working principles from hierarchy theory. United States Department of Agriculture, Forest Service, Rocky Mountain Forest and Range Experiment Station, Fort Collins, Colorado, General Technical Report RM-110

Alvarez LW, Alvarez W, Asaro F, Michel H (1980) Extraterrestrial cause for the Cretaceous-Tertiary extinction. Science 208:1095–1108

Anderson RY (1961) Solar-terrestrial climatic patterns in varved sediments. Ann N Y Acad Sci 95:424–439

Anderson RY (1984) Orbital forcing of evaporite sedimentation. In: Berger A, Imbrie J, Hays J, Kukla G, Saltzman B (eds) Milankovitch and climate: understanding the response of astronomical forcing, Part 1. D Reidel, Dordrecht, pp 147–162 (Proceedings of the NATO Advanced Research Workshop on Milankovitch and Climate, Palisades, New York, 1982. NATO ASI Series C, Mathematical and Physical Sciences, vol 126)

Andrews JA (1985) True polar wander: an analysis of Cenozoic and Mesozoic paleomagnetic poles. J Geophys Res 90:7737–7750

Angström A (1936) A coefficient of humidity of general applicability. Geogr Ann 18:245–254

Anonymous (1928) Growth of trees in the Forest of Dean in relation to rainfall. Meteorol Mag 63:29–33

Ansted DT (1871) On some phenomena of the weathering of rocks, illustrating the nature and extent of sub-aerial denudation. Trans Camb Philos Soc 11:387–395

Antevs E (1941) Arroyo cutting and filling. J Geol 60:375–385

Archer AW, Kvale EP, Johnson HR (1990) Discussion on late Precambrian tidal rhythmites in South Australia and the history of the Earth's rotation. J Geol Soc Lond 147:401–402

Arkley RJ (1963a) Relationships between plant growth and respiration. Hilgardia 34:499–584

Arkley RJ (1963b) Calculation of carbonate and water movement in soil from climatic data. Soil Sci 96:239–248

Arkley RJ (1967) Climates of some great soil groups of the western United States. Soil Sci 103:389–400

Arkley RJ, Ulrich R (1962) The use of calculated actual and potential evapotranspiration for estimating potential plant growth. Hilgardia 32:443–462

Arrhenius S (1896) On the influence of carbonic acid in the air upon the temperature of the ground. Philos Mag 41:237–275

Arrhenius S (1903) Lehrbuch der kosmischen Physik, Part 2. S Hirzel, Leipzig

Arthur MA, Bottjer DJ, Dean WE, Fischer AG, Hattin DE, Kauffman EG, Pratt LM, Scholle PA (1986) Rhythmic bedding in Upper Cretaceous pelagic carbonate sequences: varying sedimentary response to climatic forcing. Geology 14:153–156

Aurada KD (1982) Zur Anwendung des systemtheoretischen Kalküls in der Geographie. Petermanns Geogr Mitt 126:241–249

Aurada KD (1988) Raum-Zeit-Phänomene im Ostseeraum. Petermanns Geogr Mitt 132:1–14

Axelrod DI (1981) Role of volcanism in climate and evolution. Geol Soc Am Spec Pap 185:1–32

Bach W (1984) CO_2-sensitivity experiments using general circulation models. Prog Phys Geogr 8:583–609

Bailey IW, Sinnott EW (1915) A botanical index of Cretaceous and Tertiary climates. Science 41:823–832

Baillie MGL, Munro MAR (1988) Irish tree-rings, Santorini and volcanic dust veils. Nature 332:344–346

Barnett TP (1989) A solar-ocean relation: fact or fiction? Geophys Res Lett 16:803–806

Barnola JM, Raynaud D, Korotkevich YS, Lorius C (1987) Vostok ice core provides 160,000-year record of atmospheric CO_2. Nature 329:408–414

Barnosky AD (1985) Taphonomy and herd structure of the extinct Irish elk (*Megaloceras giganteus*). Science 228:340–344

Barnosky AD (1986) "Big game" extinction caused by late Pleistocene climatic change: Irish elk (*Megaloceras giganteus*) in Ireland. Quat Res 25:128–135

Barnosky AD (1989) The late Pleistocene event as a paradigm for widespread mammal extinction. In: Donovan SK (ed) Mass extinctions: processes and evidence. Belhaven, London, pp 235–254

Barrell J (1908) Relations between climate and terrestrial deposits. J Geol 16:159–190, 255–295, 363–384

Barron EJ (1983) A warm, equable Cretaceous: the nature of the problem. Earth-Sci Rev 19:305–338

Barron EJ (1984) Climatic implications of the variable obliquity explanation of Cretaceous-Paleogene high-latitude floras. Geology 12:595–598

Barron EJ (1985) Explanations of the Tertiary global cooling trend. Palaeogeogr Palaeoclimatol Palaeoecol 50:45–61

Barron EJ (1989) Studies of Cretaceous climate. In: Berger A, Dickinson RE, Kidson JW (eds) Understanding climatic change. American Geophysical Union, Washington DC, Geophysical Monograph 52, pp 149–157 (International Union of Geodesy and Geophysics, vol 7)

Barron EJ, Washington WM (1982a) Cretaceous climate: a comparison of atmospheric simulations with the geologic record. Palaeogeogr Palaeoclimatol Palaeoecol 40:103–133

Barron EJ, Washington WM (1982b) Atmospheric circulation during warm geologic periods: is the equator-to-pole surface-temperature gradient the controlling factor? Geology 10:633–636

Barron EJ, Washington WM (1984) The role of geographic variables in explaining paleoclimates: results from Cretaceous climate sensitivity studies. J Geophys Res 89:1267–1279

Barron EJ, Washington WM (1985) Warm Cretaceous climates: high atmospheric CO_2 as a plausible mechanism. In: Sundquist ET, Broecker WS (eds) The carbon cycle and atmospheric CO_2; natural variations Archean to present. American Geophysical Union, Washington DC, Geophysical Monograph 32, pp 546–553

Barron EJ, Sloan JL, Harrison CGA (1980) Potential significance of land-sea distribution and surface albedo variations as a climatic forcing factor; 180 million years to the present. Palaeogeogr Palaeoclimatol Palaeoecol 30:17–40

Barron EJ, Thompson SL, Schneider SH (1981) An ice-free Cretaceous? Results from climate model simulations. Science 212:501–508

Barron EJ, Thompson SL, Hay WW (1984) Continental distribution as a forcing factor for global-scale temperature. Nature 310:574–575

Barry JG, Johnson NM, Raza SM, Jacobs LL (1985) Neogene mammalian faunal change in southern Asia: correlations with climatic, tectonic, and eustatic events. Geology 13:637–640

Bartlein PJ, Prentice IC (1989) Orbital variations, climate, and palaeoecology. Trends Ecol Evol 4:195–199

Battiau-Queney Y (1984) The pre-glacial evolution of Wales. Earth Surface Processes and Landforms 9:229–252

Battiau-Queney Y (1987) Tertiary inheritance in the present landscape of the British Isles. In: Gardiner V (ed) International geomorphology 1986, Part II. John Wiley, Chichester, pp 979–989 (Proceedings of the First International Conference on Geomorphology)

Baumgartner A, Reichel E (1975) The world water balance: mean annual global, continental and maritime precipitation, evaporation and runoff. English translation by Richard Lee. Elsevier, Amsterdam

Baur F (1956) Physikalische-statische Regeln als Grundlagen für Wetter- und Witterungsvorher-sagen, vol 1. Akademische Verlagsgesellschaft, Frankfurt

Baur F (1959) Physikalische-statische Regeln als Grundlagen für Wetter- und Witterungsvorher-sagen, vol 2. Akademische Verlagsgesellschaft, Frankfurt

Bayer U, Seilacher A (eds) (1985) Sedimentary and evolutionary cycles. Springer, Berlin Heidelberg New York Tokyo

Bazilevich NI, Rodin LE (1967) Maps of productivity and the biological cycle in the Earth's principal vegetation types. Izv Geogr Obshchestva, Leningrad 99:190–194

Bazilevich NI, Titlyanova AA (1980) Comparative studies of ecosystem function. In: Breymeyer AI, Van Dyne GM (eds) Grasslands, systems analysis and Man. Cambridge University Press, Cambridge, pp 713–758 (International Biological Programme, vol 19)

Bazilevich NI, Rodin LE, Rozov NN (1971) Geographical aspects of biological productivity. Sov Geogr 12:293–317

Beardmore N (1862) Manual of hydrology: containing I. Hydraulic and other tables, II. Rivers, flow of water, springs, wells, and percolation, III. Tidal estuaries and tidal rivers, IV. Rainfall and evaporation. Waterlow, London

Beaty C (1978a) The causes of glaciation. Am J Sci 66:452–459

Beaty C (1978b) Ice ages and continental drift. New Sci 80:776–777

Begét JE, Hawkins DB (1989) Influence of orbital parameters on Pleistocene loess deposition in central Alaska. Nature 337:151–153

Belt T (1874a) An examination of the theories that have been proposed to account for the climate of the glacial period. Q J Sci 44:1–44

Belt T (1874b) The naturalist in Nicaragua: a narrative of a residence at the gold mines of Chontales; journeys in the savannahs and forests. With observations on plants and animals in reference to the theory of evolution of living forms. John Murray, London

Benkö F (1985) Geological and cosmogonic cycles as reflected by the new law of universal cyclicity. Akadémiai Kiadó, Budapest

Berger A (1978) Long-term variations of caloric insolation resulting from the Earth's orbital elements. Quat Res 9:139–167

Berger A, Loutre MF, Dehant V (1989) Pre-Quaternary Milankovitch frequencies. Nature 342:133

Berger WH, Killingley JS (1982) The Worthington effect and the origin of the Younger Dryas. J Mar Res 40:27–38

Berger WH, Vincent E (1986) Sporadic shutdown of North American deep water production during the Glacial-Holocene transition. Nature 324:53–55

Bergmann C (1847) Über die Verhältnisse der Wärmeökonomie der Thiere zu ihrer Grösse. Gött Stud 3:595–708

Berner EK, Berner RA (1987) The global water cycle: geochemistry and environment. Prentice-Hall, Englewood Cliffs, New Jersey

Beven K (1987) Towards a new paradigm in hydrology. In: Rodda JC, Matalas NC (eds) Water for the future: hydrology in perspective. International Association of Hydrological Sciences, Wallingford, England, Publication 164, pp 393–403

Bik MJ (1968) Morphoclimatic observations on prairie mounds. Z Geomorphol 12:409–469

Birchfield GE, Weertman J (1983) Topography, albedo-temperature feedback, and climate sensitivity. Science 219:284–285

Birkeland PW, Burke RM, Benedict JB (1989) Pedogenic gradients for iron and aluminium accumulation and phosphorus depletion in Arctic and Alpine soils as a function of time and climate. Quat Res 32:193–204

Birot P (1949) Essai sur quelques problèmes de morphologie générale. Instituto Para a Alta Cultura, Centro de Estudos Geograficos, Lisbon

Birot P (1968) The cycle of erosion in different climates. Translated by C Ian Jackson and Keith M Clayton. BT Batsford, London

Bischof KGC (1854–59) Elements of chemical and physical geology, 3 vols. Translated by Benjamin Horatio Paul and John Drummond. The Cavendish Society, London

Bishop P, Young RW, McDougall I (1985) Stream profile change and long-term landscape evolution: early Miocene and modern rivers of the east Australian highland crest, central New South Wales, Australia. J Geol 93:455–474

Blackman VH (1919) The compound interest law of plant growth. Ann Bot 33:353–360

Blodget L (1857) Climatology of the United States, and the temperate latitudes of the North American continent. Embracing a full comparison of these with the climatology of the temperate latitudes of Europe and Asia. And especially in regard to agriculture, sanitary investigations, and engineering. With isothermal and rain charts for each season, the extreme months, and the year. Including a summary of the statistics of meteorology observations in the United States, condensed from recent scientific and official publications. JB Lippincott, Philadelphia

Bloemendal J, deMenocal P (1989) Evidence for a change in the periodicity of tropical climate cycles at 2.4 Myr from whole-core magnetic susceptibility measurements. Nature 342:897–900

Bloom AL, Yonekura N (1985) Coastal terraces generated by sea-level change and tectonic uplift. In: Woldenberg MJ (ed) Models in geomorphology. Allen and Unwin, Boston, pp 139–154 (The Binghampton Symposia in Geomorphology, International Series)

Blumenstock DI, Thornthwaite CW (1941) Climate and the world pattern. In: Climate and Man. US Department of Agriculture, Yearbook of Agriculture for 1941, US Government Printing Office, Washington DC, pp 98–127

Blytt A (1899) The probable cause of the displacement of beach-lines: an attempt to compute geological epochs. Fordhanglinger Videnskaps-selskapet, Kristiana, no 1, 93 pp

Bockheim JG (1980) Solution and use of chronofunctions in studying soil development. Geoderma 24:71–85

Bolin B (1980) Climatic changes and their effects on the biosphere. World Meteorological Organization, Geneva, Switzerland, Publication no 542

Bolin B (1988) Linking terrestrial ecosystem process models to climate models. In: Rosswall T, Woodmansee RG, Risser PG (eds) Scales and global change: spatial and temporal variability in biospheric and geospheric processes. John Wiley, Chichester, pp 109–124 (SCOPE 35)

Boll J, Thewessen TJM, Meijer EL, Kroonenberg SB (1988) A simulation of the development of river terraces. Z Geomorphol 32:31–45

Bonnet RM (1985) Solar terrestrial relations. In: Malone TF, Roederer JG (eds) Global change. Published on behalf of the ICUS Press by Cambridge University Press, Cambridge, pp 397–419

Boot R, Raynal DJ, Grime JP (1986) A comparative study of the influence of drought stress on flowering in *Urtica dioica* and *U. urens*. J Ecol 74:485–495

Borinsenkov YeP, Tsvetkov AV, Agaponov SV (1983) On some characteristics of insolation changes in the past and future. Clim Change 5:237–244

Bosák P, Ford DC, Głazek J, Horáček I (eds) (1989) Paleokarst: a systematic and regional review. Elsevier, Amsterdam (Developments in Earth Surface Processes 1)

Boussingault JB (1844) Die Landwirtschaft in ihren Beziehungen zur Chemie, Physik und Meteorologie, vol 1. C Graeger, Halle

Boyce MS (1978) Climatic variability and body size variation in the muskrats (*Ondatra zibethicus*) of North America. Oecologia 36:1–19

Box EO (1981) Macroclimate and plant forms: an introduction to predictive modeling in phytogeography. W. Junk, The Hague

Bradley WH (1929) The varves and climate of the Green River epoch. US Geol Surv Washington DC Prof Pap 158, pp 87–110

Bradley WH (1931) The origin of the oil shale and its microfossils of the Green River Formation in Colorado and Utah. US Geol Surv Washington DC, Prof Pap 168, 58 pp

Bray JR (1974) Volcanism and glaciation during the past 40 millenia. Nature 252:679–680

Bray JR (1977) Volcanic dust veils and north Atlantic climatic change. Nature 268:616–617

Bremer H (1989) Allgemeine Geomorphologie: Methodik — Grundvorstellungen — Ausblick auf den Landschaftshaushalt. Gebrüder Borntraeger, Berlin

Briffa KR, Bartholin TS, Eckstein D, Jones PD, Karlén W, Schweingruber FH, Zetterberg P (1990) A 1400-year tree-ring record of summer temperatures in Fennoscandia. Nature 346:434–439

Broecker WS (1965) Isotope geochemistry and the Pleistocene climatic record. In: Wright HE Jr, Frey DG (eds) The Quaternary of the United States. Princeton University Press, Princeton, New Jersey, pp 737–753

Broecker WS, Denton GH (1989) The role of ocean-atmosphere reorganizations in glacial cycles. Geochim Cosmochim Acta 53:2465–2501

Broecker WS, Denton GH (1990) What drives glacial cycles? Sci Am 262:43–50

Broecker WS, Peng T-H (1986) Carbon cycle 1985: glacial to interglacial changes in the operation of the global carbon cycle. Radiocarbon 28:309–327

Broecker WS, van Donk J (1970) Insolation changes, ice volumes, and the O^{18} record in deep-sea cores. Rev Geophys Space Phys 8:169–197

Broecker WS, Thurber DL, Goddard J, Ku T, Matthews RK, Mesolella KJ (1968) Milankovitch hypothesis supported by precise dating of coral reefs and deep-sea sediments. Science 159:1–4

Brooks CEP (1922) The evolution of climate. With a preface by G.C. Simpson. Benn Brothers, London

Brooks CEP, Glasspoole J (1928) British droughts and floods. With an introductory note by Hugh Robert Mill. Ernest Benn, London

Brosche P, Sündermann J (eds) (1990) Earth's rotation from eons to days. Springer, Berlin Heidelberg New York Tokyo

Brown GM, John JI (1979) Solar cycle influences in tropospheric circulation. J Atmos Terrest Phys 42:43

Brown GM, Price IT (1984) Solar control of terrestrial temperatures. In: Berger AL, Nicolis C (eds) New perspectives in climate modelling. Elsevier, Amsterdam, pp 93–114

Brown JH (1981) Two decades of homage to Santa Rosalia: toward a general theory of diversity. Am Zool 21:877–888

Brown JH, Lee AK (1969) Bergmann's rule and climatic adaptation in woodrats (*Neotoma*). Evolution 23:329–338

Brückner E (1890) Klimaschwankungen seit 1700 nebst Bemerkungen über die Klimaschwankungen der Diluvialzeit. Geographische Abhandlungen herausgegeben von A Penck 4:153–484

Brunt D (1925) Periodicities in European weather. Philos Trans R Soc Lond 222A:247–302

Bryan K (1928) Historic evidence on changes in the channel of the Rio Puerco, a tributary of the Rio Grande in New Mexico. J Geol 36:265–282

Bucha V (1984) Mechanism for linking solar activity to weather-scale effects, climatic changes and glaciations in the northern hemisphere. In: Mörner N-A, Karlén W (eds) Climatic changes on a yearly to millenial basis: geological, historical and instrumental records, D Reidel, Dordrecht, pp 415–448

Büdel J (1982) Climatic geomorphology. Translated by Lenore Fischer and Detlef Busche. Princeton University Press, Princeton, New Jersey

Budyko MI (1956) Teplovoí balans zemnoí poverkhnosti. Gridometeorologicheskoe Izdatel'stvo, Leningrad

Budyko MI (1958) The heat balance of the Earth. Translated by Nina A. Stepanova. Office of Technical Services, US Department of Commerce, Washington DC

Budyko MI (1974) Climate and life. English edition edited by David H Miller. Academic Press, New York (International Geophysics Series, vol 18)

Budyko MI (1982) The Earth's climate: past and future. Translated by the author. Academic Press, New York (International Geophysics Series, vol 29)

Buffon GLL, Comte de (1749–89) Histoire naturelle, générale et particulière, avec la description du Cabinet du Roi (Buffon, Daubenton, Guéneau de Montbeillard, Bexan, and Lacépède). 44 vols. De l'Imprimerie Royale, Paris

Buffon GLL, Comte de (1799) Histoire naturelle, générale et particulière, par Leclerc de Buffon; nouvelle edition, accompagnée de notes, et dans laquelle les Supplémens sont insérés dans le premier texte, à la place qui leur convient. L'on y a ajouté l'histoire naturelle des quadrupèdes et des oiseaux découverts depuis la mort de Buffon, celle des reptiles, des poissons, des insectes et des vers; enfin, l'histoire des plantes dont ce grand naturaliste n'a pas eu le tems de s'occuper. Ouvrage formant un cours complet d'Histoire Naturelle; rédigé par CS Sonnini, membre de plusieurs sociétés savantes, vol 1. Chez Deboffe, Libraire, London

Burke K, Francis P (1985) Climatic effects of volcanic eruptions. Nature 314:136

Burnet T (1691) The theory of the Earth: containing an account of the original of the Earth, and of all the general changes which it hath already undergone, or is to undergo, till the consummation of all things, 2nd edn. Walter Kettilby, London

Burnet T (1965) The sacred theory of the Earth. A reprint of the 1691 edn, with an introduction by Basil Willey. Centaur, London

Butler EJ, Hoyle F (1979) On the effects of a sudden change in the albedo of the Earth. Astrophys Space Sci 60:505–511

Butzer KW (1962) Coastal geomorphology of Majorca. Ann Assoc Am Geogr 52:191–212

Butzer KW (1975) Pleistocene littoral-sedimentary cycles of the Mediterranean basin: a Mallorquin view. In: Butzer KW, Isaacs GLL (eds) After the Australopithecines. Mouton, The Hague, pp 25–71

Callendar GS (1938) The artificial production of carbon dioxide and its influence on temperature. Q J R Meteorol Soc 64:223–240

Campbell WH, Blechman JB, Bryson RA (1983) Long-period tidal forcing of Indian monsoon rainfall: an hypothesis. J Clim Appl Meteorol 22:289–296

Caputo MV, Crowell JC (1985) Migration of glacial centers across Gondwana during the Paleozoic Era. Bull Geol Soc Am 96:1020–1036

Carey SW (1976) The expanding Earth. Elsevier, Amsterdam

Castagnoli GC, Bonino G, Provenzale A, Serio M (1990) On the presence of regular periodicities in the thermoluminescence profile of a recent sea sediment core. Philos Trans R Soc Lond 330A:481–486

Catt JA (1988) Soils of the Plio-Pleistocene: do they distinguish types of interglacial? Philos Trans R Soc Lond 318B:539–557

Catt JA (1989) Relict properties in soils of the central and north-west European temperate region. In: Bronger A, Catt JA (eds) Paleopedology — nature and applications of paleosols. Catena Supplement 16, Catena Verlag, Cremlingen, pp 41–58

Causse C, Coque R, Fontes J-C, Gasse F, Gibert E, Ouezdou HB, Zouari K (1989) Two high levels of continental waters in the southern Tunisian chotts at about 90 and 150 ka. Geology 17:922–925

Cerling TE (1984) The stable isotopic composition of modern soil carbonate and its relationship to climate. Earth Planet Sci Lett 71:229–240

Cerling TE, Hey RL (1986) An isotopic study of paleosol carbonates from Olduvai Gorge. Quat Res 25:63–74

Chaloner WG, Creber GT (1990) Do fossil plants give a climatic signal? J Geol Soc Lond 147:343–350
Chamberlin TC (1897) A group of hypotheses bearing on climatic changes. J Geol 5:653–683
Chamberlin TC (1898) The influence of great epochs of limestone formation upon the constitution of the atmosphere. J Geol 6:609–621
Chamberlin TC (1899) An attempt to frame a working hypothesis of the cause of glacial periods on an atmospheric basis. J Geol 7:545–584
Chanin M-L, Keckhut P, Hauchecorne A, Labitzke K (1989) The solar activity — Q.B.O. effect in the lower thermosphere. Ann Geophys 7:463–470
Chappell J (1973) Astronomical theory of climatic change: status and problem. Quat Res 3:221–236
Chappellaz J, Barnola JM, Raynaud D, Korotkevich YS, Lorius C (1990) Ice-core record of atmospheric methane over the past 160,000 years. Nature 345:127–131
Chartres CJ, Pain CF (1984) A climosequence of soils on Late Quaternary volcanic ash in highland Papua New Guinea. Geoderma 32:131–155
Chernyakhovsky AG, Gradusov BP, Chizhikova NP (1976) Types of recent weathering crusts and their global distribution. Geoderma 16:235–255
Chester DK (1988) Volcanoes and climate: recent volcanological perspectives. Prog Phys Geogr 12:1–35
Chesworth W (1979) The major element geochemistry and the mineralogical evolution of granitic rocks during weathering. In: Ahrens LH (ed) Origin and distribution of the elements. Pergamon, Oxford, pp 305–313
Chesworth W (1980) The haplosoil system. Am J Sci 280:969–985
Chesworth W (1982) Late Cenozoic geology and the second oldest profession. Geosci Can 9:54–61
Chorley RJ (1957) Climate and morphometry. J Geol 65:628–638
Chorley RJ, Schumm SA, Sugden DA (1984) Geomorphology. Methuen, London
Clarke FW (1924) The data of geochemistry, 5th edn. US Geological Survey Bulletin 770, US Government Printing Office, Washington DC (1st edn 1908)
Clausen J, Keck DD, Hiesey WM (1948) Experimental studies on the nature of species. III. Environmental responses of climatic races races of *Achillea*. Carnegie Institution of Washington, Washington DC, Publication 581
Clayden B (1982) Soil classification. In: Bridges EM, Davidson DA (eds) Principles and applications of soil geography. Longman, London, pp 58–96
Clements FE, Shelford VE (1939) Bio-Ecology. John Wiley, New York
Clifton HE (1981) Progradational sequences in Miocene shoreline deposits, southeastern Caliente Range, California. J Sediment Petrol 51:165–184
CLIMAP Project Members (1976) The surface of the ice-age Earth. Science 191:1131–1137
CLIMAP Project Members (1981) Seasonal reconstruction of the Earth's surface at the Last Glacial Maximum. Geological Society of America, Map Chart Series, MC-36
Clube SVM (1986) Giant comets or ordinary comets; parent bodies or planetesimals? Proceedings of the 20th ESLAB Symposium on the Exploration of Halley's Comet, Heidelberg, ESA SP-250, pp 403–408
Clube SVM, Napier WM (1982) The role of episodic bombardment in geophysics. Earth Planet Sci Lett 57:251–262
Clube SVM, Napier WM (1984) The microstructure of terrestrial catastrophism. Mon Not R Astronom Soc 211:953–968
Clube SVM, Napier WM (1986) Giant comets and the Galaxy: implications of the terrestrial record. In: Smoluchowski R, Bahcall JN, Matthews MS (eds) The Galaxy and the Solar System. The University of Arizona Press, Tucson, Arizona, pp 260–285
Clube SVM, Napier WM (1990) The cosmic winter. Basil Blackwell, Oxford
Clube SVM, Napier WM (1991) Catastrophism now. Astronomy Now (in press)
Cogley J (1979) Albedo contrast of glaciation due to continental drift. Nature 279:712–713
Cohen TJ, Sweetser EI (1975) The "spectra" of the solar cycle and of data for Atlantic tropical cyclones. Nature 256:295–296
COHMAP Members (1988) Climatic changes of the last 18,000 years: observations and model simulations. Science 241:1043–1052
Colacino M, Rovelli A (1983) The yearly averaged air temperature in Rome from 1782 to 1975. Tellus 35A:389–397

Cole KD (1985) Solar-terrestrial physics. In: Malone TF, Roederer JG (eds) Global change. Published on behalf of the ICUS Press by Cambridge University Press, Cambridge, pp 371–396

Cole LC (1958) The ecosphere. Sci Am 198:83–96

Common R (1966) Slope failure and morphogenetic regions. In: Dury GH (ed) Essays in geomorphology. Heinemann, London

Connell JH, Orias E (1964) The ecological regulation of species diversity. Am Nat 98:399–414

Coon CS (1953) Climate and race. In: Shapley H (ed) Climatic change: evidence, causes, and effects. Harvard University Press, Cambridge, Massachusetts, pp 13–34

Coon CS (1966) The taxonomy of human variation. Ann N Y Acad Sci 134:516–523

Corbel J (1964) L'érosion terrestre, étude quantitative (méthodes-techniques-résultats). Ann Géogr 73:385–412

Cossins AR, Bowler K (1987) Temperature biology of animals. Chapman & Hall, London

Cotton CA (1942) Climatic accidents in landscape-making, 2nd edn. Whitcombe and Tombs, Christchurch, New Zealand

Cotton CA (1961) The theory of savanna planation. Geography 46:89–101

Courtillot V, Cisowski S (1987) The Cretaceous-Tertiary boundary events: external or internal causes? Eos 68:193,200

Cousins SH (1989) Letter under "Species richness and the energy theory". Nature 340:350–351

Covey C, Barron EJ (1988) The role of ocean heat transport in climatic change. Earth-Sci Rev 24:429–445

Cox A (1968) Polar wandering, continental drift, and the onset of Quaternary glaciation. In: Mitchell JM (ed) Causes of climatic change. American Meteorological Society, Boston, Massachusetts, Meteorological Monograph 8, pp 112–125

Cox NJ (1977) Climatic geomorphology and fully developed slopes: a discussion. Catena 4:229–231

Creber G, Chaloner W (1984) Climatic indications from growth rings in fossil wood. In: Benchley P (ed) Fossils and climate. John Wiley, New York, pp 49–74

Croll J (1864) On the physical cause of the change of climate during geological epochs. Philos Mag 28:121–137

Croll J (1867a) On the excentricity of the earth's orbit, and its physical relations to the glacial epoch. Philos Mag 33:119–131

Croll J (1867b) On the change of obliquity of the ecliptic; its influence on the climate of the polar regions and the level of the sea. Philos Mag 33:426–445

Croll J (1875) Climate and time in their geological relations: a theory of secular changes of the Earth's climate. Daldy, Isbister, London

Crompton E (1960) The significance of the weathering/leaching ratio in the differentiation of major soil groups with particular reference to some very strongly leached Brown Earths on the hills of Britain. Trans Seventh Int Congr Soil Sci, Madison, Wisc 4:406–412

Crowell JC (1978) Gondwana glaciation, cyclothems, continental positioning and climate change. Am J Sci 278:1345–1372

Crowell JC (1982) Continental glaciation through geologic time. In: Berger WH, Crowell JC (eds) Climate in Earth history. National Academy Press, Washington DC, pp 77–82

Crowell JC (1983) Ices ages recorded on Gondwanan continents. Trans Geol Soc S Afr 86:237–262

Crowell JC, Frakes LA (1970) Phanerozoic glaciation and the causes of ice ages. Am J Sci 268:193–224

Crowley TJ (1988) Paleoclimate modelling. In: Schlesinger ME (ed) Physically based modelling and simulation of climate and climatic change, Part II. Kluwer Academic, Dordrecht, pp 883–949

Crowley TJ, Short DA, Mengel JG, North GR (1986) Role of seasonality in the evolution of climate during the last 100 million years. Science 231:579–584

Crowley TJ, Mengel JG, Short DA (1987) Gondwanaland's seasonal climate. Nature 329:803–807

Crutzen PJ, Birks JW (1982) The atmosphere after a nuclear war: twilight at noon. Ambio 11:114–125

Cubasch U, Cess RD (1990) Processes and modelling. In: Houghton JT, Jenkins GJ, Ephraums JJ (eds) Climate change: the IPCC Scientific Assessment. Published for the Intergovernmental Panel on Climate Change by Cambridge University Press, Cambridge, pp 69–91

Curi N, Franzmeier DP (1984) Toposequence of Oxisols from the central plateau of Brazil. J Soil Sci Soc Am 48:341–346

Currie DJ (1991) Energy and large-scale patterns of animal and plant species richness. Am Nat 137:27–49

Currie DJ, Paquin V (1987) Large-scale biogeographical patterns of species richness of trees. Nature 329:326–327

Currie RG (1974) Solar cycle signal in surface air temperature. J Geophys Res 79:5657–5660

Currie RG (1979) Distribution of solar cycle signal in surface air temperature over North America. J Geophys Res 84:753–761

Currie RG (1980) Detection of the 11-year sunspot cycle signal in Earth rotation. Geophys J R Astronom Soc 61:131–139

Currie RG (1981a) Solar cycle signal in Earth rotation: nonstationary behavior. Science 211:386–389

Currie RG (1981b) Solar cycle signal in air temperature in North America: amplitude, gradient, phase and distribution. J Atmos Sci 38:808–818

Currie RG (1981c) Amplitude and phase of the 11-year term in sea level: Europe. Geophys J R Astronom Soc 67:547–556

Currie RG (1981d) Evidence for 18.6-year signal in temperature and drought condition in North America since A.D. 1800. J Geophys Res 86:11,055–11,064

Currie RG (1982) Evidence for 18.6-year term in air pressure in Japan and geophysical implications. Geophys J R Astronom Soc 69:321–327

Currie RG (1983) Detection of 18.6-year lunar nodal-induced drought in the Patagonian Andes. Geophys Res Lett 10:1089–1092

Currie RG (1984a) On bistable phasing of 18.6-year-induced flood in India. Geophys Res Lett 11:50–53

Currie RG (1984b) Evidence for 18.6-year lunar nodal drought in western North America during the past millenium. J Geophys Res 89:1295–1308

Currie RG (1984c) Periodic 18.6-year and cyclic 11-year-induced drought and flood in western North America. J Geophys Res 89:7215–7230

Currie RG (1987a) On bistable phasing of 18.6-year-induced drought and the flood in the Nile records since AD 650. J Climatol 7:373–389

Currie RG (1987b) Examples and implications of 18.6- and 11-yr terms in world weather records. In: Rampino MR, Sanders JE, Newman WS, Königsson LK (eds) Climate: history, periodicity, and predictability. Van Nostrand Reinhold, New York, pp 378–403

Currie RG (1988) Lunar tides and the wealth of nations. New Sci 120:52–55

Currie RG, Fairbridge RW (1985) Periodic 18.6-year and cyclic 11-year-induced drought and flood in northeastern China and some global implications. Quat Sci Rev 4:109–134

Currie RG, O'Brien DP (1988) Periodic 18.6-year and cyclic 10 to 11 year signals in northeastern United States precipitation data. J Climatol 8:255–281

Currie RG, O'Brien DP (1989) Morphology of bistable 180° phase switches in 18.6-year-induced rainfall over the northeastern United States of America. Int J Climatol 9:501–525

Currie RG, O'Brien DP (1990) Deterministic signals in precipitation records from the American Corn Belt. Int J Climatol 10:179–189

Curtis CD (1990) Aspects of climatic influence on the clay mineralogy and geochemistry of soils, palaeosols and clastic sedimentary rocks. J Geol Soc Lond 147:351–357

Dalquest WW (1965) New Pleistocene formation and local fauna from Hardman County, Texas. J Paleontol 39:63–79

Dalton J (1799) Experiments and observations to determine whether the quantity of rain and dew is equal to the quantity of water carried off by the rivers and raised by evaporation; with an enquiry into the origin of springs. Printed by R and W Dean, Manchester

Dana JD (1856) On American geological history. Address before the American Association for the Advancement of Science, August, 1855, by James D Dana. Am J Sci 2nd ser 22:305–334

Dansgaard W, Johnsen SJ, Clausen HB, Langway CC Jr (1971) Climatic record revealed by the Camp Century ice core. In: Turekian KL (ed) Late Cenozoic glacial ages. Yale University Press, New Haven, Connecticut, pp 37–56

Darwin GH (1880) On the secular changes in the elements of the orbit of a satellite revolving about a tidally distorted planet. Philos Trans R Soc 171:713–891

Davis WM (1895–96) A speculation in topographical climatology. Am Meteorol J 12:372–381

Davis WM (1899) The geographical cycle. Geogr J 14:481–504

Davis WM (1900) Glacial erosion in France, Switzerland and Norway. Proc Boston Soc Nat Hist 29:273–322

Davis WM (1902) Base level, grade, and peneplain. J Geol 10:77–111

Davis WM (1905) The geographical cycle in an arid climate. J Geol 13:381–407

Davis WM (1906) The sculpture of mountains by glaciers. Scott Geogr Mag 22:76–89

De Angelis M, Barkov NI, Petrov VN (1987) Aerosol concentrations over the last climatic cycle (160 kyr) from an Antarctic ice core. Nature 325:318–321

de Boer DH, Campbell IA (1989) Spatial scale dependence of sediment dynamics in a semi-arid badland drainage basin. Catena 16:277–290

de Candolle A-P (1820) Géographie botanique. In: Levrault FC (ed) Dictionnaire des Sciences Naturelles, vol 18. Levrault, Paris, pp 359–436

de Candolle A (1855) Géographie botanique raisonnée; ou, exposition des faits principaux et des lois concernant la distribution géographique des plantes de l'époque actuelle 2 vols. V Masson, Paris

Deeley RM (1915) Polar climates. Geol Mag Decade 6, 2:450–455

Deepak A (ed) (1983) Atmospheric effects and potential climatic impact of the 1980 eruptions of Mount St. Helens. NASA, Washington DC, NASA Conference Publication 2240

Defant A (1924) Die Schwankungen der atmosphärischen Zirkulation über dem nord-atlantischen Ozean im 25-jährigen Zeitraum 1881–1905. Geogr Ann 6:13–41

Degens ET, Paluska A (1979) Tectonic and climatic pulses recorded in Quaternary sediments of the Caspian-Black Sea region. Sediment Geol 23:149–163

Degens ET, Wong HK, Kempe S (1981) Factors controlling global climate of the past and future. In: Likens GE (ed) Some perspectives of the major biogeochemical cycles. John Wiley, Chichester, pp 3–24

Deirmendjian D (1973) On volcanic and other particulate turbidity anomalies. Adv Geophys 16:267–296

Delcourt PA, Delcourt HR (1987) Long-term forest dynamics of the temperate zone: a case study of late-Quaternary forests in eastern North America. (Ecological Studies: analysis and synthesis, vol 63) Springer, Berlin Heidelberg New York

de Martonne E (1909) Traité de geomorphologie physique, 1st edn. Armand Colin, Paris

de Martonne E (1913) Le climat-facteur du relief. Scientia (1913):339–355

de Martonne E (1926) Aréism et indice d'aridité. C R Acad Sci Paris 182:1395–1398

de Martonne E (1940) Problèmes morphologiques du Brésil tropical atlantique. Ann Géogr 49:1–27, 106–129

Denton GH, Hughes TJ, Karlén W (1986) Global ice-sheet system interlocked by sea level. Quat Res 26:3–26

Derbyshire E (1973) Introduction. In: Derbyshire E (ed) Climatic geomorphology. Macmillan, London, pp 11–18 (Geographical Readings Series)

Derbyshire E (1976) Geomorphology and climate: background. In: Derbyshire E (ed) Geomorphology and climate. John Wiley, London, pp 1–24

de Saussure HB (1779–96) Voyages dans les Alpes, précédes d'un essai d'histoire naturelle dans les environs de Genève, 4 vols. Samuel Fauché, Neuchâtel

de Saussure NT (1804) Recherches chimiques sur la végétation. Chez la Ve, Nyon, Paris

de Visser JP, Ebbing JHJ, Gudjonsson L, Hilgen FJ, Jorissen FJ, Verhallen PJJM, Zevenboom D (1989) The origin of rhythmic bedding in the Pliocene Trubi formation of Sicily, southern Italy. Palaeogeogr Palaeoclimatol Palaeoecol 69:45–66

Dickinson RE (1984) Modelling evapotranspiration for three-dimensional global climate models. In: Hansen JE, Takahashi T (eds) Climate processes and climate sensitivity. American Geophysical Union, Washington, DC, Geophysical Monographs 29, pp 58–72

Dole RB, Stabler H (1909) Denudation. US Geological Survey Water Supply Paper 234, US Government Printing Office, Washington DC, pp 78–93

Donn WL (1982) The enigma of high-latitude palaeoclimate. Palaeogeogr Palaeoclimatol Palaeoecol 40:199–212

Donn WL (1987) Terrestrial climate change from the Triassic to Recent. In: Rampino MR, Sanders JE, Newman WS, Königsson LK (eds) Climate: history, periodicity, and predictability. Van Nostrand Reinhold, New York, pp 343–352

Donn WL, Ewing M (1966) A theory of ice ages: III. Science 152:1706–1712

Donn WL, Ewing M (1968) The theory of an ice-free Arctic ocean. In: Mitchell JM Jr (ed) Causes of climatic change. American Meteorological Society, Boston, Massachusetts, Meteorological Monograph 8, pp 100–105

Donn WL, Shaw DM (1977) Model of climate evolution based on continental drift and polar wandering. Bull Geol Soc Am 88:390–396

Dorman JL, Sellers PJ (1989) A global climatology of albedo, roughness length and stomatal resistance for atmospheric general circulation models as represented by the simple biosphere model (SiB). J Appl Meteorol 28:833–855

Douglas I (1967) Man, vegetation and the sediment yield of rivers. Nature 215:925–928

Douglas JG, Williams GE (1982) Southern polar forests: the early Cretaceous floras of Victoria and their paleoclimatic significance. Palaeogeogr Palaeoclimatol Palaeoecol 39:171–185

Douglass AE (1909) Weather cycles in the growth of big trees. Mon Weather Rev 37:225–237

Douglass AE (1918) A method of estimating rainfall by the growth of trees. Carnegie Institution of Washington, Washington DC, Publication 192

Douglass AE (1919) Climatic cycles and tree growth: a study of annual rings of trees in relation to climate and solar activity. Carnegie Institution of Washington, Washington DC, Publication 289 (1)

Drayson AW (1871) On the cause, date, and duration of the glacial epoch of geology. Q J Geol Soc Lond 27:232–234

Drayson AW (1873) On the cause, date, and duration of the last glacial epoch of geology, and the probable antiquity of man. With an investigation and description of a new movement of the Earth. Chapman & Hall, London

Drever JI (1982) The geochemistry of natural water. Prentice-Hall, Englewood Cliffs, New Jersey

Drewry GE, Ramsay ATS, Smith AG (1974) Climatically controlled sediments, the geomagnetic field, and trade wind belts in Phanerozoic time. J Geol 82:531–553

Drude KGO (1897) Manuel de géographie botanique. Translated by Georges Poirault. P Klincksieck, Paris

Dubois E (1893) Die Klimate der geologischen Vergangenheit und ihre Beziehungen zur Entwickelungsgeschichte der Sonne. H C A Thieme, Nijegen; M Spohr, Leipzig

Dubois E (1895) The climates of the geological past and their relation to the evolution of the Sun. Swan Sonnenschein, London

Duchaufour P (1982) Pedology: pedogenesis and classification. Translated by TR Paton. George Allen & Unwin, London

Duplessy JC, Labeyrie L, Moyes J, Turon JL, Duprat J, Pujol C, de Beaulieu JL, Clerc J, Couteaux M, Pons A, Reille M, Van Campo M, Jalut G, Sabatier R (1986) The impact on Europe of large-scale climatic changes: the onset of glaciation and the last deglaciation. In: Ghazi A, Fantechi R (eds) Current issues in climate research. D. Reidel, Dordrecht, pp 28–41 (Commission of the European Communities. Proceedings of the EC Climatology Programme Symposium, Sophia Antipolis, France, 2–5 October 1984)

Dury GH (1971) Relict deep weathering and duricrusting in relation to the palaeoenvironments of middle latitudes. Geogr J 137:511–522

Ebermayer EWF (1876) Die gesamte Lehre der Waldstreu mit Rücksicht auf die chemische Statik des Waldbaues. Julius Springer, Berlin

Ebermayer EWF (1882) Naturgesetzliche Grundlagen des Wald- und Ackerbaues, vol 1. Die Bestandteile der Pflanzen, Part I, Physiologische Chemie der Pflanzen. Springer, Berlin

Eddy JA (1977a) Anomalous solar radiation during the seventeenth century. Science 198:824–829

Eddy JA (1977b) The case of the missing sunspots. Sci Am 236:80–92

Eddy JA (1977c) Climate and the changing Sun. Clim Change 1:173–190

Eddy JA (1983) The Maunder minimum: a reappraisal. Sol Phys 89:195–207

Eddy JA (1990) Some thoughts on Sun-weather relations. Philos Trans R Soc Lond 330A:543–545

Ehleringer JR (1989) Carbon isotope ratios and physiological processes in arid-land plants. In: Rundel PW, Elheringer JR, Nagy KA (eds) Application of stable isotope ratios to ecological research. Springer, Berlin Heidelberg New York Tokyo, pp 41–54

Ehrlich PR, Sagan C, Kennedy D, Roberts WO (1984) The cold and the dark: the world after nuclear war. WW Norton, New York

Einsele G, Ricken W, Seilacher A (eds) (1990) Cycles and events in stratigraphy. Springer, Berlin Heidelberg New York Tokyo

Ekholm N, Arrhenius S (1898) Über den Einfluss des Mondes auf die Polarichter und Gewitter. Handlingar Svenska Vetenskaps-Akademien, Stockholm 31:2

Eldredge N (1985) Unfinished synthesis: biological hierarchies and modern evolutionary thought. Oxford University Press, New York

Elkins N (1989) Letter under "Species richness and the energy theory". Nature 340:350

Ellsaesser HW (1986) Comments on "Surface temperature changes following the six major volcanic episodes between 1780 and 1980". J Climatol Appl Meteorol 25:1184–1185

Elton CS (1927) Animal ecology. Sidgwick & Jackson, London

Embleton BJJ, Williams GE (1986) Low palaeolatitude of deposition for late Precambrian periglacial varvites in South Australia: implications for palaeoclimatology. Earth Planet Sci Lett 79:419–430

Emiliani C, Kraus EB, Shoemaker EM (1981) Sudden death at the end of the Mesozoic. Earth Planet Sci Lett 55:317–334

Endler JA (1977) Geographic variation, speciation, and clines. Princeton University Press, Princeton, New Jersey (Monographs in Population Biology 10)

Engel AEJ, Engel CG (1964) Continental accretion and the evolution of North America. In: Subramaniam AP, Balakrishna S (eds) Advancing frontiers in geology and geophysics. Indian Geophysical Union, Hyderabad, pp 17–37e

Erman GA (1833–48) Reise um die Erde durch Nord-Asien und die Beider Ozean in den Jähren 1828, 1829 und 1830 ausgeführt von Adolph Erman. 2 vols in 5, G Reimer, Berlin

Esser G (1984) The significance of biospheric carbon pools and fluxes for the atmospheric CO_2: a proposed model structure. In: Leith H, Fantechi R, Schnitzler H (eds) Interactions between climate and biosphere. Swets and Zeitlinger, Lisse, pp 253–294 (Progress in Biometeorology, vol 3)

Esser G (1987) Sensitivity of global carbon pools and fluxes to human and potential climatic impacts. Tellus 39B:245–260

Evans J (1866) On a possible geological cause of changes in the position of the axis of the Earth's crust. Proc R Soc 82:46–54

Ewing M, Donn WL (1956) A theory of ice ages: I. Science 123:1061–1066

Ewing M, Donn WL (1958) A theory of ice ages: II. Science 127:1159–1162

Eybergen FA, Imeson AC (1989) Geomorphic processes and climatic change. Catena 16:307–319

Fabricius FH, Braune K, Funk G, Hieke W, Schmolin J (1983) Plio-Quaternary sedimentation and tectonics in the Ionian area: clues to the recent evolution of the Mediterranean. In: Stanley DJ, Wezel F-C (eds) Geographical evolution of the Mediterranean Basin. Springer, Berlin Heidelberg New York Tokyo, pp 297–310

Fabricius JC (1778) Philosophia entomologica, sistens scientiae fundamenta; adiectis definitionibus, exemplis, observationibus, adumbrationibus. CE Bohnii, Hamburg

Fairbanks RG (1989) A 17,000-year glacio-eustatic sea level record: influence of glacial melting rates on the Younger Dryas event and deep-ocean circulation. Nature 342:637–642

Fairbridge RW (1970a) South Pole reaches the Sahara. Science 168:878–881

Fairbridge RW (1971) Upper Ordovician glaciation in northwest Africa? Reply. Bull Geol Soc Am 82:269–274

Fairbridge RW (1973) Glaciation and plate migration. In: Tarling DH, Runcorn SK (eds) Implications of continental drift to the Earth sciences. Academic Press, New York, pp 503–515

Fairbridge RW (1976) Effects of Holocene climatic change on some tropical geomorphic processes. Quat Res 6:529–556

Fairbridge RW (1978) Exo- and endogenetic geomagnetic modulation of climates on decadal to galactic scale. Eos 59:269 (Abstract)

Fairbridge RW (1984) Planetary periodicities and terrestrial climate stress. In: Mörner N-A, Karlén W (eds) Climatic changes on a yearly to millenial basis: geological, historical and instrumental records. D Reidel, Dordrecht, pp 509–520

Fairbridge RW (1986) Monsoons and paleomonsoons. Episodes 9:143–149

Fairbridge RW, Sanders JE (1987) The Sun's orbit, A.D. 750–2050: basis for new perspectives on planetary dynamics and Earth-Moon linkage. In: Rampino MR, Sanders JE, Newman WS,

Königsson LK (eds) Climate: history, periodicity, and predictability. Van Nostrand Reinhold, New York, pp 446–471

Fairbridge RW, Shirley JH (1987) Prolonged minima and the 179-year cycle of the solar inertial motion. Sol Phys 110:191–220

Fallou FA (1855) Die Ackererden des Königreichs Sachsen und der angrenzenden Gegend, geognotisch nach ihren äußeren Verhältnissen un Beziehungen zum Grundgebirge, sowie nach Bestand und Gehalt untersucht und klassifiziert. W Gerhard, Leipzig

Fallou FA (1862) Pedologie; oder allgemeine und besondere Bodenkunde. G Schönfeld, Dresden

Fallou FA (1875) Die Hauptbodenarten der Nord- und Ostsee-Länder Deutschen Reiches naturwissenschaftlich wie landwirtschaftlich betrachtet. G Schönfeld, Dresden

Faure H, Leroux M (1990) Are there solar signals in the African monsoon and rainfall? Philos Trans R Soc Lond 330A:575

Feakes CR, Holland HD, Zbinden EA (1989) Ordovician paleosols at Arisaig, Nova Scotia, and the evolution of the atmosphere. In: Bronger A, Catt JA (eds) Paleopedology — nature and applications of paleosols. Catena Supplement 16, Catena Verlag, Cremlingen, pp 207–232

Feth JH (1971) Mechanisms controlling world water chemistry: evaporation-crystallization process. Science 172:870–871

Fischer AG (1965) Fossils, early life, and atmospheric history. Proc Natl Acad Sci 53:1205–1215

Fischer AG (1972) Atmosphere and the evolution of life. Main Currents in Modern Thought 28 (5) May-June: unpaginated

Fischer AG (1981) Climatic oscillations in the biosphere. In: Nitecki MH (ed) Biotic crises in ecological and evolutionary time. Academic Press, New York, pp 103–131

Fischer AG (1984a) The two Phanerozoic supercycles. In: Berggren WA, Van Couvering JA (eds) Catastrophes and Earth history: the new uniformitarianism. Princeton University Press, Princeton, New Jersey, pp 129–150

Fischer AG (1984b) Biological innovations and the sedimentary record. In: Holland HD, Trendall AF (eds) Patterns of change in Earth evolution. Springer, Berlin Heidelberg New York Tokyo, pp 145–157 (Dahlem Konferenzen 1984)

Fischer AG (1986) Climatic rhythms recorded in strata. Ann Rev Earth Planet Sci 14:351–376

Fischer AG (1988) Cyclostratigraphy. A position paper prepared for the CRER Workshop 3, Cyclostratigraphy, Perugia, Italy, September 1988, 15 pp

Fischer AG, Arthur MA (1977) Secular variations in the pelagic realm. In: Cook HE, Enos P (eds) Deep-water carbonate environments. Soc Econ Paleontol Mineral Spec Publ 25:19–50

Fischer AG, Schwartzacher W (1984) Cretaceous bedding rhythms under orbital control? In: Berger A, Imbrie J, Hays J, Kukla G, Saltzman B (eds) Milankovitch and climate: understanding the response of astronomical forcing, Part 1. D Reidel, Dordrecht, pp 163–175 (Proceedings of the NATO Advanced Research Workshop on Milankovitch and Climate, Palisades, New York, 1982. NATO ASI Series C, Mathematical and Physical Sciences, vol 126)

Fischer AG, Herbert TD, Silva IP (1985) Carbonate bedding cycles in Cretaceous pelagic and hemipelagic sequences. In: Pratt LM, Kauffman EG, Zelt FB (eds) Fine-grained deposits and biofacies of the western interior seaway: evidence of cyclic sedimentary processes. Soc Econ Paleontol Mineral Field Trip Guidebook 4:1–10

Fleming J (1822) The philosophy of zoology, 2 vols. Archibald Constable, Edinburgh

Flohn H (1951) Solare Vorgänge im Wettergeschehen. Arch Meteorol Geophys Bioklimatol 3A:303–329

Flynn LJ, Jacobs LL (1982) Effects of changing environments on Siwalik rodent faunas of northern Pakistan. Palaeogeogr Palaeoclimatol Palaeoecol 38:129–138

Fontes J-C, Gasse F (1989) On the ages of humid Holocene and late Pleistocene phases in North Africa — remarks on "Late Quaternary climatic reconstruction for the Maghreb (North Africa)" by P Rognon. Palaeogeogr Palaeoclimatol Palaeoecol 70:393–398

Forbes E (1839) Report on the distribution of pulmoniferous Mollusca in the British Isles. Reports of the British Association for the Advancement of Science, London, Part 1, pp 127–147

Forbes WTM (1931) The great glacial cycle. Science 74:294–295

Foucault A, Stanley DJ (1989) Late Quaternary palaeoclimatic oscillations in East Africa recorded as heavy minerals in the Nile Delta. Nature 339:44–46

Fournier F (1960) Climat et érosion: la relation entre l'érosion du sol par l'eau et les précipitations atmosphériques. Presses Universitaires de France, Paris

Frakes LA (1979) Climates throughout geologic time. Elsevier, Amsterdam

Frakes LA, Francis JE (1988) A guide to Phanerozoic cold polar climates from high-latitude ice-rafting in the Cretaceous. Nature 333:547–549

Frakes LA, Kemp E (1972) Influence of continental positions on early Tertiary climate. Nature 240:97–100

Frakes LA, Kemp E (1973) Palaeogene continental positions and evolution of climate. In: Tarling DH, Runcorn SK (eds) Implications of continental drift to the Earth sciences. Academic Press, London, pp 535–550

Franklin B (1789) Meteorological imaginations and conjectures. Mem Manchester Literary Philos Soc 2:373–377

Fränzle O (1965) Klimatische Schwellenwerte der Bodenbildung in Europa und der USA. Die Erde 96:84–104

Fritts HC (1976) Tree rings and climate. Academic Press, London

Fritz H (1878) Über die Beziehungen der Sonnenflecken zu den magnetischen und meteorologischen Erscheinungen der Erde. De Erven Loosjes, Haarlem

Frye JC (1959) Climate and Lester King's "Uniformitarian nature of hillslopes". J Geol 67:111–113

Gallimore RG, Kutzbach JE (1989) Effects of soil moisture on the sensitivity of a climate model to Earth orbital forcing at 9000 yr BP. Clim Change 14:175–205

Garrels RM, Mackenzie FJ (1971) Evolution of sedimentary rocks. WW Norton, New York

Gasse F, Fontes J-C (1989) Palaeoenvironments and palaeohydrology of a tropical closed lake (Lake Asal, Djibouti) since 10,000 yr BP. Palaeogeogr Palaeoclimatol Palaeoecol 69:67–102

Gasse F, Stabell B, Fourtanier E, van Iperen Y (1989) Freshwater diatom influx in intertropical Atlantic: relationships with continental records from Africa. Quat Res 32:229–243

Gasse F, Téhet R, Durand A, Gibert E, Fontes J-C (1990) The arid-humid transition in the Sahara and the Sahel region during the last deglaciation. Nature 346:141–146

Gates WL (1976a) Modeling the ice-age climate. Science 191:1138–1144

Gates WL (1976b) The numerical simulation of ice-age climate with a global general circulation model. J Atmos Sci 33:1844–1873

Gaudrea DC, Webb T III (1985) Late Quaternary pollen stratigraphy and isochrone maps for the northeastern United States. In: Bryant VM, Holloway RG (eds) Pollen records of Late Quaternary North American sediments. American Association of Stratigraphic Palynologists, Dallas, pp 247–280

Geikie A (1868a) On denudation now in progress. Geol Mag 5:249–254

Geikie A (1868b) On modern denudation. Trans Geol Soc Glasgow 3:153–190

Geikie A (1880) Rock-weathering, as illustrated in Edinburgh churchyards. Proc R Soc Edinb 10:518–532

Genthon C, Barnola JM, Raynaud D, Lorius C, Jouzel J, Barkov NI, Korotkevich YS, Kotlyakov VM (1987) Vostok ice core: climatic response to CO_2 and orbital forcing changes over the last climatic cycle. Nature 329:414–418

Gentilli J (1948) Present day volcanicity and climatic change. Geol Mag 85:172–175

Gérard J-C (1989) Aeronomy and paleoclimate. In: Berger A, Dickinson RE, Kidson JW (eds) Understanding climatic change. American Geophysical Union, Geophysical Monograph 52, pp 139–148 (International Union of Geodesy and Geophysics, vol 7)

Gérard J-C (1990) Modelling the climatic response to solar variability. Philos Trans R Soc Lond 330A:561–574

Gerson R, Grossman S (1987) Geomorphic activity on escarpments and associated fluvial systems in hot deserts. In: Rampino MR, Sanders JE, Newman WS, Königsson LK (eds) Climate: history, periodicity, and predictability. Van Nostrand Reinhold, New York, pp 300–322

Ghil M (1981) Internal climatic mechanisms participating in glaciation cycles. In: Berger A (ed) Climatic variations and variability: facts and theories. D Reidel, Dordrecht, pp 539–557 (NATO Advanced Study Institute First Course of the International School of Climatology, Ettore Majorana Center for Scientific Culture, Erice, Italy, March 9–21, 1980. NATO Advanced Study Institute Series C-Mathematical and Physical Sciences, vol 72)

Gibbs RJ (1970) Mechanisms controlling world water chemistry. Science 170:1088–1090

Gibbs RJ (1973) Mechanisms controlling world water chemistry: evaporation-crystallization process. Science 172:871–872

Gilbert GK (1877) Geology of the Henry Mountains (Utah). US Geographical and Geological Survey of the Rocky Mountains Region. US Government Printing Office, Washington DC

Gilbert GK (1890) Lake Bonneville. US Geological Survey, Washington DC, Monograph 1, pp 340–345

Gilbert GK (1895) Sedimentary measurement of geologic time. J Geol 3:121–127

Gilbert GK (1900) Rhythms in geologic time. Proceedings of the Am Assoc Adv Sci 49:1–19

Gilliland RL (1981) Solar radius variations over the last 265 years. Astrophys J 248:1144–1155

Gilliland RL (1982) Solar, volcanic and CO_2 forcing of recent climatic changes. Clim Change 4:111–131

Gilman DL (1982) The nature of climatic variability. In: Eddy JA (ed) Solar variability, weather, and climate. National Academy Press, Washington DC, pp 53–63

Giorgi F (1989) Simulation of regional climate using a limited-area model nested in a general circulation model. Eos 70:1011 (Abstract)

Gledhill JA (1985) Dinosaur extinction and volcanic activity. Eos 66:153

Gleick J (1988) Chaos: making a new science. Cardinal (Sphere Books), Harmondsworth, Middlesex

Gleissberg W (1955) The 80-year sunspot cycle. J Br Astronom Assoc 68:148–152

Gleissberg W (1965) The eighty-year cycle in auroral frequency numbers. J Br Astronom Assoc 75:227–231

Glinka KD (1914) Die Typen der Bodenbildung, ihre Klassifikation und geographische Verbreitung. Gebrüder Borntraeger, Berlin

Glinka KD (1927) The great soil groups of the world and their development. Translated from the German by CF Marbut. Edwards Brothers, Ann Arbor, Michigan

Glinka KD (1928) Dokuchaev's ideas in the development of pedology and cognate sciences. Proceedings and Papers of the First Int Congr Soil Sci 1:116–136

Gloger CWL (1833) Das Abändern der Vögel durch Einfluss des Klimas: nach zoologischen, zunächst von den europäischen Landvögeln entnommenen Beobachtungen dargestellt, mit den entsprechenden Erfahrungen bei den europäischen Säugethieren verglichen, und durch Tatsachen aus dem Gebiete der Physiologie, der Physik und der physischen Geographie erläutert. A Schultz, Breslau

Godard A (1965) Recherches de géomorphologie en Écosse de Nord-Ouest. Les Belles Lettres, Paris (Publications de la Faculté des Lettres de l'Université de Strasbourg, Fondation Baulig 1)

Gold T (1955) Instability of the Earth's axis of rotation. Nature 175:526–529

Goldhammer RK, Dunn DA, Hardie LA (1987) High-frequency glacio-eustatic sea-level oscillations with Milankovitch characteristics recorded in Middle Triassic platform carbonates in northern Italy. Am J Sci 287:853–892

Goodchild JG (1890) Notes on some observed rates of weathering on limestones. Geol Mag Decade 3, 7:463–466

Goodwin PW, Anderson EJ (1985) Punctuated aggradational cycles: a general hypothesis of episodic stratigraphic accumulation. J Geol 93:515–533

Gordon RG (1987) Polar wandering and paleomagnetism. Annu Rev Earth Planet Sci 15:567–593

Gordon WA (1975) Distribution by latitude of Phanerozoic evaporite deposits. J Geol 83:671–684

Goudie AS (1985) Duricrusts and landforms. In: Richards KS, Arnett RR, Ellis S (eds) Geomorphology and soils. George Allen & Unwin, London, pp 37–57

Gould SJ (1974) The origin and function of "bizarre" structures: antler size and skull size in the "Irish elk", *Megaloceras giganteus*. Evolution 28:191–220

Gow AJ, Williamson T (1971) Volcanic ash in the Atlantic ice sheet and its possible climatic implications. Earth Planet Sci Lett 13:210–218

Graham RW (1976) Late Wisconsin mammal faunas and environmental gradients of the eastern United States. Paleobiology 2:343–350

Graham RW (1979) Paleoclimates and late Pleistocene faunal provinces in North America. In: Humphrey RL, Stanford DJ (eds) Pre-Llano cultures of the Americas: paradoxes and possibilities. Anthropological Society of Washington, Washington DC, pp 46–69

Graham RW, Lundelius EL Jr (1984) Coevolutionary disequilibrium and Pleistocene extinctions. In: Martin PS, Klein RG (eds) Quaternary extinctions: a prehistoric revolution. The University of Arizona Press, Tucson, Arizona, pp 223–249

Graham RW, Mead JI (1987) Environmental fluctuations and evolution of mammalian faunas during the last deglaciation. In: Ruddiman WF, Wright HE Jr (eds) North American and adjacent oceans during the last deglaciation. The geology of North America, vol K-3. The Geological Society of America, Boulder, Colorado, pp 371–402

Graham RW, Semken HA Jr, Graham MA (eds) (1987) Late Quaternary mammalian biogeography and communities of the Great Plains and prairies. Illinois State Museum Scientific Papers, vol 22, Springfield, Ill

Greenwood G (1857) Rain and rivers; or, Hutton and Playfair against Lyell and all comers. Longman, Brown, Green, Longmans, & Roberts, London

Gribbin J (1978) Astronomical influences: long-term effects. In: Gribbin J (ed) Climatic change. Cambridge University Press, Cambridge, pp 133–138

Grime JP (1989) The stress debate: symptom of impending synthesis? Biol J Linn Soc 37:3–17

Grisebach AHR (1872) Der Vegetation der Erde nach ihrer klimatischen Anordnung: ein Abriss der vergleichenden Geographie der Pflanzen, 2 vols. W Engelmann, Leipzig

Grotch SL (1988) Regional intercomparisons of general circulation model predictions and historical climate data. US Department of Energy, Oak Ridge, Tennessee, DOE/NBB-0084

Grove JM (1972) The incidence of landslides, avalanches and floods in western Norway during the Little Ice Age. Arct Alp Res 4:131–138

Grove JM (1988) The Little Ice Age. Methuen, London

Guilday JE (1984) Pleistocene extinction and environmental change, a case study of the Appalachians. In: Martin PS, Klein RG (eds) Quaternary extinctions: a prehistoric revolution. The University of Arizona Press, Tucson, Arizona, pp 250–258

Guilday JE, Martin PS, McCrady AD (1964) New Paris No. 4: a Pleistocene cave deposit in Bedford County, Pennsylvania. Bull Nat Speleol Soc 26:121–194

Gupta BL, Singh NN (1981) Some findings of rainfall characteristics in Rajasthan. Geogr Rev India 43:326–341

Gupta VK, Waymire E, Rodriguez-Iturbe I (1986) On scales, gravity and network structure in basin runoff. In: Gupta VK, Rodriguez-Iturbe I, Wood EF (eds) Scale problems in hydrology. D Reidel, Dordrecht, pp 159–184

Guthrie RD (1984) Mosaics, allelochemics, and nutrients: an ecological theory of late Pleistocene megafaunal extinctions. In: Martin PS, Klein RG (eds) Quaternary extinctions: a prehistoric revolution. The University of Arizona Press, Tucson, Arizona, pp 259–298

Haeckel E (1866) Generelle Morphologie der Organismen. Allgemeine Grundzüge der organischen Formen-wissenschaft, mechanisch begründet durch die von Charles Darwin reformierte Descendenz-Theorie, 2 vols. G Reimer, Berlin

Haigh MJ (1987) The holon: hierarchy theory and landscape research. In: Ahnert F (ed) Geomorphological models: theoretical and empirical aspects. Catena Supplement 10, Catena Verlag, Cremlingen, pp 181–192

Hale GE (1924) Sunspots as magnets and the periodic reversal of their polarity. Nature 113:105

Hall AM (1985) Cenozoic weathering covers in Buchan, Scotland, and their significance. Nature 315:392–395

Hallam A (1977) Secular changes in marine inundation of USSR and North America through the Phanerozoic. Nature 269:769–772

Hambrey MJ, Harland WB (1981) The evolution of climates. In: Cocks LRM (ed) The evolving Earth. British Museum (Natural History), Cambridge University Press, Cambridge, pp 137–152

Hameed S (1984) Fourier analysis of Nile flood level. Geophys Res Lett 11:843–845

Hameed S, Yeh W-M, Li M-T, Cess RD, Wang W-C (1983) An analysis of periodicities in the 1470 to 1974 Beijing precipitation record. Geophys Res Lett 10:436–439

Hamilton W (1968) Cenozoic climatic change and its cause. Meteorol Monogr 8:128–133

Hammer CU, Clausen HB, Dansgaard W (1980) Greenland ice sheet evidence of post-glacial volcanism and its climatic impact. Nature 288:230–235

Hammer CU, Clausen HB, Friedrich WL, Tauber H (1987) The Minoan eruption of Santorini in Greece dated to 1645 BC? Nature 328:517–519

Hansen JE, Wand WC, Lacis AA (1978) Mount Agung eruption provides a test of global climatic perturbations. Science 199:1065–1068

Hardie LA, Bosellini A, Goldhammer RK (1986) Repeated subaerial exposure of subtidal carbonate platforms, Triassic, Northern Italy: evidence for high-frequency sea level oscillations on a 10^4 year time scale. Paleoceanography 1:447–457

Hart MB (1987) Orbitally induced cycles in the chalk facies of the United Kingdom. Cretaceous Res 8:335–348

Hart MH (1978) The evolution of the atmosphere of the Earth. Icarus 33:23–29

Harvey LDD (1980) Solar variability as a contributing factor to Holocene climatic change. Prog Phys Geogr 4:487–530

Harvey LDD (1988a) Climatic impact of ice-age aerosols. Nature 334:333–335

Harvey LDD (1988b) A semi-analytic energy balance climate model with explicit sea ice and snow physics. J Clim 1:1065–1085

Harvey LDD (1988c) Development of a sea ice model for use in zonally averaged energy balance climate models. J Clim 1:1211–1238

Harvey LDD (1989a) An energy balance climate model study of radiative forcing and temperature response at 18 Ka. J Geophys Res 94D:12,873–12,884

Harvey LDD (1989b) Modelling the Younger Dryas. Quat Sci Rev 8:137–149

Harvey LDD (1989c) Milankovitch forcing, vegetation, and North Atlantic deep-water formation. J Clim 2:800–815

Harvey LDD (1989d) Effect of model structure on the response of terrestrial biosphere models to CO_2 increases and temperature increases. Glob Biogeochem Cycles 3:137–153

Harvey LDD (1989e) Transient climatic response to an increase of greenhouse gases. Clim Change 15:15–30

Harvey LDD (1989f) Managing atmospheric CO_2. Clim Change 15:343–381

Harvey LDD (1990) Managing atmospheric CO_2: policy implications. Energy 15:91–104

Hayes MO (1967) Relationship between coastal climate and bottom sediment type on the inner continental shelf. Mar Geol 5:11–132

Hays JD, Imbrie J, Shackleton NJ (1976) Variations in the Earth's orbit: pacemaker of the ice ages. Science 194:1121–1132

Heer O (1868–83) Flora fossilis Arctica: die fossile Flora der Polarländer, 7 vols. J Wurster, Zürich

Hennessy H (1859) Terrestrial climate as influenced by the distribution of land and water at different geological epochs. Am J Sci 2nd Ser 27:316–328

Hennessy H (1860) Change of climate. The Athenaeum 1717:384–386

Herbert TD, Fischer AG (1986) Milankovitch climatic origin of mid-Cretaceous black shale rhythms in central Italy. Nature 321:739–743

Herbert TD, Stallard RF, Fischer AG (1986) Anoxic events, productivity rhythms, and orbital signature in a mid-Cretaceous deep-sea sequence from central Italy. Paleoceanography 1:495–506

Herder JG von (1827) Aus den Ideen zur Philosophie der Geschichte der Menschheit. Verlag des Bibliographischen Instituts, Gotha

Herschel JFW (1935) On the astronomical causes which may influence geological phaenomena. Trans Geol Soc Lond 2nd Ser 3:293–299

Hesse R (1924) Tiergeographie auf ökologischer Grundlage. Gustav Fischer, Jena

Hesse R, Allee WC, Schmidt KP (1937) Ecological animal geography. An authorized, rewritten edition based on Tiergeographie auf ökologischer Grundlage by Richard Hesse. Prepared by WC Allee and Karl P Schmidt. John Wiley, New York; Chapman & Hall, London

Hibbard CW (1960) Pliocene and Pleistocene climates in North America. Annu Rep Mich Acad Sci Arts Lett 62:5–30

Hilgard EW (1892) A report on the relation of soils to climate. US Dep Agric Weather Bur Bull 3

Hilgard EW (1906) Soils. Macmillan, New York

Hill RS, Read J, Busby JR (1988) The temperature-dependence of photosynthesis of some Australian temperate rainforest trees and its biogeographical significance. J Biogeogr 15:431–449

Hinds RB (1842) The physical agents of temperature, humidity, light, and soil, considered as developing climate, and in connexion with geographic botany. Ann Mag Nat Hist 9:169–189, 311–333, 469–475, 521–527

Hinds RB (1843) The regions of vegetation; being an analysis of the distribution of vegetable forms over the surface of the globe in connexion with climate and physical agents. GJ Palmer, London

Hirschboeck KK (1980) A new worldwide chronology of volcanic eruptions. Palaeogeogr Palaeoclimatol Palaeoecol 29:223–241

Hirth P (1926) Die Isonotiden. Petermanns Geogr Mitt 72:145–149

Holdridge LR (1947) Determination of world plant formations from simple climatic data. Science 105:367–368

Holdridge LR (1959) Simple method for determining potential evapotranspiration from temperature data. Science 130:572

Holdridge LR (1967) Life zone ecology, revised edn. Tropical Science Center, San José, Costa Rica

Holeman JN (1968) The sediment yield of major rivers of the world. Water Resour Res 4:737–747

Holland HD, Feakes CR (1989) Paleosols and their relevance to Precambrian atmospheric composition: a discussion. J Geol 97:761–762

Holland HD, Zbinden EA (1988) Paleosols and the evolution of the atmosphere. Part I. In: Lerman A, Meybeck M (eds) Physical and chemical weathering in geochemical cycles. D Reidel, Dordrecht, pp 61–82

Holland HD, Feakes CR, Zbinden EA (1989) The Flin Flon paleosol and the composition of the atmosphere 1.8 by BP. Am J Sci 289:362–389

Holliday VT (1988) Mt Blanco revisited: soil-geomorphic implications for the ages of the upper Cenozoic Blanco and Blackwater Draw Formations. Geology 16:505–508

Holliday VT (1989a) The Blackwater Draw Formation (Quaternary): a 1.4-plus-my record of eolian sedimentation and soil formation in the Southern High Plains. Bull Geol Soc Am 101:1598–1607

Holliday VT (1989b) Middle Holocene drought on the Southern High Plains. Quat Res 31:74–82

Hooke R (1705) Lectures and discourses of earthquakes, and subterraneous eruptions. Explicating the causes of the rugged and uneven face of the Earth; and what reasons may be given for the frequent finding of shells and other sea and land petrified substances, scattered over the whole terrestrial superficies. In: Waller R (ed) The posthumous works of Robert Hooke. Containing his Cutlerian lectures, and other discourses, read at the meetings of the illustrious Royal Society. Richard Waller for the Royal Society, London, Part V

Hooke R (1978) Lectures and discourses of earthquakes, and subterraneous eruptions. Explicating the causes of the rugged and uneven face of the Earth; and what reasons may be given for the frequent finding of shells and other sea and land petrified substances, scattered over the whole terrestrial superficies. In: Waller R (ed) The posthumous works of Robert Hooke. Containing his Cutlerian lectures, and other discourses, read at the meetings of the illustrious Royal Society. 2nd facsimile edn, with a new introduction by TM Brown of Princeton University. Frank Cass, London, Part V

Horton RE (1945) Erosional development of streams and their drainage basins: hydrophysical approach to quantitative morphology. Bull Geol Soc Am 56:275–370

House MR (1985) A new approach to an absolute timescale from measurements of orbital cycles and sedimentary micro-rhythms. Nature 315:721–725

Hoyle F (1981) Ice. Hutchinson, London

Hoyle F, Lyttleton RA (1939) The effect of interstellar matter on climatic variation. Proceedings of the Cambridge Philos Soc 35:405–415

Hoyle F, Lyttleton RA (1950) Variations in solar radiation and the cause of ice ages. J Glaciol 1:453–455

Hoyle F, Wickramasinghe C (1978) Comets, ice ages, and ecological catastrophes. Astrophys Space Sci 53:523–526

Hoyle F, Wickramasinghe NC (1990) Sunspots and influenza. Nature 343:304

Huggett RJ (1982) Models and spatial patterns of soils. In: Bridges EM, Davidson DA (eds) Principles and applications of soil geography. Longman, London, pp 132–170

Huggett RJ (1985) Earth surface systems. Springer, Berlin Heidelberg New York Tokyo (Springer Series in Physical Environment, vol 1)

Huggett RJ (1989a) Cataclysms and Earth history: the development of diluvialism. Clarendon, Oxford

Huggett RJ (1989b) Drayson's hypothesis: the Earth's tilt cycle. Chronol Catastroph Rev 11:12–20

Huggett RJ (1990) Catastrophism: systems of Earth history. Edward Arnold, London

Humboldt A von (1820–21) On isothermal lines, and the distribution of heat over the globe. Edinb Philos J 3(1820):1–20, 256–274; 4(1820–21):23–37, 262–281; 5(1821):28–39

Humboldt A von (1849) Cosmos: a sketch of a physical description of the universe, vol. i. Translated from the German by EC Otté. Henry G Bohn, London

Humboldt A von, Bonpland A (1807) Essai sur la géographie des plantes; accompagné d'un tableau physique des régions equinoxiales, fondé sur les mesures exécutées, depuis le dixième degré de latitude boréale jusqu'au dixième degré de latitude australe, pendant les années 1799, 1800, 1801, 1802 and 1803. Fr Schoell, Paris

Humphreys WJ (1940) Physics of the air. McGraw-Hill, New York

Hundeshagen J (1830–32) Forstliche Berichte und Miscellen, 2 vols. H Laupp, Tübingen

Hunt BG (1979) The effects of past variations of the Earth's rotation rate on climate. Nature 281:188–191

Hunt BG (1982) The impact of large variations of the Earth's obliquity on the climate. J Meteorol Soc Jpn 60:309–318

Hunt BG (1984) Polar glaciation and the genesis of ice ages. Nature 308:48–51

Huntington E (1907) The pulse of Asia. Houghton Mifflin, Boston

Huntington E (1911) Palestine and its transformation. Houghton Mifflin, Boston

Huntington E (1914a) The solar hypothesis of climatic changes. Bull Geol Soc Am 25:477–590

Huntington E (1914b) The climatic factor as illustrated in arid America. With contributions by Charles Schuchert, AE Douglas, and CJ Kullmer. Carnegie Institution of Washington, Washington DC, Publication 192

Huntington E (1917) The geographical work of Dr MA Veeder. Geogr Rev 3:188–211

Huntington E, Visher SS (1922) Climatic changes: their nature and causes. Yale University Press, New Haven; Oxford University Press, Humphrey Milford, London

Huntley B, Webb T III (eds) (1988) Vegetation history. (Handbook of vegetation science, vol 7) Kluwer, Dordrecht

Hurt PR, Libby LM, Pandolphi LJ, Levine LH, Van Engel WA (1979) Periodicites in blue crab population of Chesapeake Bay. Clim Change 2:75–78

Hutchinson GE (1959) Homage to Santa Rosalia, or why are there so many kinds of animals? Am Nat 93:145–159

Huxley JS (1942) Evolution: the modern synthesis. George Allen & Unwin, London

Imbrie J (1985) A theoretical framework for the Pleistocene ice ages. J Geol Soc Lond 142:417–432

Imbrie J, Imbrie KP (1986) Ice ages: solving the mystery. Harvard University Press, Cambridge

Imbrie J, Hays JD, Martinson DG, McIntyre A, Mix AG, Morley JJ, Pisias NG, Prell WL, Shackleton NJ (1984) The orbital theory of Pleistocene climate: support from a revised chronology of the marine $\delta^{18}O$ record. In: Berger A, Imbrie J, Hays J, Kukla G, Saltzman B (eds) Milankovitch and climate: understanding the response of astronomical forcing. Part 1. D Reidel, Dordrecht, pp 269–305 (Proceedings of the NATO Advanced Research Workshop on Milankovitch and Climate, Palisades, New York, 1982. NATO ASI Series. Series C, Mathematical and Physical Sciences, vol 126)

Isaac KP (1981) Tertiary weathering profiles in plateau deposits of East Devon. Proc Geol Assoc 92:159–168

Isaac KP (1983) Tertiary lateritic weathering in Devon, England, and the Palaeogene continental environment of South West England. Proc Geol Assoc 94:105–114

James FC (1970) Geographical size variation in birds and its relationship to climate. Ecology 51:365–390

James H (1860) On the change of climate in different regions of the Earth. The Athenaeum, 1713:256–257

Janacek TR, Rea DK (1984) Pleistocene fluctuations in northern hemisphere tradewinds and westerlies. In: Berger A, Imbrie J, Hays J, Kukla G, Saltzman B (eds) Milankovitch and climate: understanding the response of astronomical forcing, Part 1. D Reidel, Dordrecht, pp 331–347 (Proceedings of the NATO Advanced Research Workshop on Milankovitch and Climate, Palisades, New York, 1982. NATO ASI Series. Series C, Mathematical and Physical Sciences, vol 126)

Jansen E, Veum T (1990) Evidence for a two-step deglaciation and its impact on North Atlantic deep-water circulation. Nature 343:612–616

Jansen JML, Painter RB (1974) Predicting sediment yield from climate and topography. J Hydrol 21D:371–380

Jansson MB (1982) Land erosion by water in different climates. University of Uppsala, Department of Physical Geography, Uppsala, Sweden, Uppsala Universitet Naturgeografiska Institutionen Rapport No 57

Jansson MB (1988) A global survey of sediment yield. Geogr Ann 70A:81–98

Jenny H (1930) A study of the influence of climate upon the nitrogen and organic content of the soil. MO Agric Exp Stn Res Bull 152

Jenny H (1935) The clay content of the soil as related to climatic factors, particularly temperature. Soil Sci 40:111–128

Jenny H (1941) Factors of soil formation: a system of quantitative pedology. McGraw-Hill, New York

Jenny H (1980) The soil resource: origin and behaviour. Springer, Berlin Heidelberg New York (Ecological Studies, vol 37)

Jenny H, Leonard CD (1934) Functional relationships between soil properties and rainfall. Soil Sci 38:363–381

Jessen O (1936) Reisen und Forschungen in Angola. D Reimer, Berlin

Joffe JS (1949) Pedology, 2nd edn. With an introduction by Curtis F Marbut. Pedology Publications, New Brunswick, New Jersey

Johnson DL (1985) Soil thickness processes. In: Jungerius PD (ed) Soils and geomorphology. Catena Supplement 6, Catena Verlag, Cremlingen, pp 29–40

Johnson DL, Watson-Stegner D (1987) Evolution model of pedogenesis. Soil Sci 143:349–366

Johnson DL, Kell EA, Rockwell TK (1990) Dynamic pedogenesis: new views on some key soil concepts, and a model for interpreting Quaternary soils. Quat Res 33:306–319

Johnson WD (1901) The High Plains and their utilization. US Geological Survey, Washington DC, 11th Annual Report, part iv, pp 601–741

Johnston RF, Selander RK (1971) Evolution in the house sparrow. II. Adaptive differentiation in North American populations. Evolution 25:1–28

Jones MDH, Henderson-Sellers A (1990) History of the greenhouse effect. Prog Phys Geogr 14:1–18

Jordan CF (1971) A world pattern in plant energetics. Am Sci 59:425–433

Jukes JB (1862) On the mode of formation of some river-valleys in the south of Ireland. Q J Geol Soc Lond 18:378–403

Kachanoski RG (1988) Processes in soils — from pedon to landscapes. In: Rosswall T, Woodmansee RG, Risser PG (eds) Scales and global change: spatial and temporal variability in biospheric and geospheric processes. John Wiley, Chichester, pp 153–177 (SCOPE 35)

Kaiser E (1926) Die Diamantenwüste Südwest-Afrikas, zugleich Erläuterung zu einer geologischen Spezialkarte der Südlichen Diamantfelder 1:25000, aufgenommen von W. Beetz und E. Kaiser mit Beiträgen von W. Beetz, J. Böhm, R. Martin [et al.], 2 vols. D Reimer, Berlin

Karlstrom ET (1987) Stratigraphy and genesis of five superposed paleosols in pre-Wisconsin drift on Mokowan Butte, southwestern Alberta. Can J Earth Sci 24:2235–2253

Karlstrom ET (1988) Multiple paleosols in pre-Wisconsin drift, northwestern Montana and south-western Alberta. Catena 15:147–178

Karrasch H (1972) The planetary and hypsometric variation of valley asymmetry. In: Adams WP, Helleiner FM (eds) International Geography 1972, vol 1. Toronto University Press, Toronto, pp 31–34

Kasting JF (1989) Long-term stability of the Earth's climate. Palaeogeogr Palaeoclimatol Palaeoecol (Global and Planetary Change Section) 75:83–95

Keller G (1989) Extended periods of extinctions across the Cretaceous/Tertiary boundary in planktonic foraminifera of continental shelf sections: implications for impact and volcanic theories. Bull Geol Soc Am 101:1408–1419

Keller G, d'Hondt SL, Orth CJ, Gilmore JS, Oliver PQ, Shoemaker EM, Molina E (1987) Late Eocene impact microspherules: stratigraphy, age and geochemistry. Meteoritics 22:25–60

Kendeigh SC (1969) Tolerance to cold and Bergmann's rule. Auk 86:13–25

Kennedy BA (1976) Valley-side slopes and climate. In: Derbyshire E (ed) Geomorphology and climate. John Wiley, London, pp 171–201

Kennett JP, Thunell RC (1975) Global increase in Quaternary explosive volcanism. Science 187:497–503

Kennett JP, Thunell RC (1977) On explosive Cenozoic volcanism and climatic implications. Science 196:1231–1234

Kennett JP, von der Borch CC (1986) Southwest Pacific Cenozoic paleoceanography. Deep Sea Drill Proj Init Rep 90:1493–1517

Kerr RA (1981) Mount St Helens and a climate quandary. Science 211:371–374

Kerr RA (1984) The Moon influences western US drought. Science 224:587

Kerr RA (1988) Sunspot-weather link holding up. Science 242:1124–1125

Kimura T, Sekido S (1975) *Nilssoniocladus* n. gen. (Nilssoniaceae n fam) newly found from the early Lower Cretaceous in Japan. Palaeontographica 153B:111–118

Kinahan GH (1866) The effects of weathering on rocks. Geol Mag 3:86–88

King JE, Saunders JJ (1984) Environmental insularity and the extinction of the American mastodon. In: Martin PS, Klein RG (eds) Quaternary extinctions: a prehistoric revolution. The University of Arizona Press, Tucson, Arizona, pp 315–344

King LC (1957) The uniformitarian nature of hillslopes. Trans Edinb Geol Soc 17:81–102

King LC (1983) Wandering continents and spreading sea floors on an expanding Earth. John Wiley, Chichester

Kirkby MJ (1989) A model to estimate the impact of climatic change on hillslope and regolith form. Catena 16:321–341

Klein RG (1986) Carnivore size and Quaternary climatic change in southern Africa. Quat Res 26:153–170

Klüpfel W (1917) Über die Sedimente der Flachsee im Lothringer Jura. Geol Rundschau 7:98–109

Knoll AH (1984) Patterns of extinction in the fossil record of vascular plants. In: Nitecki MH (ed) Extinctions. The University of Chicago Press, Chicago, pp 21–68

Knox JC (1984) Responses of river systems to Holocene climates. In: Wright HE Jr (ed) Late-Quaternary environments of the United States, vol 2. The Holocene. Longman, London, pp 26–41

Koch PL (1986) Clinal geographic variation in mammals: implications for the study of chronoclines. Paleobiology 12:269–281

Koerner RM, Fisher DA (1990) A record of Holocene summer climate from a Canadian high-Arctic ice core. Nature 343:630–631

Koestler A (1967) The ghost in the machine. Hutchinson, London

Koestler A (1978) Janus: a summing up. Hutchinson, London

Kondratyev KYa (1988) Climatic shocks: natural and anthropogenic. Translated from the Russian by AP Kostrova. John Wiley, New York

Köppen W (1873) Über mehrjährige Perioden der Witterung insbesondere über die 11-jährige Periode der Temperaturen. Meteorol Z (Österreichische Gesellschaft für Meteorologie; Deutsche Meteorologische Gesellschaft) 8:241–248, 257–267

Köppen W (1931) Grundriss der Klimakunde. Zweite, Verbesserte Auflage der Klimate der Erde. Walter de Gruyter, Berlin

Köppen W, Wegener AL (1924) Die Klimate der geologischen Vorzeit. Gebrüder Borntraeger, Berlin

Krause GCL (1832) Über Gemeinheitsheilung und landwirthschaftliche Ablösungen; oder, Entwicklung der Gesetze für die Gemeinheitsheilungen und Ablösung der passiven Berechtigungen des Landbaues, sowie der Principien und des Geschäftsganges zur Ausführung derselben und der Abschätzungen zu den verschiedenen Zwecken, nach den Forderungen erwachsender Bedürfnisse der Gesellschaft, der fortschreitenden Industrie und den Grundsätzen der rationellen Landwirtschaft. Vol. 2. Bodenkunde und Klassifikation des Bodens nach seinen physischen und chemischen Eigenschaften, Bestandtheilen und Kulturverhältnissen. Flinzer, Gotha

Kreichgauer PD (1902) Die Äquatorfrage in der Geologie, 2nd edn. Missionsdruckerei, Steyl

Kronberg BI, Nesbitt HW (1981) Quantification of weathering, soil geochemistry and soil fertility. J Soil Sci 32:453–459

Kröner A (1977) Non-synchroneity of late Precambrian glaciations in Africa. J Geol 85:289–300

Krook M (1953) Interstellar matter and the solar constant. In: Shapley H (ed) Climatic change: evidences, causes, and effects. Harvard University Press, Cambridge, Massachusetts, pp 143–146

Kuhn WR, Walker JCG, Marshall HG (1989) The effect on the Earth's surface temperature from variations in rotation rate, continent formation, solar luminosity, and carbon dioxide. J Geophys Res 94D:11,129–11,136

Kukla GJ (1968) "Comment". Curr Anthropol 9:37–39

Kukla GJ (1975) Loess stratigraphy of central Europe. In: Butzer KW, Isaac GL (eds) After the Australopithecines. Mouton, The Hague, pp 99–188

Kukla GJ (1977) Pleistocene land-sea correlations. 1. Europe. Earth-Sci Rev 13:307–374

Kullmer CJ (1914) The shift of the storm track. In: Huntington E (ed) The climatic factor as illustrated in arid America. Carnegie Institution of Washington, Washington DC, Publication 192, pp 193–205

Kullmer CJ (1933) The latitude shift of the storm track in the 11-year solar period. Smithsonian Institution, Washington DC, Smithsonian Miscellaneous Collections 89, no 2

Kump LR (1988a) Alternative modeling approaches to the geochemical cycles of carbon, sulfur, and strontium isotopes. Am J Sci 289:390–410

Kump LR (1988b) Terrestrial feedback in atmospheric oxygen regulation by fire and phosphorus. Nature 335:152–154

Kump LR (1989) Chemical stability of the atmosphere and ocean. Palaeogeogr Palaeoclimatol Palaeoecol (Global and Planetary Change Section) 75:123–136

Kump LR, Garrels RM (1986) Modeling atmospheric O_2 in the global sedimentary redox cycle. Am J Sci 286:337–360

Kurtén B, Anderson E (1980) Pleistocene mammals of North America. Columbia University Press, New York

Kutzbach JE (1981) Monsoon climate of the early Holocene: climatic experiment using the Earth's orbital parameters for 9000 years ago. Science 214:59–61

Kutzbach JE (1987) Model simulations of the climatic patterns during the deglaciation of North America. In: Ruddiman WF, Wright HE Jr (eds) North America and adjacent oceans during the last deglaciation. The Geology of North America, vol K-3. Boulder, Colorado, The Geological Society of America, pp 425–446

Kutzbach JE, Gallimore RG (1988) Sensitivity of a coupled atmosphere/mixed-layer ocean model to changes in orbital forcing at 9000 yr BP. J Geophys Res 93D:803–821

Kutzbach JE, Gallimore RG (1989) Pangaean climates: megamonsoons of the megacontinent. J Geophys Res 94:3341–3357

Kutzbach JE, Guetter PJ (1986) The influence of changing orbital parameters and surface boundary conditions on climate simulations for the past 18 000 years. J Atmos Sci 43:726–1759

Kutzbach JE, Otto-Bliesner BL (1982) The sensitivity of the African-Asian monsoonal climate to orbital parameter changes for 9000 years BP in a low-resolution general circulation model. J Atmos Sci 39:1177–1188

Kutzbach JE, Street-Perrott FA (1985) Milankovitch forcing of fluctuations in the level of tropical lakes from 18 to 0 kyr BP. Nature 317:130–134

Kutzbach JE, Guetter PJ, Ruddiman WF, Prell WL (1989) Sensitivity of climate to late Cenozoic uplift in southern Asia and the American west: numerical experiments. J Geophys Res 94D:18,393–18,407

Kyle PR, Jezek PA, Mosley-Thompson E, Thompson LG (1981) Tephra layers in the Byrd station ice core and the Dome C ice core, Antarctica and their climatic importance. J Volcanol Geotherm Res 11:29–39

Labitzke K, Chanin M-L (1988) Changes in the middle of the atmosphere in winter related to the 11-year solar cycle. Ann Geophys 6:643–644

Labitzke K, van Loon H (1988) Associations between the 11-year solar cycle, the QBO and the atmosphere. Part I: The troposphere and stratosphere in the northern hemisphere winter. J Atmos Terrest Phys 50:197–206

Labitzke K, van Loon H (1989a) Association between the 11-year solar cycle, the QBO, and the atmosphere. Part III: Aspects of the association. J Clim 2:554–565

Labitzke K, van Loon H (1989b) The 11-year solar cycle in the stratosphere in the northern summer. Ann Geophys 7:595–598

Labitzke K, van Loon H (1989c) Recent work correlating the 11-year solar cycle with atmospheric elements grouped according to the phase of the quasi-biennial oscillation. Space Sci Rev 49:239–258

Labitzke K, van Loon H (1990) Sonnenflecken und Wetter. Gibt es doch einen Zusammenhang? Geowissenschaften 8:1–6

LaMarche VC Jr, Hirschboeck KK (1984) Frost rings in trees as records of major volcanic eruptions. Nature 307:121–126

Lamb HH (1961) Climatic change within historic time as seen in circulation maps and diagrams. Ann N Y Acad Sci 95:124–161

Lamb HH (1965) The early Medieval warm epoch and its sequel. Palaeogeogr Palaeoclimatol Palaeoecol 1:13–37

Lamb HH (1970) Volcanic dust in the atmosphere; with a chronology and assessment of its meteorological significance. Philos Trans R Soc Lond 266A:425–533

Lamb HH (1971) Volcanic activity and climate. Palaeogeogr Palaeoclimatol Palaeoecol 10:203–230

Lamb HH (1972) Climate: present, past and future, 2 vols. Methuen, London

Lamb HH (1982) Climate, history and the modern world. Methuen, London

Landsberg HE (1964) Early stages of climatology in the United States. Bull Am Meteorol Soc 45:268–275

Landsberg HE (1980) Variable solar emissions, the "Maunder Minimum" and climatic temperature fluctuations. Arch Meteorol Geophys Bioklimatol 28B:181–191

Landsberg HE, Albert JM (1974) The summer of 1816 and volcanism. Weatherwise 27:63–66

Landsberg HE, Mitchell JM Jr, Crutcher HI (1959) Power spectrum analysis of climatological data for Woodstock College, Maryland. Mon Weather Rev 87:283–298

Landscheidt T (1987) Long-range forecasts of solar cycles and climatic change. In: Rampino MR, Sanders JE, Newman WS, Königsson LK (eds) Climate: history, periodicity, and predictability. Van Nostrand Reinhold, New York, pp 421–445

Lane RP, Marshall JE (1981) Geographical variation, races and subspecies. In: Forey PL (ed) The evolving biosphere. British Museum (Natural History), London and Cambridge University Press, Cambridge, pp 9–19

Lang R (1915) Versuch einer exakten Klassifikation der Böden in klimatischer und geologischer Hinsicht. Int Mitt Bodenk 5:312–346

Lang R (1920) Verwitterung und Bodenbildung als Einführung in die Bodenkunde. E Schweizerbart, Stuttgart

Lean JL (1984) Solar ultraviolet irradiance variations and the Earth's atmosphere. In: Mörner N-A, Karlén W (eds) Climatic changes on a yearly to millenial basis: geological, historical and instrumental records. D Reidel, Dordrecht, pp 449–471

Leckie D, Fox C, Tarnocai C (1989) Multiple paleosols of the late Albian Boulder Creek Formation, British Columbia, Canada. Sedimentology 36:307–323

Le Conte J (1882) Elements of geology: a text-book for colleages and the general reader. D Appleton, New York

Lehmann H (1957) Klimageomorphologie-Beobachtungen in der Serra de Mantiqueria und im Paraiba-Tal (Brasilien). Abh Geogr Inst Freien Univ Berl 5:67–72

Leopold LB, Wolman MG, Miller JP (1964) Fluvial processes in geomorphology. WH Freeman, San Francisco, California

Lettau H (1969) Evapotranspiration climatonomy I. Mon Weather Rev 97:691–699

Lettau H (1973) Evapotranspiration climatonomy II. Mon Weather Rev 101:636–649

Lézine A-M (1988a) New pollen data from the Sahel, Senegal. Rev Palaeobot Palynol 55:141–154

Lézine A-M (1988b) Les variations de la couverture forestière mésophile d'Afrique occidentale au cours de l'Holocène. C R Acad Sci Paris Ser II 307:439–445

Lézine A-M (1989) Late Quaternary vegetation and climate of the Sahel. Quat Res 32:317–334

Lézine A-M, Casanova J (1989) Pollen and hydrological evidence for the interpretation of past climates in tropical West Africa during the Holocene. Quat Sci Rev 8:45–55

Libby LM (1983) Past climates: tree thermometers, commodities, and people. The University of Texas Press, Austin, Texas

Libby LM (1987) Evaluation of historic climate and prediction of near-future climate from stable-isotope variations in tree rings. In: Rampino MR, Sanders JE, Newman WS, Königsson LK (eds) Climate: history, periodicity, and predictability. Van Nostrand Reinhold, New York, pp 81–89

Lidmar-Bergstrom K (1985) Exhumed Mesozoic landforms in south Sweden. In: Spencer T (ed) Abstracts of papers for the first international geomorphology conference. Department of Geography, University of Manchester, Manchester, p 363

Liebig J von (1862) Der chemische Prozess der Ernährung der Vegetabilien. F Viehweg, Brunswick

Lieth HFH (1961) La production de sustancia organica por la capa vegetal y sus problemas. Acta Sci Venezol 12:107–114

Lieth HFH (1972) Modelling the primary productivity of the world. Nat Resour 8:5–10

Lieth HFH (1973) Primary production: terrestrial ecosystems. J Human Ecol 1:303–332

Lieth HFH (1975a) Editor's comments on papers 1 through 11. In: Lieth HFH (ed) Patterns of primary production in the biosphere. Dowden, Hutchinson & Ross, Stroudsberg, Pennsylvania, pp 4–21 (Benchmark Papers in Ecology, vol 8)

Lieth HFH (1975b) Historical survey of primary productivity research. In: Lieth HFH, Whittaker RH (eds) Primary productivity of the biosphere. Springer, Berlin Heidelberg New York, pp 7–16 (Ecological Studies 14)

Lieth HFH (1975c) Primary productivity in ecosystems: comparative analysis of global patterns. In: van Dobben WH, Lowe-McConnell RH (eds) Unifying concepts in ecology. W Junk, The Hague; Centre for Agricultural Publishing and Documentation, Wageningen, pp 67–88 (Report of the Plenary Sessions, First International Congress of Ecology, The Hague, Netherlands, 1974)

Lieth HFH, Box E (1972) Evapotranspiration and primary productivity; CW Thornthwaite memorial model. In: Mather JR (ed) Papers on selected topics in climatology, vol 2. Thornthwaite Associates, Elmer, New Jersey, pp 37–46

Lindeman RL (1942) The trophic-dynamic aspects of ecology. Ecology 23:399–418

Lindsay JF, Srnka LJ (1975) Galactica dust lanes and lunar soil. Nature 257:776–778

Lindsey CC (1966) Body sizes of poikilotherm vertebrates at different latitudes. Evolution 20:456–465

Link F (1968) The 400-year cycle. J Br Astronom Assoc 78:195–205

Linnaeus C (1751) Specimen academicem de oeconomia naturae. Amoenitat Acad 2:1–58

Linton DL (1955) The problem of tors. Geogr J 121:289–291

Littmann T (1989) Spatial patterns and frequency distribution of late Quaternary water budget tendencies in Africa. Catena 16:163–188

Littmann T, Schmidt K-H (1989) The response of different relief units to climatic change in an arid environment (southern Morocco). Catena 16:343–355

Livingstone DA (1963) Chemical composition of rivers and lakes. US Geological Survey Professional Paper 440G, US Government Printing Office, Washington DC, pp G1-G64

Livingstone EB, Shreve F (1921) The distribution of vegetation in the United States as related to climatic conditions. The Carnegie Institution of Washington, Washington DC, Publication 284

Lockyer JN, Hunter WW (1877) Sunspots and famines. Nineteenth Century 2:583–602

Loder JW, Garrett C (1978) The 18.6-year cycle of sea surface temperature in shallow seas due to variations in tidal mixing. J Geophys Res 83:1967–1970

Löffelholz von Colberg C, Freiherr (1886) Die Drehung der Erdkruste in geologischen Zeiträumen: eine neue geologisch-astronomische Hypothese. JA Finsterlin, Munich

Loper DE, McCartney K, Buzyna G (1988) A model of correlated episodicity in magnetic-field reversals, climate, and mass extinctions. J Geol 96:1–15

Lorius C (1990) Environmental records from polar ice cores. Philos Trans R Soc 330A:459–462

Louis H (1957) Rumpfflächen-Problem, Erosionszyklus und Klimageomorphologie. Petermanns Geogr Mitt 262:9–26

Louis H (1973) The problem of erosion surfaces, cycles of erosion and climatic geomorphology. In: Derbyshire E (ed) Climatic geomorphology. Methuen, London, pp 153–170 (Geographical Readings Series) (Translation of 1957 paper by Roger S. Mays and Edward Derbyshire)

Lovelock JE (1972) Gaia as seen through the atmosphere. Atmos Environ 6:579–580

Lovelock JE (1979) Gaia: a new look at life on Earth. Oxford University Press, Oxford

Lovelock JE (1988) The Ages of Gaia: a biography of our living Earth. Oxford University Press, Oxford

Lovelock JE (1989) Geophysiology. Trans R Soc Edinb 80:169–175

Lovelock JE, Margulis L (1974a) Atmospheric homeostasis by and for the biosphere: the Gaia hypothesis. Tellus 26:1–10

Lovelock JE, Margulis L (1974b) Homeostatic tendencies of the Earth's atmosphere. Origin Life 1:12–22

Lovenburg MF, Dell CI, Johnson MJS (1972) Effect of a shorter day upon biotic diversity. Bull Geol Soc Am 83:3529–3530

Lubbock J (1848) On change of climate resulting from a change in the Earth's axis of rotation. Q J Geol Soc Lond 4:4–7

Lundelius EL Jr (1976) Vertebrate paleontology of the Pleistocene: an overview. In: West RC (ed) Geoscience and Man. Louisiana State University, Baton Rouge, pp 49–59

Lundelius EL Jr, Graham RW, Anderson E, Guilday J, Holman JA, Steadman D, Webb SD (1983) Terrestrial vertebrate faunas. In: Porter SC (ed) Late-Quaternary environments of the United States, vol 1, The Late Pleistocene. Longman, London, pp 311–353

Lundelius EL Jr, Downs T, Lindsay E, Semken HA Jr, Zakrzewski RA (1987) The North American Quaternary sequence. In: Woodburne MO (ed) Cenozoic mammals of North America: geochronology and biostratigraphy. University of California Press, Berkeley, pp 211–235

Lungershausen GF (1957) Periodic changes in climate and the major glaciations of the globe (some problems of historical palaeogeography and absolute chronology). Sov Geol 59:88–115 (in Russian)

Luyendyk B, Forsyth D, Phillips J (1972) An experimental approach to the paleocirculation of oceanic surface waters. Bull Geol Soc Am 83:2649–2664

Lyell C (1830–33) Principles of geology, being an attempt to explain the former changes of the Earth's surface, by reference to causes now in operation, 3 vols. John Murray, London

Lyell C (1863) The geological evidences of the antiquity of Man with remarks on the origin of species by variation. John Murray, London

Magaritz M, Kaufman A, Yaalon DH (1981) Calcium carbonate nodules in soils: $^{18}O/^{16}O$ and $^{13}C/^{12}C$ ratios and ^{14}C contents. Geoderma 25:157–172

Major J (1963) A climatic index to vascular plant activity. Ecology 44:1–9

Manabe S, Broccoli AJ (1985) The influence of continental ice sheets on the climate of an ice age. J Geophys Res 90:2167–2190

Manabe S, Hahn DG (1977) Simulation of the tropical climate of an ice age. J Geophys Res 82:3889–3911

Manabe S, Wetherald RT (1967) Thermal equilibrium of the atmosphere with a given distribution of relative humidity. J Atmos Sci 24:241–259

Manabe S, Wetherald RT (1975) The effects of doubling the CO_2 concentration on the climate of a general circulation model. J Atmos Sci 32:3–15

Manabe S, Wetherald RT (1980) On the distribution of climatic change resulting from an increase in CO_2 content of the atmosphere. J Atmos Sci 37:99–118

Manley G (1961) Late and postglacial climatic fluctuations and their relationship to those shown by the instrumental record of the last 300 years. Ann N Y Acad Sci 95:162–172

Manley G (1971) Interpreting the meteorology of the late and post-glacial. Palaeogeogr Palaeoclimatol Palaeoecol 10:163–175

Manley G (1974) Central England temperatures: monthly means 1659–1973. Q J R Meteorol Soc 100:389–405

Marbut CF (1927) A scheme for soil classification. Proc Pap First Int Congr Soil Sci 4:1–31

Marbut CF (1935) Soils of the United States. US Department of Agriculture, Atlas of American Agriculture, Part III, Advance Sheets No 8, US Government Printing Office, Washington DC

Marianatos S (1939) The volcanic destruction of Minoan Crete. Antiquity 13:425–439

Marion GM (1989) Correlation between long-term pedogenic $CaCO_3$ formation rate and modern precipitation in deserts of the American Southwest. Quat Res 32:291–295

Mariotté E (1686) Traité du mouvement des eaux et des autres corps fluides. E Michallet, Paris

Martin J-M, Meybeck M (1979) Elemental mass-balance of material carried by major world rivers. Mar Chem 7:173–206

Martin PS (1984a) Prehistoric overkill: the global model. In: Martin PS, Klein RG (eds) Quaternary extinctions: a prehistoric revolution. The University of Arizona Press, Tucson, Arizona, pp 354–403

Martin PS (1984b) Catastrophic extinctions and late Pleistocene blitzkreig: two radiocarbon tests. In: Nitecki MH (ed) Extinctions. The University of Chicago Press, Chicago, pp 153–190

Matthews E (1983) Global vegetation and land use: new high-resolution data bases for climate studies. J Clim Appl Meteorol 22:474–487

Matthews E (1984) Global inventory of pre-agricultural and present biomass. Prog Biometeorol 3:237–246

Matthews E (1986) Data needs and data bases for climate studies. European Space Agency SP-248 (May 1986), pp 191–203 (Proceedings of ISLSCP Conference, Rome, Italy, 2–6 December 1985)

Matthews E, Rossow WB (1987) Regional and seasonal variations of surface reflectance from satellite observations at 0.6 μm. J Clim Appl Meteorol 26:170–202

May BR, Hitch TJ (1989) Periodic variations in extreme hourly rainfalls in the United Kingdom. Meteorol Mag 118:45–50

May RM (1973) Stability and complexity in model ecosystems. Princeton University Press, Princeton, New Jersey

May RM (1981) Models for single populations. In: May RM (ed) Theoretical ecology: principles and applications, 2nd edn. Blackwell Scientific Publications, Oxford, pp 5–29

Mayr E (1942) Systematics and the origin of species. Columbia University Press, New York

McCrea WH (1975) Ice ages and the galaxy. Nature 255:607–609

McCrea WH (1981) Long time-scale fluctuations in the evolution of the Earth. Proc R Soc Lond 375:1–41

McFarlane MJ (1976) Laterite and landscape. Academic Press, London

McLean DM (1981) A test of terminal Mesozoic "catastrophe". Earth Planet Sci Lett 53:103–108

McLean DM (1985) Deccan traps mantle degassing in the terminal Cretaceous marine extinctions. Cretac Res 6:235–259

McNab BK (1971) On the ecological significance of Bergmann's rule. Ecology 52:845–854

Melosh HJ, Scheider NM, Zahnle KJ, Lathan D (1990) Ignition of global wildfires at the Cretaceous/Tertiary boundary. Nature 343:251–254

Melton MA (1957) An analysis of the relations among elements of climate, surface, properties, and geomorphology. Columbia University, New York, Department of Geology Technical Report 11, 102 pp

Mengel RM, Jackson JA (1977) Geographic variations of the red-cockaded woodpecker. Condor 79:349–355

Merriam CH (1893) The geographic distribution of life in North America, with special reference to the *Mammalia*. Smithsonian Institution, Washington DC, Annual Report 1891, pp 365–415

Merriam CH (1894) Laws of temperature control of the geographic distribution of terrestrial animals and plants. Natl Geogr Mag 6:229–238

Merriam CH, Stejneger LH (1890) Results of a biological survey of the San Francisco mountain region and desert of the Little Colorado, Arizona. North American Fauna, No 3. US Government Printing Office, Washington DC

Merrill GP (1897) A treatise on rocks, rock weathering and soils. Macmillan, London (New edn 1906)

Mesolella KJ, Matthews RK, Broecker WS, Thurber DL (1969) The astronomical theory of climatic change: Barbados data. J Geol 77:250–274

Meybeck M (1979) Concentrations des eaux fluviales en éléments majeurs et apports en solution aux océans. Rev Géol Dynam Géogr Phys 21:215–246

Meybeck M (1987a) Global chemical weathering of surficial rocks estimated from river-dissolved loads. Am J Sci 287:401–428

Meybeck M (1987b) Les transports fluviaux en solution dans les sciences de la terre. Inst Géol Bassin d'Aquitaine, Bordeaux 41:19–36

Meyer A (1926) Über einige Zusammenhänge zwischen Klima und Boden in Europa. Chem Erde 2:209–347

Meyer HLF (1916) Klimazonen der Verwitterung und ihre Bedeutung für die jüngste geologische Geschichte Deutschlands. Geol Rundschau 7:193–248

Middleton WEK, Spilhaus AF (1953) Invention of meteorological instruments. University of Toronto Press, Toronto

Milankovitch M (1920) Théorie mathématique des phénomènes thermiques produits par la radiation solaire. Gauthier-Villars, Paris

Milankovitch M (1930) Mathematische Klimalehre und astronomische Theorie der Klimaschwankungen. In: Köppen W, Geiger R (eds) Handbuch der Klimatologie, I(A). Gebrüder Bornträger, Berlin, pp 1–176

Milankovitch M (1938) Astronomische Mittel zur Erforschung der erdgeschichtlichen Klimate. Handb Geophys 9:593–698

Milliman JD (1980) Transfer of river-borne particulate material to the oceans. In: Martin J-M, Burton JD, Eisma D (eds) River inputs to ocean systems. FAO, Rome, SCOR/UNEP/UNESCO Review and Workshop, pp 5–12

Milliman JD, Meade RH (1983) World-wide delivery of sediment to the oceans. J Geol 91:1–21

Mitchell GF (1980) The search for Tertiary Ireland. J Earth Sci 3:13–33

Mix AC (1987) Hundred-kiloyear cycle queried. Nature 327:370

Mix AC, Ruddiman WF (1985) Structure and timing of the last deglaciation: oxygen-isotope evidence. Quat Sci Rev 4:59–108

Mohr ECJ, van Baren FA (1954) Tropical soils: a critical study of soil genesis as related to climate, rock and vegetation. Wiley Interscience, New York

Molfino B, Heusser LH, Woillard GM (1984) Frequency components of a Grand Pile pollen record: evidence of precessional orbital forcing. In: Berger A, Imbrie J, Hays J, Kukla G, Saltzman B (eds) Milankovitch and climate: understanding the response of astronomical forcing, Part I. D Reidel, Dordrecht, pp 391–404 (Proceedings of the NATO Advanced Research Workshop on Milankovitch and Climate, Palisades, New York, 1982. NATO ASI Series C, Mathematical and Physical Sciences, vol 126)

Möller F (1963) On the influence of changes in the CO_2 concentration in the air on the radiation balance at the Earth's surface and on climate. J Geophys Res 68:3877–3886

Molnar P, England P (1990) Late Cenozoic uplift of mountain ranges and global climatic change: chicken or egg? Nature 346:29–34

Monin AS (1986) An introduction to the theory of climate. D Reidel, Dordrecht

Monteith JL (1965) Light distribution and photosynthesis in field crops. Ann Bot 29:17–37

Moore B III, Bolin B, Björkström A, Holmén K, Ringo C (1989) Ocean carbon models and inverse methods. In: Anderson DLT, Willebrand J (eds) Ocean circulation models: combining data and dynamics. Kluwer Academic Publishers, Dordrecht, pp 409–449

Mörner N-A (1981) Eustasy, paleoglaciation and paleoclimatology. Geol Rundschau 70:691–702

Mörner N-A (1984a) Planetary, solar, atmospheric, hydrospheric and endogene processes as origin of climatic changes on Earth. In: Mörner N-A, Karlén W (eds) Climatic changes on a yearly to millenial basis: geological, historical and instrumental records. D Reidel, Dordrecht, pp 483–507

Mörner N-A (1984b) Eustacy, geoid changes, and multiple geophysical interaction. In: Berggren WA, Van Couvering JA (eds) Catastrophes and Earth history: the new uniformitarianism. Princeton University Press, Princeton, New Jersey, pp 395–415

Mörner N-A (1984c) Low sea levels, droughts, and mammalian extinctions. In: Berggren WA, Van Couvering JA (eds) Catastrophes and Earth history: the new uniformitarianism. Princeton University Press, Princeton, New Jersey, pp 397–393

Mörner N-A (1987) Short-term paleoclimatic changes: observational data and a novel causation model. In: Rampino MR, Sanders JE, Newman WS, Königsson LK (eds) Climate: history, periodicity, and predictability. Van Nostrand Reinhold, New York, pp 256–269

Mörth HT, Schlamminger L (1979) Planetary motion, sunspots and climate. In: McCormac BM, Seliga TA (eds) Solar-terrestrial influence on weather and climate, D Reidel, Dordrecht, pp 193–207

Morton AG (1981) History of botanical science. Academic Press, London

Murphy EC (1985) Bergmann's rule, seasonality, and geographical variation in body size of house sparrows. Evolution 39:1327–1334

Naeser N, Westgate J, Hughes O, Péwé T (1982) Fission-track ages of late Cenozoic distal tephra beds in the Yukon Territory and Alaska. Can J Earth Sci 19:2167–2178

Nairn AEM (1964) Problems in palaeoclimatology. Wiley Interscience, London (Proceedings of the NATO Palaeoclimate Conference held at the University of Newcastle upon Tyne January 7–12, 1963)

Nance RD, Worsley TR, Moody JB (1988) The supercontinent cycle. Sci Am 259:44–51

Napier WM, Clube SVM (1979) A theory of terrestrial catastrophism. Nature 282:455–459

Nettleton WD, Gamble EE, Allen BL, Borst G, Peterson FF (1989) Relict soils of subtropical regions of the United States. In: Bronger A, Catt JA (eds) Paleopedology — nature and applications of paleosols, Catena Supplement 16, Cremlingen, pp 59–93

Neubauer L (1983) The sun-weather connection — sudden stratospheric warmings correlated with sudden commencements and solar proton events. In: McCormac BM (ed) Weather and climate responses to solar variations. Colorado Associated University Press, Boulder, Colorado, pp 395–397

Nevo E (1986) Mechanisms of adaptive speciation at the molecular and organismal levels. In: Karlin S, Nevo E (eds) Evolutionary processes and theory. Academic Press, New York, pp 439–474

Newell NE, Newell RE, Hsiung J, Zhongxiang W (1989) Global marine temperature variation and the solar magnetic cycle. Geophys Res Lett 16:311–314

Newell RE (1970) Stratospheric temperature change from the Mount Agung volcanic eruption of 1963. J Atmos Sci 27:977–978

Newell RE (1981) Further studies of the atmospheric temperature change from the Mount Agung volcanic eruption of 1963. J Volcanol Geotherm Res 11:61–66

Newhall CG, Self S (1982) The Volcanic Exposivity Index (VEI): an estimate of explosive magnitude for historical volcanism. J Geophys Res 87C:1231–1238

Newkirk G Jr (1980) Solar variability on time scales of 10^5 to 10^9 years. In: Pepin R, Eddy J, Merrill R (eds) Proceedings of the conference on the ancient Sun. Pergamon Press, New York, pp 293–320

Newton I (1729) The mathematical principles of natural philosophy, 2 vols. Translated by Andrew Motte. Benjamin Motte, London

Ninkovich D, Down WL (1976) Explosive Cenozoic volcanism and climatic implications. Science 194:899–906

Nix HA (1981) The environment of *Terra Australis*. In: Keast A (ed) Ecological biogeography of Australia. W Junk, The Hague, pp 103–133

Noddack W (1937) Der Kohlenstoff im Haushalt der Natur. Angew Chem 50:505–510

Noddack W (1975) Carbon in the household of Nature. In: Lieth HFH (ed) Patterns of primary production in the biosphere. Dowden, Hutchinson & Ross, Stroudsberg, Pennsylvania, p 293 (Benchmark Papers in Ecology, vol 8) (Translation of part of 1937 paper)

Nölke F (1909) Die Entstehung der Eiszeiten. Dtsch Geogr Blätter Bremen 32:1–30

North GR, Crowley TJ (1985) Application of a seasonal climate model to Cenozoic glaciation. J Geol Soc Lond 142:475–482

North GR, Mengel JG, Short DA (1983) Simple energy balance model resolving the seasons and the continents: application to the astronomical theory of ice ages. J Geophys Res 88:6576–6586

Oeschger H, Beer J (1990) The past 5000 years history of solar modulation of cosmic radiation from ^{10}Be and ^{14}C studies. Philos Trans R Soc Lond 330A:471–480

Officer CB, Hallam A, Drake CL, Devine JD (1987) Late Cretaceous and paroxysmal Cretaceous/Tertiary extinctions. Nature 326:143–149

Oliver MA, Webster R (1986) Combining nested and linear sampling for determining the scale and form of spatial variation of regionalized variables. Geogr Anal 18:227–242

Ollier CD (1979) Evolutionary geomorphology of Australia and New Guinea. Trans Inst Br Geogr New Ser 4:516–539

Ollier CD (1981) Tectonics and landforms. Longman, London (Geomorphology Texts, vol 6)

Olsen PE (1984) Periodicity of lake-level cycles in the Late Triassic Lockatong Formation of the Newark Basin (Newark Supergroup, New Jersey and Pennsylvania). In: Berger A, Imbrie J, Hays J, Kukla G, Saltzman B (eds) Milankovitch and climate: understanding the response of astronomical forcing, Part I. D Reidel, Dordrecht, pp 129–146 (Proceedings of the NATO Advanced Research Workshop on Milankovitch and Climate, Palisades, New York, 1982. NATO ASI Series C, Mathematical and Physical Sciences, vol 126)

Olsen PE (1986) A 40-million-year lake record of early Mesozoic orbital climatic forcing. Science 234:842–848

Olson JS (1963) Energy storage and the balance of producers and decomposers in ecological systems. Ecology 44:322–331

O'Neill RV (1988) Hierarchy theory and global change. In: Rosswall T, Woodmansee RG, Risser PG (eds) Scales and global change: spatial and temporal variability in biospheric and geospheric processes. John Wiley, Chichester, pp 29–45 (SCOPE 35)

O'Neill RV, De Angelis DL, Waide JB, Allen TFH (1986) A hierarchical concept of ecosystems. Princeton University Press, Princeton, New Jersey

O'Neill RV, Johnson AR, King AW (1989) A hierarchical framework for the analysis of scale. Landscape Ecology 3:193–205

Öpik EJ (1950) Secular changes of stellar structure and the ice ages. Armagh Observatory, Northern Ireland, Contribution no 5

Öpik EJ (1953) A climatological and astronomical interpretation of the ice ages and of past variations of terrestrial climate. Armagh Observatory, Northern Ireland, Contribution no 9

Öpik EJ (1958a) Climate and the changing Sun. Sci Am 198:85–92

Öpik EJ (1958b) Solar variability and palaeoclimatic changes. Ir Astronom J 5:97–109

Overpeck JT, Peterson LC, Kipp N, Imbrie J, Rind D (1989) Climate change in the circum-North Atlantic region during the last deglaciation. Nature 338:553–557

Ovid (edn 1955) Metamorphoses. Translated by Mary M. Innes. Penguin Books, Harmondsworth

Owen JG (1988) On productivity as a predictor of rodent and carnivore diversity. Ecology 69:1161–1165

Owen R (1846) A history of British fossil mammals and birds. John van Voorst, London

Owen-Smith N (1987) Pleistocene extinctions: the pivotal role of megaherbivores. Paleobiology 13:351–362

Paepe R, Mariolakos I, Mettos A, Sabot V, Thorez J, Livaditis G, Van Overloop E, Hatziotis ME, Hus J, Lin J-X, Vanhoorne R (1986) Paleoclimatic reconstruction in Belgium and in Greece based on Quaternary lithostratigraphic sequences. In: Ghazi A, Fantechi R (eds) Current issues in climate research. D Reidel, Dordrecht, pp 113–129 (Commission of the European Communities. Proceedings of the EC Climatology Programme Symposium, Sophia Antipolis, France, 2–5 October 1984)

Palais JM, Sigurdsson H (1989) Petrologic evidence of volatile emissions from major historic and pre-historic volcanic eruptions. In: Berger A, Dickinson RE, Kidson JW (eds) Understanding climate change. American Geophysical Union, Washington DC, Geophysical Monograph 52, pp 31–53 (International Union of Geodesy and Geophysics, vol 7)

Palmer JA, Phillips GN, McCarthy TS (1989) Paleosols and their relevance to Precambrian atmospheric composition. J Geol 97:77–92

Paluska A, Degens ET (1979) Climatic and tectonic events controlling the Quaternary in the Black Sea region. Geol Rundshau 68:284–301

Pang KD, Chou H-H (1985) Three very large volcanic eruptions in antiquity and their effects on the climate of the ancient world. Eos 67:880–881 (Abstract)

Pantič N, Stefanovič (1984) Complex interaction of cosmic and geological events that affect the variation of Earth climate through the geological history. In: Berger A, Imbrie J, Hays J, Kukla G, Saltzman B (eds) Milankovitch and climate: understanding the response of astronomical forcing, Part 1. D Reidel, Dordrecht, pp 251–26 (Proceedings of the NATO Advanced Research Workshop on Milankovitch and Climate, Palisades, New York, 1982. NATO ASI Series. Series C, Mathematical and Physical Sciences, vol 126)

Paracelsus PA (1951) Paracelsus: selected writings. Edited with an introduction by Jolande Jacobi. Translated by Norbert Guterman. Princeton University Press, Princeton, New Jersey (Bollingen Series, vol 23)

Passarge S (1926) Morphologie der Klimazonen oder Morphologie der Landschaftsgürtel? Petermanns Geogr Mitt 72:173–175

Passarge S (1973) Morphology of climatic zones or morphology of landscape belts? In: Derbyshire E (ed) Climatic geomorphology. Geographical Readings Series. Methuen, London, pp 91–95 (Translation of 1926 paper by Roger S Mays and Edward Derbyshire)

Paterson SS (1956) The forest area of the world and its potential productivity. Ph D Dissertation, The Royal University of Göteborg, Sweden

Paterson SS (1975) The forest area of the world and its potential productivity. In: Lieth HFH (ed) Patterns of primary production in the biosphere. Dowden, Hutchinson & Ross, Stroudsberg, Pennsylvania, pp 240–247 (Benchmark Papers in Ecology Volume 8) (Partial reprint of 1956 thesis)

Pearson R (1978) Climate and evolution. Academic Press, London

Pecker J-C, Runcorn SK (eds) (1990) The Earth's climate and variability of the Sun over recent millenia: geophysical, astronomical and archaeological aspects. The Royal Society, London (Proceedings of a Royal Society and Académie des Sciences Discussion Meeting held on 15 and 16 February 1989) (First published in Philos Trans R Soc Lond 330A (no 1615):395–687)

Pedro G (1966) Essai sur la caractérisation des différents processus zonaux résultant de l'altération des roches superficielles (cycle aluminosilicique). C R Acad Sci Paris 262D:1828–1831

Pedro G (1968) Distribution des principaux types d'altération chimique à la surface du globe: présentation d'une esquisse géographique. Rev Géogr Phys Géol Dynam 10:457–470

Pedro G (1979) Caractérisation générale des processus de l'altération hydrolitique. Sci Sol 2:93–105

Pedro G (1983) Structuring of some basic pedological processes. Geoderma 31:289–299

Peltier LC (1950) The geographic cycle in periglacial regions as it is related to climatic geomorphology. Ann Assoc Am Geogr 40:214–236

Peltier LC (1975) The concept of climatic geomorphology. In: Melhorn WN, Flemal RC (eds) Theories of landform development. George Allen and Unwin, London, pp 129–143

Penck A (1905) Climatic features in the land surface. Am J Sci 4th Ser 19:165–174

Penck A (1910) Versuch einer Klimaklassifikation auf physiographischer Grundlage. Sitzungsber Preuss Akad Wiss Berlin phys math Klasse 1:36–246

Penck A (1914a) Die Formen der Landoberfläche und Verschiebungen der Klimagürtel. Sitzungsber Preuss Akad Wiss Berlin phys math Klasse 4:77–97

Penck A (1914b) The shifting of the climatic belts. Scott Geogr Mag 30:281–293

Penck A (1973) Attempt at a classification of climate on a physiographic basis. In: Derbyshire E (ed) Climatic geomorphology. Geographical Readings Series. Methuen, London, pp 51–60 (Translation of 1910 paper Roger S. Mays and Edward Derbyshire)

Perrault P (1678) De l'origine des fontaines. Chez Pierre le Petit, Paris

Pestiaux P, Duplessy JC, Berger A (1987) Paleoclimatic variability at frequencies ranging from 10^{-4} cycles per year to 10^{-3} cycles per year — evidence for nonlinear behavior of the climate system. In: Rampino MR, Sanders JE, Newman WS, Königsson LK (eds) Climate: history, periodicity, and predictability. Van Nostrand Reinhold, New York, pp 285–299

Petersen GH (1984) Energy-flow budgets in aquatic ecosystems and the conflict between biology and geophysics about Earth-tilt axis. In: Mörner N-A, Karlén W (eds) Climatic changes on a yearly to millenial basis: geological, historical and instrumental records. D Reidel, Dordrecht, pp 621–633

Petit JR, Mournier L, Jouzel J, Korotkevich YS, Kotlyakov VI, Lorius C (1990) Palaeoclimatological and chronological implications of the Vostok core dust record. Nature 343:56–58

Phillips J (1831) On some effects of the atmosphere in wasting the surfaces of buildings and rocks. Proc Geol Soc Lond 1:323–324

Phillips JD (1986a) Sediment storage, sediment yield, and time scales in landscape denudation studies. Geogr Anal 18:161–167

Phillips JD (1986b) Spatial analysis of shoreline erosion, Delaware Bay, New Jersey. Ann Assoc Am Geogr 76:50–62

Phillips JD (1987) Choosing the level of detail for depicting two-variable spatial relationships. Math Geol 19:539–547

Phillips JD (1988a) The role of spatial scale in geomorphic systems. Geogr Anal 20:308–317

Phillips JD (1988b) Nonpoint source pollution and spatial aspects of risk assessment. Ann Assoc Am Geogr 78:611–623

Phillips JD (1989a) An evaluation of the state factor model of soil ecosystems. Ecol Model 45:165–177

Phillips JD (1989b) Erosion and planform irregularity of an estuarine shoreline. Z Geomorphol Suppl 73:59–71

Phillips JD (1990) Relative importance of factors influencing fluvial loss at the global scale. Am J Sci 290:547–568

Pisias NG, Imbrie J (1986) Orbital geometry, CO_2, and Pleistocene climate. Oceanus 29:43–49

Pisias NG, Shackleton NJ (1984) Modelling the global climate response to orbital forcing and atmospheric carbon dioxide changes. Nature 30:757–759

Pitty AF (1982) The nature of geomorphology. Methuen, London

Plass GN (1956) The carbon dioxide theory of climatic change. Tellus 8:140–153

Pokras EM, Mix AC (1985) Eolian evidence for spatial variability of late Quaternary climates in tropical Africa. Quat Res 24:137–149

Pollack JB, Toon OB, Ackermann TP, McKay TP, Turco RP (1983) Environmental effects of an impact-generated dust cloud: implications for the Cretaceous-Tertiary extinctions. Science 219:287–289

Polunin N, Grinevald J (1988) Vernadsky and biospheral ecology. Environ Conserv 15:117–122

Polynov BB (1937) The cycle of weathering. Translated from the Russian by Alexander Muir. With a foreword by WG Ogg. Thomas Murby, London

Porter SC (1981) Recent glacier variations and volcanic eruptions. Nature 291:139–142

Potter PE (1978) Petrology and chemistry of modern big-river sands. J Geol 86:423–449

Powell CM, Veevers JJ (1987) Namurian uplift in Australia and South America triggered by the main Gondwanan glaciation. Nature 326:177–179

Prell WL (1984) Monsoonal climate of the Arabian Sea during the late Quaternary: a response to changing solar radiation. In: Berger A, Imbrie J, Hays J, Kukla G, Saltzman B (eds) Milankovitch and climate: understanding the response of astronomical forcing, Part 1. D Reidel, Dordrecht, pp 349–366 (Proceedings of the NATO Advanced Research Workshop on Milankovitch and Climate, Palisades, New York, 1982. NATO ASI Series C, Mathematical and Physical Sciences, vol 126)

Prell WL, Kutzbach JE (1987) Monsoon variability over the past 150,000 years. J Geophys Res 92D:8411–8425

Prescott JA (1934) Single value climatic factors. Trans R Soc S Aust 58:48–61

Prinn RG, Fegley B Jr (1987) Bolide impacts, acid rain, and biospheric traumas at the Cretaceous-Tertiary boundary. Earth Planet Sci Lett 83:1–15

Purdue JR (1980) Clinal variations of some mammals during the Holocene in Missouri. Quat Res 13:242–258

Purdue JR (1986) The size of the white-tailed deer (*Odocoileus virginianus*) during the archaic period in central Illinois. In: Neusius SW (ed) Foraging, collecting, and harvesting: archiac period subsistence and settlement in the eastern woodlands. Southern Illinois University at Carbondale, Center for Archaeological Investigations Occasional Paper No 6, pp 65–95

Purdue Jr (1989) Changes during the Holocene in the size of white-tailed deer (*Odocoileus virginianus*) from central Illinois. Quat Res 32:307–316

Quade J, Cerling TE, Bowman JR (1989a) Systematic variations in the carbon and oxygen isotopic composition of pedogenic carbonate along elevation transects in the southern Great Basin, United States. Bull Geol Soc Am 101:464–475

Quade J, Cerling TE, Bowman JR (1989b) Development of Asian monsoon revealed by marked ecological shift during the latest Miocene in northern Pakistan. Nature 342:163–166

Rabinovitch E (1971) An unfolding discovery. Proc Natl Acad Sci USA 68:2875–2876

Raffi S, Stanley SM, Marasti R (1985) Biogeographic patterns and Plio-Pleistocene extinction of Bivalvia in the Mediterranean and southern North Sea. Paleobiology 11:368–388

Raisbeck GM, Yiou F, Jouzel J, Petit JR (1990) [10]Be and δ^{2}H in polar ice cores as a probe of the solar variability's influence on climate. Philos Trans R Soc Lond 330A:463–470

Ramann E (1911) Bodenkunde. Springer, Berlin

Ramann E (1918) Bodenbildung und Bodeneinteilung (System der Böden). Springer, Berlin

Ramann E (1928) The evolution and classification of soils. Translated by CL Whittles. W Heffer, Cambridge

Rambler MB, Margulis L, Fester R (eds) (1989) Global ecology: towards a science of the biosphere. Academic Press, London

Rampino MR (1987) Impact cratering and flood basalt volcanism. Nature 327:468

Rampino MR, Stothers RB (1986) Periodic flood-basalt eruptions, mass extinctions, and comet impacts. Eos 67:1247 (Abstract)

Rampino MR, Volk T (1988) Mass extinctions, atmospheric sulphur and climatic warming at the K/T boundary. Nature 332:63–65

Rampino MR, Self S, Fairbridge RW (1979) Can rapid climatic change cause volcanic eruptions? Science 206:826–829

Rampino MR, Stothers RB, Self S (1985) Climatic effects of volcanic eruptions. Nature 313:272

Rampino MR, Self S, Stothers RB (1988) Volcanic winters. Annu Rev Earth Planet Sci 16:73–99

Ramsay W (1924) The probable solution of the climate problem in geology. Geol Mag 61:152–163

Rawson HE (1907) Anticyclones as aids to long distance forecasts. Q J R Meteorol Soc 33:309–310

Rawson HE (1908) The anticyclonic belt of the Southern Hemisphere. Q J R Meteorol Soc 34:165–188

Rawson HE (1909) The anticyclonic belt of the Northern Hemisphere. Q J R Meteorol Soc 35:233–248

Ray C (1960) The application of Bergmann's and Allen's rules to poikilotherms. J Morphol 106:85–108

Read J, Hill RS (1988) Comparative responses to temperature of the major canopy species of Tasmanian cool temperate rainforest and their ecological significance. I. Foliar frost resistance. Aust J Bot 36:131–143

Read J, Hill RS (1989) The response of some Australian temperate rain forest tree species to freezing temperatures and its biogeographical significance. J Biogeogr 16:21–27

Read J, Hope GS (1989) Foliar frost resistance of some evergreen tropical and extratropical Australasian *Nothofagus* species. Aust J Bot 37:361–373

Reader JR (1973) Phenological investigation in Eastern North America. Ph D Thesis, University of North Carolina, Chapel Hill, North Carolina

Rehbock PF (1983) The philosophical naturalists: themes in early nineteenth-century British biology. The University of Wisconsin Press, Madison, Wisconsin (Wisconsin Publications in the History of Science and Medicine, no 3)

Reinhardt J, Sigleo WR (eds) (1988) Paleosols and weathering through geologic time: principles an applications. Geol Soc Am Spec Pap 206

Renoir (1839) Note sur les glaciers qui ont recouvert anciennement la partie méridionale de la chaîne des Vosges. Bull Soc Géol Fr 11:53–66

Renoir (1840) Sur la cause probable de l'ancienne existence des glaces générales. Bull Soc Géol Fr 11:148–155

Rensch B (1932) Über die Abhängigkeit der Grösse, des relativen Gewichtes und der Oberflächenstruktur der Landschneckenschalen von den Umweltsfaktoren. Z Morphol Ökol 25:757–807

Rensch B (1937–38) Some problems of geographical variations and species formation. Proc Linn Soc Lond 150th session:275–285

Retallack GJ (1986) Reappraisal of a 2200-Ma-old paleosol near Waterval Onder, South Africa. Precambrian Res 32:195–232

Retallack GJ (1990) Soils of the past: an introduction to paleopedology. Unwin Hyman, Boston

Rhodes RS II (1984) Paleoecological and regional paleoclimatic implications of the Farmdalian Craigmile and Woodfordian Waubonsie mammalian local faunas, southwestern Iowa. Ill State Mus Rep Invest 40:1–51

Richardson LF (1922) Weather prediction by numerical process. Cambridge University Press, Cambridge

Richthofen FPW, Freiherr von (1886) Führer für Forschungsreisende. Anleitung zu Beobachtungen über Gegendstände der physischen Geographie und Geologie. R Oppenheim, Berlin

Rind D (1986) The dynamics of warm and cold climates. J Atmos Sci 43:3–24

Rind D, Peteet D, Kukla G (1989) Can Milankovitch orbital variations initiate the growth of ice sheets in a general circulation model? J Geophys Res 94D:12,851–12,871

Risser PG, Rosswall T, Woodmansee RG (1988) Spatial and temporal variability of biospheric and geospheric processes: a summary. In: Rosswall T, Woodmansee RG, Risser PG (eds) Scales and global change: spatial and temporal variability in biospheric and geospheric processes. John Wiley, Chichester, pp 1–10 (SCOPE 35)

Roberts DF (1973) Climate and human variability. Benjamin-Cummings, Menlo Park, California

Roberts N (1990) Ups and downs of African lakes. Nature 346:107

Rodin LE, Basilevich NI (1968) Production and mineral cycling in terrestrial vegetation. English translation by GE Fogg. Oliver & Boyd, Edinburgh

Rodin LE, Bazilevich NI, Rozov NN (1975) Productivity of the world's main ecosystems. In: Reichle DE, Franklin JF, Goodall DW (eds) Productivity of world ecosystems. National Academy of Science, Washington DC, pp 13–26

Rognon P (1987) Late Quaternary climatic reconstruction for the Maghreb (North Africa). Palaeogeogr Palaeoclimatol Palaeoecol 58:11–34

Rooth CGH (1990) Meltwater Younger Dryas upheld. Nature 343:702

Rosenzweig ML (1968a) The strategy of body size in mammalian carnivores. Am Midl Nat 80:299–315

Rosenzweig ML (1968b) Net primary productivity of terrestrial communities: prediction from climatological data. Am Nat 102:67–74

Rossignol-Strick M (1983) African monsoons, an immediate climate response to orbital insolation. Nature 303:46–49

Rossignol-Strick M, Planchais N (1989) Climate patterns revealed by pollen and oxygen isotope records of a Tyrrhenian sea core. Nature 342:413–416

Röthlisberger F (1986) 10,000 Jahre Gletschergeschichte der Erde. Verlag Sauerländer, Arau

Ruddiman WF (1971) Pleistocene sedimentation in the equatorial Atlantic: stratigraphy and faunal climatology. Bull Geol Soc Am 82:283–302

Ruddiman WF, Kutzbach JE (1989) Forcing of late Cenozoic Northern Hemisphere climate by plateau uplift in southern Asia and the American west. J Geophys Res 94D:18,409–18,427

Ruddiman WF, Raymo ME (1988) Northern Hemisphere climatic regimes during the last 3 Ma: possible tectonic connections. Philos Trans R Soc Lond 318B:411–430

Ruddiman WF, Prell WL, Raymo ME (1989) Late Cenozoic uplift in southern Asia and the American west: rationale for general circulation modeling experiments. J Geophys Res 94D:18,379–18,391

Ruedemann R (1939) Climates of the past in North America. In: Ruedemann R, Balk R (eds) Geology of North America, vol 1: Introductory chapters, and geology of the stable areas. Gebrüder Borntraeger, Berlin, pp 88–99 (Geologie der Erde)

Ruhe RV (1975) Climatic geomorphology and fully developed slopes. Catena 2:309–320

Ruhe RV (1977) Mr. Cox's comments on the paper by Ruhe. Catena 4:232

Ruhe RV (1983) Aspects of Holocene pedology in the United States. In: Wright HE Jr (ed) Late-Quaternary environments of the United States, vol 2, The Holocene. Longman, London, pp 12–25

Ruhe RV (1984) Loess-derived soils, Mississippi Valley region: II. Soil-climate system. Soil Sci Soc Am J 48:864–867

Runcorn SK, Suess HE (1982) Investigating solar activity. Science 218:842

Russell DA (1979) The enigma of the extinction of the dinosaurs. Annu Rev Earth Planet Sci 7:163–182

Sabadini R, Yuen DA (1989) Mantle stratification and long-term polar wander. Nature 339:373–375

Sabadini R, Yuen DA, Boschi E (1982) Interaction of cryospheric forcings with rotational dynamics has consequences for ice ages. Nature 296:338–341

Sabadini R, Yuen DA, Boschi E (1983) Dynamic effects from mantle phase transitions on true polar wander during ice ages. Nature 303:694–696

Saltzman B (1987a) Carbon dioxide and the $\delta^{18}O$ record of late-Quaternary climatic change: a global model. Clim Dynam 1:77–85

Saltzman B (1987b) Modeling the $\delta^{18}O$-derived record of Quaternary climatic change with low order dynamical systems. In: Nicolis C, Nicolis G (eds) Irreversible phenomena and dynamical systems analysis in geosciences. D Reidel, Dordrecht, pp 355–380

Saltzman B (1988) Modelling the slow climatic attractor. In: Schlesinger ME (ed) Physically-based modelling and simulation of climate and climatic change. Part II. Kluwer Academic Publishers, Dordrecht, pp 737–754

Saltzman B, Maasch KA (1988) Carbon cycle instability as a cause of the late Pleistocene ice age oscillations: modeling the asymmetric response. Glob Biogeochem Cycles 2:177–185

Saltzman B, Maasch KA (1990) A first-order global model of late Cenozoic climatic change. Trans R Soc Edinb Earth Sci 81:315–325

Saltzman B, Sutera A (1984) A model of the internal feedback system involved in late Quaternary climatic variations. J Atmos Sci 41:736–745

Saltzman B, Sutera A (1987) The Mid-Quaternary climatic transition as a free response of a three-variable dynamical model. J Atmos Sci 44:236–241

Saltzman B, Hansen AR, Maasch KA (1984) The late Quaternary glaciations as the response of a three-component feedback system to earth-orbital forcing. J Atmos Sci 41:3380–3389

Sanders JE, Fairbridge RW (1987) Selected bibliography on Sun-Earth relationships and cycles having periods of less than 10,000 years. In: Rampino MR, Sanders JE, Newman WS, Königsson LK (eds) Climate: history, periodicity, and predictability. Van Nostrand Reinhold, New York, pp 475–541

Sapper K (1935) Geomorphologie der feuchten Tropen. BG Teubner, Leipzig (Geographische Schriften, herausgegeben von A Hettner, vol 7)

Sartorius von Waltershausen W (1865) Untersuchungen über die Klimate der Gegenwart und Vorwelt, mit besonderer Berücksichtigung der Gletscher-Erscheinungen in der Diluvialzeit. De Erven Loosjes, Haarlem

Sato N, Sellers PJ, Randall DA, Schneider EK, Shuckla J, Kinter JL III, Hou Y-T, Albertazzi E (1989) Effects of implementing the simple biosphere model in a general circulation model. J Atmos Sci 46:2757–2782

Sauer CO (1925) The morphology of landscapes. Univ Calif Publ Geogr 2:19–53

Saull WD (1848) An elucidation of the successive changes of temperature and the levels of ocean waters upon the Earth's surface, in harmony with geological evidences. Q J Geol Soc Lond 4:7

Saunders I, Young A (1983) Rates of surface processes on slopes, slope retreat and denudation. Earth Surface Processes and Landforms 8:473–501

Schaffer WM (1981) Ecological abstraction: the consequences of reduced dimensionality in ecological models. Ecol Monogr 51:383–401

Scheidegger AE (1990) Theoretical geomorphology, 3rd edn. Springer, Berlin Heidelberg New York Tokyo

Schidlowski M (1987) Evolution of the early sulphur cycle. In: Rodriguez-Clemente R, Tardy Y (eds) Proceedings of the International Meeting "Geochemistry of the Earth Surface and Processes of Mineral Formation" held in Granada (Spain) 16–22 March 1986. Consejo Superior de Investigaciones Cientificas, Madrid, pp 29–49

Schidlowski M (1988) A 3,800-million-year isotopic record of life from carbon in sedimentary rocks. Nature 333:313–318

Schimper AFW (1898) Pflanzengeographie auf physiologischer Grundlage. Gustav Fischer, Jena

Schimper AFW (1903) Plant geography upon a physiological basis. Revised and edited by Percy Groom and Isaac Bayley Balfour. Clarendon, Oxford

Schneider SH, Londer R (1984) The coevolution of climate and life. Sierra Club Books, San Francisco, California

Schneider SH, Mass C (1975) Volcanic dust, sunspots, and temperature trends. Science 190:741–746

Schneider SH, Thompson SL (1988) Simulating the climatic effects of nuclear war. Nature 333:221–227

Schneider SH, Thompson SL, Barron EJ (1985) Mid-Cretaceous continental surface temperatures: are high CO_2 concentrations needed to simulate above-freezing winter conditions? In: Sundquist ET, Broecker WS (eds) The carbon cycle and atmospheric CO_2: natural variations, Archean to present. American Geophysical Union, Washington DC, Geophysical Monograph 32, pp 554–559

Schove DJ (1954) Summer temperatures and tree-rings in north Scandinavia. Geogr Ann 36:40–80

Schove DJ (1955) The sunspot cycle 649 BC-AD 2000. J Br Astronom Assoc 66:59–61

Schove DJ (1961) Tree rings and climatic chronology. Ann N Y Acad Sci 95:605–622

Schove DJ (1983) Sunspot cycles. Van Nostrand Reinhold, New York (Benchmark Papers in Geology Series, vol 68)

Schove DJ (1987) Sunspot cycles and weather history. In: Rampino MR, Sanders JE, Newman WS, Königsson LK (eds) Climate: history, periodicity, and predictability. Van Nostrand Reinhold, New York, pp 355–377

Schroeder H (1919) Die jährliche Gesamtproduktion der grünen Pflanzendecke der Erde. Die Naturwissenschaften 7:8–12, 23–29

Schroeder H (1975) The annual primary production of the green vegetation cover of the Earth. In: Lieth HFH (ed) Patterns of primary production in the biosphere. Dowden, Hutchinson & Ross, Stroudsberg, Pennsylvania, pp 291–292 (Benchmark Papers in Ecology, vol 8) (Translation of part of 1919 paper)

Schumm SA (1977) The fluvial system. John Wiley, New York

Schumm SA (1988) Variability of the fluvial system in space and time. In: Rosswall T, Woodmansee RG, Risser PG (eds) Scales and global change: spatial and temporal variability in biospheric and geospheric processes. John Wiley, Chichester, pp 225–250 (SCOPE 35)

Schumm SA, Lichty RW (1965) Time, space, and causality in geomorphology. Am J Sci 263:110–119

Schwabe H (1838) Über die Flecken der Sonne. Astronomische Nachrichten 15:243–248

Schwabe H (1843) Die Sonne. Astronomische Nachrichten 20:213–286

Schwabe H (1844) Solar observations during 1843. Astronomische Nachrichten 21:233–236

Schwartzacher W (1975) Sedimentation models and quantitative stratigraphy. Elsevier, Amsterdam (Developments in Sedimentology, vol 19)

Schwartzacher W, Fischer AG (1982) Limestone-shale bedding and perturbations of the Earth's orbit. In: Einsele G, Seilacher A (eds) Cyclic and event stratification. Springer, Berlin Heidelberg New York, pp 72–95

Schwartzbach M (1963) Climates of the past: an introduction to paleoclimatology. Translated and edited by Richard O Muir. D van Nostrand, London (The University Series in Geology, edited by Rhodes W Fairbridge)

Sellers A, Meadows AJ (1975) Long-term variations in the albedo and surface temperature of the Earth. Nature 254:44

Sellers PJ, Dorman JL (1987) Testing the simple biosphere model (SiB) using point micrometeorological and biophysical data. J Clim Appl Meteorol 26:622–651

Sellers PJ, Mintz Y, Sud YC, Dalcher A (1986) A simple biosphere model (SiB) for use with general circulation models. J Atmos Sci 43:505–531

Sellers PJ, Shuttleworth WJ, Dorman JL, Dalcher A, Roberts JM (1989) Calibrating the simple biosphere model for Amazonian tropical forest using field and remote sensing data. Part I: average calibration with field data. J Appl Meteorol 28:727–759

Semken HA (1984) Holocene mammalian biogeography and climatic change in eastern and central United States. In: Wright HE Jr (ed) Late-Quaternary environments in the United States, vol 2: The Holocene. Longman, London, pp 182–207

Semper KG (1881) Animal life as affected by natural conditions of existence. D Appleton, New York

Shackleton NJ, Opdyke ND (1973) Oxygen isotope analysis and paleomagnetic stratigraphy of equatorial Pacific core, V 28–238: oxygen isotope temperature and ice volumes on a 10^5-year and 10^6-year scale. Quat Res 3:39–55

Shantz HL (1923) The natural vegetation of the Great Plains region. Ann Assoc Am Geogr 8:81–107

Shapley H (1921) Note on a possible factor in changes of geological climate. J Geol 29:502–504

Shapley H (1949) Galactic rotation and cosmic seasons. Sky and Telescope 9:36–37

Sharma HS (1987) Climate and drainage basin morphometric properties — a case study of Rajasthan. In: Gardiner V (ed) International geomorphology 1986, Part II. John Wiley, Chichester, pp 69–87 (Proceedings of the First International Conference on Geomorphology)

Shaw CF (1930) Potent factors in soil formation. Ecology 11:239–245

Shaw HR (1987) The periodic structure of the natural record, and nonlinear dynamics. Eos 1651–1663,1655

Shaw HR, Moore JG (1988) Magmatic heat and the El Niño cycle. Eos 69:1552,1564–1565

Shelford VE (1911) Physiological animal geography. J Morphol 22:551–618

Shelford VE (1913) Animal communities in temperate America. Williams & Wilkins, Baltimore

Shelford VE (1932) Life zones, modern ecology, and the failure of temperature summing. Wilson Bull 44:144–157

Shelford VE (1945) The relative merits of the life zone and biome concepts. Wilson Bull 57:248–252

Sherman GD (1952) The genesis and morphology of alumina-rich laterite clays. In: Problems in clay and laterite genesis. American Institute of Mineral and Metallurgical Engineering, New York, pp 154–161 (St. Louis Symposium)

Simpson GC (1929) Past climates. Mem Manchester Liter Philos Soc 74:1–34

Simpson GC (1934) World climate during the Quaternary period. Q J R Meteorol Soc 60:425–478

Simpson GC (1959) World temperatures during the Pleistocene. Q J R Meteorol Soc 85:332–349

Simpson GG (1953) The major features of evolution. Columbia University Press, New York

Singer A (1980) The paleoclimatic interpretation of clay minerals in soils and weathering profiles. Earth-Sci Rev 15:303–326

Sleep NH, Zahnle KJ, Kasting JF, Morowitz HJ (1989) Annihilation of ecosystems by large asteroid impacts on the early Earth. Nature 342:139–142

Slingerland R (1981) Qualitative stability analysis of geologic systems, with an example from river hydraulic geometry. Geology 9:491–493

Smith EA, Voner Haar TH, Hickey JR, Maschhoff R (1983) The nature of the short period fluctuations in solar irradiance received by the Earth. Clim Change 5:211–235

Soil Survey Staff (1975) Soil taxonomy: a basic system of soil classification for making and interpreting soil surveys. US Department of Agriculture Handbook 436. US Government Printing Office, Washington DC

Sonett CP, Finney SA (1990) The spectrum of radiocarbon. Philos Trans R Soc Lond 330A:413–426

Sonett CP, Williams GE (1985) Solar periodicities expressed in varves from glacial Skilak Lake, southern Alaska. J Geophys Res 90:12,019–12,026

Sonett CP, Williams GE (1987) Frequency modulation and stochastic variability of the Elatina varve record: a proxy for solar activity? Sol Phys 110:397–410

Sonett CP, Finney SA, Williams CR (1988) The lunar orbit in the late Precambrian and the Elatina sandstone laminae. Nature 335:806–808

Southward AJ, Butler EI, Pennycuick L (1975) Recent cyclic changes of climate and in abundance of marine life. Nature 253:714–717

Spicer RA (1987) The significance of the Cretaceous flora of northern Alaska for the reconstruction of the climate of the Cretaceous. Geol Jahrb 96A:265–291

Spicer RA (1989) Physiological characteristics of land plants in relation to environment through time. Trans R Soc Edinb Earth Sci 80:321–329

Spicer RA, Parrish JT (1986) Paleobotanical evidence for cool North Polar climates in middle Cretaceous (Albian-Cenomanian) time. Geology 14:703–706

Spicer RA, Parrish JT (1990a) Late Cretaceous woods of the central North Slope, Alaska. Palaeontology 33:225–242

Spicer RA, Parrish JT (1990b) Late Cretaceous-early Tertiary palaeoclimates of northern high latitudes: a quantitative review. J Geol Soc Lond 147:329–341

Sprengel K (1844) Die Bodenkunde oder Lehre vom Boden nebst einer vollständigen Anleitung zur chemischen Analyse der Ackererde und den Resultaten von 180 chemisch untersuchten Bodenarten aus Deutschland, Belgien, England, Frankreich, der Schweiz, Ungarn, Russland, Schweden, Ostindien, Westindien und Nordamerika. Ein Handbuch für Landwirthe, Forstmänner, Gärtner, Boniteure und Theilungscommissäre, 2 vols. I Müller, Leipzig

Stallard RF, Edmond JM (1983) Geochemistry of the Amazon 2: the influence of geology and weathering environment on the dissolved load. J Geophys Res 88:9671–9688

Stanley DJ, Sheng H (1986) Volcanic shards from Santorini (Upper Minoan ash) in the Nile delta, Egypt. Nature 320:733–735

Stanley SM (1984) Temperature and biotic crises in the marine realm. Geology 12:205–208

Stanley SM (1986) Anatomy of a regional mass extinction: Plio-Pleistocene decimation of the western Atlantic bivalve fauna. Palaios 1:17–36

Stanley SM (1988a) Paleozoic mass extinctions: shared patterns suggest global cooling as a common cause. Am J Sci 288:334–352

Stanley SM (1988b) Climatic cooling and mass extinction of Paleozoic reef communities. Palaios 3:228–232

Starkel L (1987) Long-term and short-term rhythmicity in terrestrial landforms and deposits. In: Rampino MR, Sanders JE, Newman WS, Königsson LK (eds) Climate: history, periodicity, and predictability. Van Nostrand Reinhold, New York, pp 323–332

Steiner J (1967) The sequence of geological events and the dynamics of the Milky Way Galaxy. J Geol Soc Aust 14:99–131

Steiner J (1973) Possible galactic causes for synchronous sedimentation sequences of the North American and eastern European cratons. Geology 1:89–92

Steiner J (1978) Lead isotope events of the Canadian Shield, ad hoc solar galactic orbits and glaciations. Precambrian Res 6:269–274

Steiner J (1979) Regularities of the revised Phanerozoic time scale and the Precambrian time scale. Geol Rundschau 68:825–831

Steiner J, Grillmair E (1973) Possible galactic causes for periodic and episodic glaciations. Bull Geol Soc Am 84:1003–1018

Stetson HT (1937) Sunspots and their effects. McGraw-Hill, New York

Stoddart DR (1969a) Climatic geomorphology: review and re-assessment. Prog Phys Geogr 1:160–222

Stoddart DR (1969b) Climatic geomorphology. In: Chorley RJ (ed) Water, Earth, and Man: a synthesis of hydrology, geomorphology, and socio-economic geography. Methuen, London, pp 473–485

Stothers RB, Wolff JA, Self S, Rampino MR (1986) Basaltic fissure eruptions, plume heights, and atmospheric aerosols. Geophys Res Lett 13:725–728

Strakhov NM (1967) Principles of lithogenesis, vol 1. Translated by JP Fitzsimmons, SI Tomkieff, JE Hemingway. Oliver & Boyd, Edinburgh

Street-Perrott FA, Harrison SP (1984) Temporal variations in lake levels since 30,000 yr B.P. — an index of the global hydrological cycle. In: Hansen JE, Takahashi T (eds) Climate processes and climate sensitivity. American Geophysical Union, Washington DC, Geophysical Monograph 29, pp 118–129 (Maurice Ewing Volume 5)

Street-Perrott FA, Perrott RA (1990) Abrupt climate fluctuations in the tropics: the influence of Atlantic Ocean circulation. Nature 343:607–612

Street-Perrott FA, Roberts N (1983) Fluctuations in closed-basin lakes as an indicator of past atmospheric circulation pattern. In: Street-Perrott FA, Beran M, Ratcliffe RAS (eds) Variations in the global water budget. D Reidel, Dordrecht, pp 331–345

Stuiver M (1983) The AD record of climatic and carbon isotope change. Radiocarbon 25:221

Sturm M, Benson C, MacKeith P (1986) Effects of 1966–68 eruptions of Mount Redoubt on the flow of Drift Glacier, Alaska, USA. J Glaciol 32:355–362

Suess E (1875) Die Entstehung der Alpen. W Braunmüller, Vienna

Suess HE, Linick TW (1990) The ^{14}C record in bristlecone pine wood of the past 8000 years based on the dendrochronology of the late CW Ferguson. Philos Trans R Soc Lond 330A:403–412

Summerfield MA (1983) Silcrete as a palaeoclimatic indicator: evidence from southern Africa. Palaeogeogr Palaeoclimatol Palaeoecol 41:65–79

Summerfield MA, Thomas MF (1987) Long-term landform development: editorial comment. In: Gardiner V (ed) International geomorphology 1986, Part II. John Wiley, Chichester, pp 927–933 (Proceedings of the First International Conference on Geomorphology)

Swindale LD, Jackson ML (1956) Genetic processes in some residual podzolized soils of New Zealand. Trans Sixth Int Congr Soil Sci Paris 5:233–239

Symons G (ed) (1888) The eruption of Krakatoa and subsequent phenomena: report of the Krakatoa Committee. The Royal Society, London

Tamrazyan GP (1967) The global historical and geological regularities of the Earth's development as a reflection of its cosmic origin (as a sequence of interaction in the course of galactic movement of the solar system). Trans Inst Mining Metall Ostrava 13:5–24 (in Russian)

Tanner WF (1961) An alternate approach to morphogenetic climates. Southeast Geol 2:251–257

Tansley AG (1935) The use and abuse of vegetational concepts and terms. Ecology 16:284–307

Tarling DH (1978) The geological-geophysical framework of ice ages. In: Gribbin J (ed) Climatic change. Cambridge University Press, Cambridge, pp 3–24

Tarnocai C, Valentine KWG (1989) Relict soil properties of the Arctic and Subarctic regions of Canada. In: Bronger A, Catt JA (eds) Paleopedology — nature and applications of paleosols. Catena Supplement 16, Catena Verlag, Cremlingen, pp 9–39

Terjung WH (1976) Climatology for geographers. Ann Assoc Am Geogr 66:199–222

Theophrastus (1916) Enquiry into plants and minor works on odours and weather signs. Translated by Sir Arthur Hort, 2 vols. William Heinemann, London

Thienemann A (1926) Limnologie: eine Einführung in die biologischen Probleme der Süsswasserforschung. Ferdinand Hirt, Breslau

Thienemann A (1939) Grundzüge einer allgemeinen Oekologie. Arch Hydrobiol 35:267–285

Thomas MF (1978) Denudation in the tropics and the interpretation of the tropical legacy in higher latitudes — a view of the British experience. In: Embleton C (ed) Geomorphology: present problems, future prospects. Oxford University Press, Oxford, pp 185–202

Thomas MF, Thorp MB (1985) Environmental change and episodic etchplanation in the humid tropics of Sierra Leone: the Koidu etchplain. In: Douglas I, Spencer T (eds) Environmental change and tropical geomorphology. George Allen & Unwin, London, pp 239–267

Thompson WF (1990) Climate-related landscapes in world mountains: criteria and maps. Z Geomorphol Suppl 78

Thomson DJ (1990) Time series analysis of Holocene climate data. Philos Trans R Soc Lond 330A:601–616

Thomson W (Lord Kelvin) (1862) On the secular cooling of the Earth. Trans R Soc Edinb 23:157–170

Thorbecke F (1927a) Der Formenschatz im periodisch trockenen Tropenklima mit überwiegender Regenzeit. In: Thorbecke F (ed) Düsseldorfer Vorträge und Erörterungen, vol 3. Ferdinand Hirt, Breslau, pp 10–17

Thorbecke F (ed) (1927b) Morphologie der Klimazonen. Düsseldorfer Vorträge und Erörterungen, vol 3. Ferdinand Hirt, Breslau

Thorbecke F (1973) Landforms of the savanna zone with a short dry season. In: Derbyshire E (ed) Climatic geomorphology. Geographical Readings Series. Methuen, London, pp 96–103 (Translation of 1927 paper by Roger S Mays and Edward Derbyshire)

Thornbury WD (1954) Principles of geomorphology. John Wiley, New York; Chapman & Hall, London

Thornes J (1990) Big rills have little rills. Nature 345:764–765

Thornthwaite CW (1931) The climates of North America. Geogr Rev 21:633–654

Thornthwaite CW (1948) An approach toward a rational classification of climate. Geogr Rev 38:55–94

Thornthwaite CW, Mather JR (1957) Instructions and tables for computing potential evapotranspiration and the water balance. Drexel Institute of Technology, Laboratory of Climatology, Centreton, New Jersey, Publications in Climatology 10, pp 181–311

Thornthwaite CW, Sharpe CFS, Dosch EF (1942) Climate and accelerated erosion in the arid and semi-arid southwest with special reference to the Polacca Wash drainage basin, Arizona. US Department of Agriculture Technical Bulletin 808, US Government Printing Office, Washington DC, 134 pp

Tinsley B (1988) The solar cycle and the QBO influences on the latitude of storm tracks in the North Atlantic. Geophys Res Lett 15:409

Toon OB, Pollack JB, Ackerman TP, Turco RP, McKay CP, Liu MS (1982) Evolution of an impact-generated dust cloud and its effect on the atmosphere. In: Silver LT, Schultz PH (eds) Geological implications of the impact of large asteroids and comets on the Earth. Geol Soc Am Spec Pap 190, pp 187–200

Transeau EN (1905) Forest centers of central America. Am Nat 39:875–889

Transeau EN (1926) The accumulation of energy by plants. Ohio J Sci 26:1–10

Treviranus GR (1802–22) Biologie; oder, Philosophie der lebenden Natur für Naturforscher und Aerzte, 6 vols in 3. JF Röwer, Göttingen

Tricart J (1963) Géomorphologie des régions froides. Presses Universitaires de France, Paris

Tricart J (1985) Evidence of Upper Pleistocene dry climates in northern South America. In: Douglas I, Spencer T (eds) Environmental change and tropical geomorphology. George Allen & Unwin, London, pp 197–217

Tricart J, Cailleux A (1962) Le modèle glaciare et nival. Traité de géomorphologie, vol 3. Société d'Édition d'Enseignement Supérieure, Paris

Tricart J, Cailleux A (1965a) Le modèle des régions chaudes, forêts et savanes. Traité de géomorphologie, vol 5. Société d'Édition d'Enseignement Supérieure, Paris

Tricart J, Cailleux A (1965b) Introduction à la géomorphologie climatique. Traité de géomorphologie, vol 1. Société d'Édition d'Einseignement Supérieure, Paris

Tricart J, Cailleux A (1967) Le modèle des régions periglaciares. Traité de géomorphologie, vol 2. Société d'Édition d'Enseignement Supérieure, Paris

Tricart J, Cailleux A (1969) Le modèle des régions sèches. Traité de géomorphologie, vol 4. Société d'Édition d'Enseignement Supérieure, Paris

Tricart J, Cailleux A (1972) Introduction to climatic geomorphology. Translated from the French by Conrad J Kiewiet de Jonge. Longman, London

Troll C (1944) Strukturböden, Solifluktion und Frostklimate der Erde. Geol Rundschau 34:545–694

Tucker ME, Benton MJ (1982) Triassic environments, climates and reptile evolution. Palaeogeogr Palaeoclimatol Palaeoecol 40:361–379

Turesson G (1922) The species and the variety as ecological units. Hereditas 3:100–113

Turesson G (1925) The plant species in relation to habitat and climate. Hereditas 6:147–236

Turesson G (1930) The selective effect of climate upon the plant species. Hereditas 14:99–152

Turner JRG, Lennon JJ (1989) Reply under "Species richness and the energy theory". Nature 340:351

Turner JRG, Gatehouse CM, Corey CA (1987) Does solar energy control organic diversity? Butterflies, moths and the British climate. Oikos 48:195–205

Turner JRG, Lennon JJ, Lawrenson JA (1988) British bird species distributions and the energy theory. Nature 335:539–541

Tüxen R (1931–32) Die Pflanzensoziologie in ihren Beziehungen zu den Nachbarwissenschaften. Der Biologe 1:180–187

Twidale CR (1976) On the survival of paleoforms. Am J Sci 276:77–95

Twidale CR, Bourne JA, Smith DM (1974) Reinforcement and stabilisation mechanisms in landform development. Rév Géomorphol Dynam 28:81–95

Tyndall J (1861) On the absorption and radiation of heat by gases and vapours, and on the physical connection, of radiation absorption, and conduction. Philos Mag 22:169–194, 273–285

Umbgrove JHF (1939) On the rhythms in the history of the Earth. Geol Mag 82:237–244

Umbgrove JHF (1940) Periodicity in terrestrial processes. Am J Sci 238:573–576

Umbgrove JHF (1942) The pulse of the Earth, 1st edn. Nijhoff, The Hague

Umbgrove JHF (1947) The pulse of the Earth, 2nd edn. Nijhoff, The Hague

Upchurch GR, Wolfe JA (1987) Mid-Cretaceous to early Tertiary vegetation and climate: evidence from fossil leaves and wood. In: Friis EM, Chaloner WG, Crane PR (eds) The origin of angiosperms and their biological consequences. Cambridge, Cambridge University Press, pp 75–104

Urey HC (1973) Cometary collisions and geological periods. Nature 242:32–33

Useinova I (1989) The astounding continental factor. Geogr Mag 61:24–26

Vail PR, Mitchum RM, Thompson S (1977) Seismic stratigraphy and global changes of sea level. Part 4, Global cycles of relative changes of sea level. In: Peyton CE (ed) Seismic stratigraphy: applications to hydrocarbon exploration. American Association of Petroleum Geologists, Tulsa, Oklahoma, Memoir 26, pp 83–97

Valentine KWG, King RH, Dormaar JF, Vreeken WJ, Tarnocai C, de Kimpe CR, Harris SA (1987) Some aspects of Quaternary soils in Canada. Can J Soil Sci 67:221–247

van Campo E, Duplessy JC, Rossignol-Strick M (1982) Climatic conditions deduced from a 150-kyr oxygen isotope-pollen record from the Arabian Sea. Nature 296:56–59

van den Broeck E (1881) Mémoire sur les phénomènes d'altération des dépôts superficiels par l'infiltration des eaux météoriques étudiés dans leurs rapports avec la géologie stratigraphique. F Hayez, l'Imprimeur de l'Académie Royale de Belgique

van der Hammen T (1961) Upper Cretaceous and Tertiary climatic periodicities and their causes. Ann New Acad Sci 95:440–448

van Houten FW (1964) Cyclic lacustrine sedimentation, Upper Triassic Lockatong Formation, central New Jersey and adjacent Pennsylvania. Kansas State Geological Survey Lawrence, Kansas, Bulletin 169, pp 497–531

van Loon H, Labitzke K (1988a) Association between the 11-year solar cycle, the QBO, and the atmosphere. Part II: Surface and 700 mb on the northern hemisphere in winter. J Clim 1:905–920

van Loon H, Labitzke K (1988b) When the wind blows. New Sci 119:58–60

van Loon H, Labitzke K (1990) Association between the 11-year solar cycle and the atmosphere. Part IV: The stratosphere, not grouped by phase of the QBO. J Clim 3:827–837

van Tassel J (1987) Upper Devonian Catskill delta margin cyclic sedimentation: Brallier, Scherr and Foreknobs Formation of Virginia and West Virginia. Bull Geol Soc Am 99:414–426

van't Hoff JH (1884) Études de dynamique chimique. F Muller, Amsterdam

Vernadksy VI (1924) La géochimie. Félix Alcan, Paris

Vernadsky VI (1926) Biosfera. Nauchnoe Khimikotekhnicheskoe Izdatelstvo, Leningrad

Vernadsky VI (1929) La biosphère. Nouvelle Collection Scientifiques. Félix Alcan, Paris

Vernadsky VI (1944) Problems of biochemistry, II: The fundamental matter-energy difference between the living and inert natural bodies of the biosphere. Translated by George Vernadsky, edited and condensed by GE Hutchinson. Trans Conn Acad Arts Sci 35:483–517

Vernadsky VI (1945a) The biosphere and the noosphere. Am Sci 33:1–12

Vernadsky VI (1945b) La biogéochimie. Scientia 77–78:77–84

Vines RG (1982) Rainfall patterns in the western United States. J Geophys Res 87:7303–7311

Vines RG (1984) Rainfall patterns in the eastern United States. Clim Change 6:79–98

Visher SS (1941) Climate and geomorphology: some comparisons between regions. J Geomorphol 5:54–64

Visher SS (1945) Climatic maps of geologic interest. Bull Geol Soc Am 56:713–736

von Bertalanffy L (1951) Theoretische Biologie, vol 2, 2nd edn. A Francke, Berne

von Bertalanffy L (1973) General system theory: foundations, development, applications. Penguin Books, Harmondsworth, Middlesex

Vörösmarty CJ, Moore B III (1991) Modeling basin-scale hydrology in support of physical climate and global biogeochemical studies: an example using the Zambezi river. In: Wood EF (ed) Land surface — atmospheric interactions: parameterization and analysis for climate modeling. Kluwer Academic Publishers, Dordrecht, pp 271–311 (Surveys in Geophysics, vol 12)

Vörösmarty CJ, Moore B III, Grace AL, Gildea MP, Melillo JM, Peterson BJ, Rastetter EB, Steudler PA (1989) Continental scale models of water balance and fluvial transport: an application to South America. Glob Biogeochem Cycles 3:241–265

Vrba ES (1984) Evolutionary pattern and process in the sister group Alcelaphini-Aepycerotini (Mammalia: Bovidae). In: Eldredge N, Stanley SM (eds) Living fossils. Springer, Berlin Heidelberg New York, pp 62–79

Vrba ES, Gould SJ (1986) The hierarchical expansion of sorting and selection: sorting and selection cannot be equated. Paleobiology 12:217–228

Vreeken (1975a) Variability of depth to carbonates in fingertip loess watersheds in Iowa. Catena 2:321–336

Vreeken WJ (1975b) Principal kinds of chronosequences and their significance in soil history. J Soil Sci 26:378–394

Wagner A (1940) Klimaänderungen und Klimaschwankungen. Friedrich Vieweg, Brunswick

Wallace AR (1878) Tropical nature and other essays. Macmillan, London

Walling DE (1987) Rainfall, runoff and erosion of the land: a global view. In: Gregory KJ (ed) Energetics of the physical environment: energetic approaches to physical geography. John Wiley, Chichester, pp 89–117

Walling DE, Webb BW (1983) The dissolved load of rivers: a global overview. In: Webb BW (ed) Dissolved load of rivers and surface water quantity/quality relationships. Int Assoc Hydrol Sci Publ 141:3–20

Walter H (1985) Vegetation of the Earth and ecological systems of the geo-biosphere, 3rd revised and enlarged edn. Translated from the 5th revised German edn by Owen Muise. Springer, Berlin Heidelberg New York Tokyo. (Heidelberg Science Library)

Walter H, Lieth HFH (1961–66) Climate diagram world atlas. Gustav Fischer, Jena

Wanless HR, Shepard FP (1936) Sea level and climatic changes related to Paleozoic cycles. Bull Geol Soc Am 47:1177–1206

Wanless HR, Weller JM (1932) Correlation and extent of Pennsylvanian cyclothems. Bull Geol Soc Am 43:1003–1016

Warming E (1895) Plantesamfund: grundtraek af den økologiske plantegeografi. PG Philipsen, Copenhagen

Warming E (1909) Oecology of plants: an introduction to the study of plant communities. Assisted by Martin Vahl. Prepared for publication in English by Percy Groom and Isaac Bayley Balfour. Clarendon, Oxford

Watkins JR (1967) The relationship between climate and the development of landforms in Cainozoic rocks of Queensland. J Geol Soc Aust 14:153–168

Watson HC (1835) Remarks on the geographical distribution of British plants; chiefly in connection with latitude, elevation and climate. Longman, Rees, Orme, Brown, Green, and Longman, London (2nd edn of Outlines of the geographical distribution of British plants. Privately published, Edinburgh)

Watts WA (1983) Vegetational history of the eastern United States 25,000 to 10,000 years ago. In: Porter SC (ed) Late-Quaternary environments of the United States, vol 1, the late Pleistocene. Longman, London, pp 294–310

Watts WA (1986) Stages of climatic change from full glacial to Holocene in northwest Spain, southern France, and Italy: a comparison of the Atlantic coast and the Mediterranean Basin. In: Ghazi A, Fantechi R (eds) Current Issues in Climate Research. D Reidel, Dordrecht, pp 101–112 (Commission of the European Communities. Proceedings of the EC Climatology Programme Symposium, Sophia Antipolis, France, 2–5 October 1984)

Webb SD (1983) The rise and fall of the late Miocene ungulate fauna in North America. In: Nitecki MH (ed) Coevolution. The University of Chicago Press, Chicago, pp 267–306

Webb SD (1984) Ten million years of mammal extinctions in North America. In: Martin PS, Klein RG (eds) Quaternary extinctions: a prehistoric revolution. The University of Arizona Press, Tucson, Arizona, pp 189–210

Webb SD (1985) Late Cenozoic mammal dispersals between the Americas. In: Stehli FG, Webb SD (eds) The great American biotic interchange. Plenum, New York, pp 357–386

Webb SD, Barnosky AD (1989) Faunal dynamics of Pleistocene mammals. Annu Rev Earth Planet Sci 17:413–438

Webb T III (1985) Holocene palynology and climate. In: Hecht AD (ed) Paleoclimate analysis and modeling. John Wiley, New York, pp 163–195

Webb T III (1986a) Is vegetation in equilibrium with climate? How to interpret late-Quaternary pollen data. Vegetatio 67:75–91

Webb T III (1986b) Vegetation change in eastern North America from 18,000 to 500 yr B.P. In: Rosenzweig C, Dickinson R (eds) Climate-vegetation interactions. Report OIES-2, Office for interdisciplinary Earth studies and University Corporation for Atmospheric Research, Boulder, Colorado, pp 63–69

Webb T III (1987) The appearance and disappearance of major vegetational assemblages: long-term vegetational dynamics in eastern North America. Vegetatio 69:177–187

Webb T III, Cushing EJ, Wright HE Jr (1983) Holocene changes in the vegetation of the midwest. In: Wright HE Jr (ed) Late Quaternary environments of the United States, vol 2, The Holocene. Longman, London, pp 142–165

Webb T III, Bartlein PJ, Kutzbach JE (1987) Climatic change in eastern North America during the past 18,000 years; comparisons of pollen data with model results. In: Ruddiman WF, Wright HE Jr (eds) North America and adjacent oceans during the last glaciation. Decade of North American Geology, vol K-3. Geological Society of America, Boulder, Colorado, pp 447–462

Wegener AL (1915) Die Entstehung der Kontinente und Ozeane. Friedrich Vieweg, Brunswick

Wegener AL (1929) Die Entstehung der Kontinente und Ozeane. 4th edn. Friedrich Vieweg, Brunswick

Wegener AL (1966) The origin of the continents and oceans. Translated by J Biram, with an introduction by BC King. Methuen, London

Weinert HH (1961) Climate and weathered Karroo dolerites. Nature 191:325–329

Weinert HH (1965) Climatic factors affecting the weathering of igneous rocks. Agric Meteorol 2:27–42

Wells JW (1963) Coral growth and geochronometry. Nature 197:948–950

Wendland WM (1978) Holocene man in North America: the ecological setting and climatic background. Plains Anthropol 23:273–287

Wendland WM, Benn A, Semken HA Jr (1987) Evaluation of climatic changes on the North American Great Plains determined from faunal evidence. In: Graham RW, Semken HA Jr, Graham MA (eds) Late Quaternary mammalian biogeography and environments of the Great Plains and prairies. Illinois State Museum, Springfield, Il, Scientific Papers 22, pp 460–473

Wexler H (1951) Spread of Krakatoa volcanic dust cloud as related to the high level circulation. Bull Am Meteorol Soc 32:48–51

Wexler H (1952) Volcanoes and climate. Sci Am 186:74–80

Whitehouse KW (1940) Studies in the late geological history of Queensland. Paper, Department of Geology, University of Queensland, St Lucia, Brisbane, No 2, pp 2–22

Whitehurst J (1778) An inquiry into the original state and formation of the Earth; deduced from facts and the laws of Nature. To which is added an appendix, containing some general observations on the strata in Derbyshire. With sections of them, representing their arrangement, affinities, and the mutations they have suffered at different periods of time, intended to illustrate the preceding enquiries, and as a specimen of subterraneous geography. Printed for the author by J Cooper, London

Whittaker RH (1962) Classification of natural communities. Bot Rev 28:1–239

Whittaker RH (1970) Communities and ecosystems. 1st edn. Macmillan, London (Current concepts in biology series)

Whyte MA (1977) Turning points in Phanerozoic history. Nature 267:679–682

Wigley TML, Kelly PM (1990) Holocene climatic change, ^{14}C wiggles and variations in solar irradiance. Philos Trans R Soc Lond 330A:547–560

Wilhelmy H (1958) Klimageomorphologie der Massengesteine. Georg Westermann, Braunschweig

Willdenow KL (1792) Grundriss der Kräuterkunde. Haude und Spener, Berlin

Willdenow KL (1805) Principles of botany, and of vegetable physiology. William Blackwood, Edinburgh

Willett HC (1949) Solar variability as a factor in the fluctuations of climate during geological time. Geogr Ann 31:295–315

Willett HC (1961) The pattern of solar climatic relationships. Ann N Y Acad Sci 95:89–106

Willett HC (1962) The relationship of total atmospheric ozone to the sunspot cycle. J Geophys Res 67:661–670

Willett HC (1980) Solar prediction of climatic change. Phys Geogr 1:95–117

Willett HC (1987) Climatic responses to variable solar activity — past, present, and predicted. In: Rampino MR, Sanders JE, Newman WS, Königsson LK (eds) Climate: history, periodicity, and predictability. Van Nostrand Reinhold, New York, pp 404–414

Williams GE (1972) Geological evidence relating to the origin and secular rotation of the Solar System. Mod Geol 3:165–181

Williams GE (1975a) Possible relation between periodic glaciation and the flexure of the Galaxy. Earth Planet Sci Lett 26:361–369

Williams GE (1975b) Late Precambrian glacial climate and the Earth's obliquity. Geol Mag 112:441–544

Williams GE (1980) Effects of the Earth's rotation rate on climate. Nature 286:309

Williams GE (ed) (1981a) Megacycles: long-term episodicity in Earth and planetary history. Dowden, Hutchinson & Ross, Stroudsberg, Pennsylvania (Benchmark Papers in Geology 57)

Williams GE (1981b) Sunspot periods in the late Precambrian glacial climate and solar-planetary relations. Nature 291:624–628

Williams GE (1985) Solar affinity of sedimentary cycles in the late Precambrian Elatina Formation. Aust J Phys 38:1027–1043

Williams GE (1986) Precambrian permafrost horizons as indicators of palaeoclimate. Precambrian Res 32:233–242

Williams GE (1988) Cyclicity in the late Precambrian Elatina Formation, south Australia: solar or tidal signature? Clim Change 13:117–128

Williams GE (1989) Late Precambrian tidal rhythmites in South Australia and the history of the Earth's rotation. J Geol Soc Lond 146:97–111

Williams GE (1990a) Discussion on Late Precambrian tidal rhythmites in South Australia and the history of the Earth's rotation. J Geol Soc Lond 147:402–407

Williams GE (1990b) Precambrian cyclic rhythmites: solar-climatic or tidal signatures? Philos Trans R Soc 330A:445–458

Williams GE, Sonett CP (1985) Solar signature in sedimentary cycles from the late Precambrian Elatina Formation, Australia. Nature 318:523–527

Wolbach WS, Lewis RS, Anders E (1985) Cretaceous extinctions: evidence for wildfires and search for meteoritic material. Science 230:167–170

Wolbach WS, Gilmour I, Anders E, Orth C, Brooks RR (1988) Global fire at the Cretaceous-Tertiary boundary. Nature 334:665–669

Wolf JR (1858) Mitteilungen über die Sonnenflecken. Astronom Mitt 6:127–143

Wolfe JA (1977) Paleogene floras from the Gulf of Alaska region. US Geol Surv Prof Pap 997

Wolfe JA (1978) A paleobotanical interpretation of Tertiary climates in the Northern Hemisphere. Am Sci 66:694–703

Wolfe JA (1979) Temperature parameters of humid to mesic forests of eastern Asia and relation to forests of other regions of the Northern Hemisphere and Australasia. US Geol Surv Prof Pap 1106

Wolfe JA (1980) Tertiary climates and floristic relationships at high latitudes in the northern hemisphere. Palaeogeogr Palaeoclimatol Palaeoecol 30:313–323

Wolfe JA (1985) Distribution of major vegetational types during the Tertiary. In: Sundquist KET, Broecker WS (eds) The carbon cycle and atmospheric CO_2: natural variations Archean to present. American Geophysical Union, Washington DC, Geophysical Monograph 32, pp 357–375

Wolfe JA, Poore RZ (1981) Tertiary marine and nonmarine climatic trends. In: Berger WH, Crowell JC (eds) Climate in Earth history. National Academy of Sciences, Washington DC, pp 154–158 (Studies in Geophysics)

Wolfe JA, Upchurch GR Jr (1986) Vegetation and floral changes at the Cretaceous-Tertiary boundary. Nature 324:148–152

Wolfe JA, Upchurch GR Jr (1987a) Leaf assemblages across the Cretaceous-Tertiary boundary in the Raton Basin, New Mexico and Colorado. Proc Natl Acad Sci USA 84:5096–5100

Wolfe JA, Upchurch GR Jr (1987b) North American nonmarine climates and vegetation during the late Cretaceous. Palaeogeogr Palaeoclimatol Palaeoecol 61:33–77

Woodward FI (1987) Climate and plant distribution. Cambridge University Press, Cambridge

Woodward FI (1989) Plants in the greenhouse world. New Sci 122:1–4

Woodward FI (1991) Review of the effects of climate on vegetation: ranges, competition and composition. In: Peters RL, Lovejoy TE (eds) Consequences of the greenhouse effect for biological diversity. Yale University Press, New Haven (in press) (Proceedings of the World Wildlife Fund Conference)

Woodward FI, McKee IF (1991) Vegetation and climate. Environ Int (in press)

Worster D (1985) Nature's economy: a history of ecological ideas. Cambridge University Press, Cambridge

Wright DH (1983) Species-energy theory: an extension of species-area theory. Oikos 41:496–506

Wright HE Jr (1987) Synthesis: land south of the ice sheets. In: Ruddiman WF, Wright HE Jr (eds) North America and adjacent oceans during the last deglaciation. The Geology of North America, vol K-3. The Geological Society of America, Boulder, Colorado, pp 470–488

Wright VP (ed) (1986) Palaeosols: their recognition and interpretation. Blackwell Scientific Publishers, Oxford

Wright VP (1990) Equatorial aridity and climatic oscillations during the early Carboniferous, southern Britain. J Geol Soc Lond 147:359–363

Wright WB (1937) The Quaternary ice age. Macmillan, London

Wundt W (1944) Die Mitwirkung der Erdbahnelemente bei der Entstehung der Eiszeiten. Geol Rundschau 34:713–747

Wyrwoll K-H, McConchie D (1986) Accelerated plate motion and rates of volcanicity as controls on Archaean climates. Clim Change 8:257–265

Xu Q-Q (1979) On the causes of ice ages. Sci Geol Sin 7:252–263 (in Chinese with English summary)

Xu Q-Q (1980) Climatic variation and the obliquity. Vertebr PalAsiat 18:334–343 (in Chinese with English summary)

Yaalon DH (1975) Conceptual models in pedogenesis: can soil-forming functions be solved? Geoderma 14:189–205

Yamamoto T (1967) On the climatic change along the current of historical times in Japan and its surroundings. Jpn Philos Mag 76:115–141

Yates J (1830–31) Remarks on the formation of alluvial deposits. Proc Geol Soc Lond 1:237–239

Yatsu E (1988) The nature of weathering: an introduction. Sozosha, Tokyo

Yeh T-C, Fu C-B (1985) Climatic change — a global and multidisciplinary theme. In: Malone TF, Roederer JG (eds) Global change. Published on behalf of the ICUS Press by Cambridge University Press, Cambridge, pp 127–145

Young A (1974) The rate of slope retreat. In: Brown EH, Waters RS (eds) Progress in geomorphology: papers in honour of David L Linton. Special Publication No 7, The Institute of British Geographers, London, pp 65–78

Young KR (1989) The tropical Andes as a morphoclimatic zone. Prog Phys Geogr 13:13–22

Young RW (1983) The tempo of geomorphological change: evidence from southeastern Australia. J Geol 91:221–230

Yule GU (1927) On a method of investigating periodicities in disturbed series, with special reference to Wolf's sunspot numbers. Philos Trans R Soc Lond 226A:267–298

Zbinden EA, Holland HD, Feakes CR, Dobos SK (1988) The Sturgeon Falls paleosol and the composition of the atmosphere 1.1 Ga BP. Precambrian Res 42:141–163

Zeuner FE (1952) Dating the past, 3rd edn. Methuen, London

Zimmerman EAW von (1777) Specimen zoologicae geographicae, quadrupedum domicilia et migrationes sistens. Dedit tabulamque mundi zoographicam adjunxit Eberh Aug Guilielm Zimmerman. T Haak, Leyden

Glossary

The definitions in the following glossary are cross-referenced, all words in italics appearing elsewhere as main entries. Terms which are defined in the body of the text are not included.

absolute time: geological time measured in years; specifically, time as an actual age determined by the radioactive decay of elements (cf. *relative time*)

acid igneous rock: an *igneous rock* containing more than 60% *silica*

actual evaporation: the amount of water vapour actually released from a land or water surface (cf. *potential evaporation*)

adiabatic warming: a rise in temperature of a parcel of air caused by compression, as for example when air flows to a lower level

aeolian: of, or referring to, the wind

aerosol: solid particles or liquid droplets dispersed in the air

aggradation: the building upwards of the Earth's surface owing to the accumulation of sediment deposited by geomorphological agencies such as wind, wave, and water; in the case of accumulation in a river system, *alluviation*

aggradational process: any Earth surface process associated with the action of wind, wave, and water which leads to the accumulation of sediment

albedo: a measure of the proportion of light reflected by a surface to the amount falling upon it: a pure white body has a fractional albedo of 1.0 (or 100%); a pure black body has a fractional albedo of 0.0 (or 0%)

alcrete: a *duricrust* rich in aluminium, commonly in the form of hardened *bauxite*

alfisol: relatively young, basic or slightly acid *soils* with a *clay*-enriched B *horizon*. Associated with deciduous forests in humid, sub-humid, temperate, and sub-tropical climates. One of the ten soil orders in the United States Department of Agriculture's Comprehensive Soil Classification System, Seventh Approximation

allitic crust: a surficial weathered layer rich in *alumina* (cf. *alcrete*)

allitization: the loss of *silica* and concentration of *sesquioxides* in the *soil*, with the formation of *gibbsite*, and with or without the formation of *laterite*; more or less synonymous with *soluviation*, *ferrallitization*, *laterization*, and *latosolization*

allocyclic: the *deposition* of a *cyclothem* owing to changes in the energy or mass received by a sedimentary system induced by changes in factors outside the system (*driving variables*) such as uplift, subsidence, climatic change, or sea-level change (*eustasy*)

allometric growth: the growth of a morphological, physiological, or chemical property of an organism relative to the growth of the entire organism or another of its properties

alluvial: of, or pertaining to, *alluvium*

alluvial fill: the deposit of sediment laid down by flowing water in river channels

alluvial terrace: a river terrace composed of *alluvium* and created by renewed downcutting of the flood plain or valley floor (which leaves *alluvial* deposits stranded on the valley sides), or by the covering of an old river terrace with *alluvium*

alluviation: the formation of *alluvium* by the action of running water; more generally, the process of *aggradation*

alluvium: an unconsolidated, stratified deposit laid down by running water; sometimes applied only to fine sediments (*silt* and *clay*), but more generally used to include *sands* and *gravels*, too

alumina: aluminium oxide; occurs in various forms, for example in *bauxite*

ammonite: an animal belonging to the order Ammonitida, a group of organisms characterized by a thick shell of rich ornamentation with sutures having finely divided lobes and saddles

amphibole: a mineral of the amphibole group which consists of dark, rock-forming silicate minerals rich in iron and magnesium (hydrous ferromagnesian silicates); on occasions, used as a synonym for hornblende, the best-known member of the group

andesite: a dark-coloured, fine-grained, extrusive rock

angiosperm: a *vascular plant* possessing true flowers with seeds in ovaries

angiosperms: the flowering plants

antediluvian world: the Earth before Noah's flood

aphelion: the point on the orbit of the Earth (or any other body of the Solar System) which is farthest from the Sun. For the Earth at present, aphelion occurs on 4 July

Archaean aeon: the oldest commonly used slice of geological time between 4 to 2.5 billion years ago

argillaceous: referring to, containing, or composed of *clay*. Used to describe rocks containing *clay*-sized material and *clay minerals*

arid cycle (see *geographical cycle*)

aridisol: any member of a group of saline and alkaline mineral *soils* (little organic matter present) found in deserts. One of the ten soil orders in the United States Department of Agriculture's Comprehensive Soil Classification System, Seventh Approximation

artiodactyl: a member of the order Artiodactyla (the "even-toed" ungulates) which includes pigs, peccaries, hippopotamuses, camels, deer, cows, sheep, goats, and antelope

astronomical pole shift: a change in the *obliquity* (tilt) of the Earth's rotatory axis

asymmetrical valley: a river valley or glacial valley with one side steeper than the other

atmosphere: the gaseous envelope of the Earth, retained by the Earth's gravitational field

atmospheric general circulation model (see *general circulation model*)

backslope: a relatively straight element of a *hillslope* linking a convex *shoulder* with a concave *footslope*; also called the midslope, transport slope, and constant slope

banded iron formation: *Precambrian* sedimentary deposits in which thin iron-rich beds alternate with thin silica-rich (chert) beds

barred bassinal system: a restricted basin: a depression in the ocean floor in which the circulation of water is restricted by the submarine *topography* and, consequently, in which oxygen depletion is commonplace

base level: the lower limit to the action of fluvial processes; the level below which land cannot be eroded. Sea level is a general base below which continents cannot be eroded but in any landscape there may be many local base levels

bauxite: a rock, chiefly associated with the *clay* deposits of the *weathering* zones in tropical regions, composed of an admixture of various amorphous and crystalline *hydrous* aluminium oxides and hydroxides along with free *silica, silt,* iron hydroxides, and *clay minerals*; a *laterite* rich in aluminium

bedload: sediment transported by water along, or close to, the stream bed

bedrock: consolidated, unweathered rock exposed at the land surface, or lying beneath the *soil* and unconsolidated surficial sediments

biogeochemical cycle: the cycling of a mineral or organic chemical constituent through the *biosphere*; for example, the carbon cycle

biogeochemistry: a branch of geochemistry focussing on the effects of life on the storage, transformation, and transfer of mineral and organic chemical constituents in the *biosphere*

biomass: the mass (or weight) of living material in a specified group of animals, or plants, or in a community, or in a unit area

biome: a community of animals and plants occupying a climatically uniform area on a continental scale

biosphere: life and life-supporting systems — all living beings, *atmosphere, hydrosphere,* pedosphere (*soil* sphere), surficial sediments; used synonymously with "world climate system" in this book

biota: all the animals (*fauna*) and plants (*flora*) living in an area or region

biotic crisis: a time when the *extinction* rate greatly exceeds the *speciation* rate; a mass extinction

biotite: dark mica, a member of the mica group of minerals

bisiallitization: the formation in the *soil* of secondary 2:1 *clay* minerals such as *smectite* and *vermiculite*

black savanna soil: a *soil* associated with basic *igneous rocks* (*silica* content between about 45 and 52%) or freshly weathered rocks in topographic depressions in tropical grasslands where leached colloidal *silica* and bases accumulate

black shale: a dark, thinly layered, *carbonaceous* variety of *mudstone*, containing much organic matter and iron sulphide. Formed by the partial decay of organic matter buried in a still water, reducing environment, such as a stagnant marine basin

blocking circulation pattern: an extreme form of the tropospheric circulation of air in which large-amplitude, stationary waves occur, so that air flow is mainly *meridional* and blocking high pressure cells develop

boreal forest: a plant *formation type* associated with cold-temperate climates (cool summers and long winters); also called taiga and coniferous evergreen forest. Spruces, firs, larches, and pines are the dominant plants

bovid: a member of the family Bovidae which includes bison, sheep, goats, and cattle

braiding: the tendency of some rivers to form several interlaced branches which separate and rejoin

broad-leaved deciduous forest: a plant *formation type* associated with cool temperate climates with a short period of frost; also called nemoral forests. Dominated by broad-leaved deciduous species such as oak, beech, maple, and ash

broad-leaved evergreen forest: a plant formation type associated with warm temperate climates. Forest trees in this *formation type* tend to be sensitive to frost

brown forest soil: any member of a group of intrazonal *soils* characterized by a mull (base-rich *humus*) *horizon* and a lack of *horizons* of *clay* or iron accumulation. Formed on calcium-rich *parent material* under a temperate climate and deciduous forest. Called brown earth in the United Kingdom, braunerde in Germany, and *sol brun acide* in France

brown podzolic soil: any member of a group of *zonal soils* characterized by light or greyish brown A *horizons*, devoid of carbonates and deficient of iron oxides and *clay*, and with signs of iron illuviation in thick, reddish or coloured B *horizons* with iron oxides coating the original *soil* material; similar to a *podzol* but lacking a true eluvial (*leached*) horizon

brown soil: any member of a group of *zonal soils* characterized by a brown A *horizon* surmounting a light-coloured B *horizon* (sometimes argillic and red), and a C *horizon* with calcium carbonate present. Formed under Mediterranean climates (cool, wet winters and hot, dry summers) (cf. *fersiallic soil*)

brunification: the development of brown colours in a *soil*

C3 plant: any plant which uses the C_3 biochemical pathway for fixing carbon dioxide which involves the enzyme ribulose-diphosphate carboxylase

C4 plant: any plant which uses the C_4 biochemical pathway for fixing carbon dioxide which involves the enzyme phospho-enol-pyruvate carboxylase

cala: a short, narrow ria (a long, narrow inlet of the sea) formed in a *limestone* coast

calcareous: rich in calcium carbonate

calcification: processes which lead to the accumulation of calcium carbonate in a *soil* body

calcrete: a terrestrial material consisting chiefly of calcium carbonate in a variety of states: powder and nodules to massive, indurated (hardened) *horizons*

caliche: a layer of calcium carbonate accumulation observed in the *soils* of many dry regions

Cambrian period: a slice of geological time between 590 to 505 million years ago; the oldest unit of the *Palaeozoic era*

Camp Century ice core: a core of ice extracted from the Greenland ice sheet at Camp Century

carbonate compensation depth: in the ocean, the level below which the solution rate of calcium carbonate is more than its *deposition* rate

carbonate rock: *limestone*; any bedded sedimentary rock composed chiefly of the mineral calcite (calcium carbonate). Includes dolomitic limestone, which consists of the minerals calcite and dolomite (calcium-magnesium carbonate), dolomite (*dolostone*), which consists chiefly of the mineral dolomite, and *chalk*

Carboniferous period: a slice of geological time between 360 to 286 million years ago; it is divided into two subperiods — the Mississippian (360 to 320 million years ago) and the Pennsylvanian (320 to 286 million years ago)

carnassial tooth: either the last upper premolar or the first lower molar in carnivorous mammals adapted as specialized shearing blades for tearing flesh asunder

Cenozoic era: a slice of geological time between 65 million years ago to the present; the youngest unit of the geological eras

cervid: a member of the family Cervidae which includes deer, elk, caribou, and moose

chalk: a soft, white, pure, fine-grained *limestone* consisting of very fine grains of calcite (calcium carbonate) mixed with the remains of microscopic calcareous fossils

cheluviation: the movement of the heavy cations of aluminium and iron (and sometimes alkaline earths derived from surface litter) as organo-metal complexes or chelates; complexolysis

chemical denudation: the laying bare of the land surface by the decomposition of rocks and the removal of soluble materials so formed by flowing water

chemical weathering: the decay or decomposition of minerals in rocks owing to chemical attack

chemolithotrophic: pertaining to organisms which manufacture their food from a low potential chemical energy source such as sulphur

Chenopodiaceae: a family of perennial herbs (and a few shrubs and trees) which are halophytic (adapted to *soils* with a high concentration of inorganic salts). Includes sugar beet, beetroot, and spinach

chernozem: any member of a group of *zonal soils* characterized by a dark, usually deep, *humic* A *horizon* saturated with cations of calcium and magnesium (a mollic epipedon), thin, brown B *horizons* with *clay* accumulation, and a *calcareous* subsoil. Formed under a temperate to cool subhumid climate associated with long-grass steppe

chert: a cryptocrystalline form of silica, a variety of chalcedony, found as nodules in *limestone*

chestnut soil: any member of a group of *zonal soils* characterized by less distinct *horizons* than a *chernozem*, a dark brown surface *horizon* lying on lighter-coloured subsoil *horizons* which show signs of *clay* accumulation, often specks of gypsum (calcium sulphate), and lime (calcium carbonate) accumulation. Associated with short-grass steppe

cirque: a hollow, backed by a bow-shaped, cliffed headwall and open down-valley; also called cwm and corrie

clay: particles of *soil* or sediment with a diameter less than 0.002 mm; a deposit formed chiefly of clay-sized particles

1:1 clay: any *clay* consisting of a *tetrahedral layer* of *silica* (in which four oxygen ions are arranged around a silicon ion) paired with an octahedral layer (in which oxygen and hydroxyl ions are grouped around metal cations such as aluminium); in other words, any clay having one *tetrahedral layer* for each octahedral layer; an example is *kaolinite*

2:1 clay: any *clay* consisting of octahedral layers (see *1:1 clay*) sandwiched between two *tetrahedral layers*; in other words, a clay having two *tetrahedral layers* for each octahedral layer; an example is *illite*

clay mineral: a member of a group of related *hydrous* aluminosilicates, the chief constituents of *clay* and *mud*

cliff face: a steep element of a *hillslope* in which *bedrock* is exposed; variously called a free face, gravity slope, derivation slope, and more colloquially, bluff, cliff, scar, and scarp

climatic geomorphology: the subject exploring the possibility that different climatic *zones* (temperature, arid, tropical, and so forth) lead to the creation of distinct assemblages of landforms (*morphoclimatic regions*)

climax vegetation: vegetation which, if left undisturbed, will eventually attain a *steady state* constrained by prevailing conditions in the environment (climate, *soil*, animals, *relief, parent material*)

climofunction: an empirical relationship between a *soil* property and a climatic variable

cline: a gradual and essentially continuous change of a character in a series of contiguous populations; a character gradient

complexolysis (see *cheluviation*)

condensation nucleus: a focal point around which water vapour may condense to form liquid water droplets; ice crystals and fine dust are examples

coniferous evergreen forest (see *boreal forest*)

continental drift: the slow movement of landmasses relative to one another across the surface of the globe

corestone: an ellipsoidal or broadly rectangular joint block of granite formed by subsurface *weathering* and entirely separated from *bedrock*

Coriolis force: geostrophic force; the apparent deflective component of the centrifugal force created by the rotation of the Earth

cosmic rays: high-energy particles, chiefly photons, emitted by stars

cosmogenic: of, or produced by, *cosmic rays*

cosmopolitan: a worldwide, or nearly worldwide, distribution

Cretaceous period: a slice of geological time between 144 to 65 million years ago; the youngest unit of the *Mesozoic era*

cryosphere: the frozen waters of the Earth; the zone where ice and frozen ground are formed

cycad: any gymnosperm of the family Cycadaceae looking like a palm tree but topped with compound, fern-like leaves

cycle of erosion (see *geographical cycle*)

cyclostratigraphy: the study of rhythmic sedimentary sequences

cyclothem: a series of beds deposited during a single cycle of sedimentation of the sort which occurred during the Pennsylvanian subperiod

debris face: a *backslope* formed of scree; also called debris slope and *talus* slope

decalcification: chemical reactions which remove calcium carbonate from one or more *soil horizons*

deep-sea core (marine core): a core of sediment extracted from the bottom of the open ocean

deglaciation: the process, usually caused by climatic amelioration, by which *ice sheets* and *glaciers* withdraw from a region so uncovering the land beneath

degradation (of clays): the formation of *secondary clays* by the considerable *transformation* of primary minerals. Associated mainly with polar and temperate climates

dendrochronology: the study of annual growth rings of trees. Used as a means of dating events over the last millenium or so

denudation: the wearing away and progressive lowering of the Earth's surface by several natural agencies including *weathering, erosion*, mass wasting, and transport

deposition: the laying down of any material; specifically, the constructive process of accumulation into beds, veins, or irregular masses of any kind of sediment by any natural agency

desert soil: any member of a group of *zonal soils* characterized by a light-coloured surface *horizon* overlying *calcareous* materials and, commonly, a hardpan. Formed in warm to cool arid climates

detritus: a collective term for loose mineral and rock that is broken or worn off by mechanical means, as by disintegration and abrasion

Devonian period: a slice of geological time between 408 to 360 million years ago

diatom: a microscopic, unicellular, marine (*planktonic*) alga with a skeleton composed of *hydrous* opaline *silica*

dimorphic: having two distinct morphs (morphological types or forms) in a single population

diploid chromosome number: the normal chromosome number of the cells (save mature germ cells) in any individual grown from a fertilized egg. The term "diploid" refers to the fact that the chromosomes are a double set (2n)

dipole model: the idea that the Earth's magnetic field could be explained by a powerful bar magnet (a dipole) placed near the middle of the Earth along the axis joining the north and south geomagnetic poles

disharmonious community: a community of animals and plants adapted to a climate which has no modern counterpart. An instance is the "boreal grassland" biome that existed in the north-central United States from about 18 000 to 12 000 years ago in which modern grassland and modern deciduous woodland species lived side by side

disharmonious fauna (see *disharmonious community*)

disharmonious flora (see *disharmonious community*)

dispersal: the spread of a *taxon* into new areas

dissection index: a measure of the degree of incision and erosion of a landscape by rivers; commonly expressed as the number of contour crenulations per unit area

dissolved load: all material, organic and inorganic, carried in solution by a river

dolostone: another name for the rock dolomite, a variety of limestone rich in magnesium carbonate (see *carbonate rock*)

drainage basin: the catchment area of a river

drainage basin relief: the difference in altitude between the highest and lowest points in a *drainage basin*

drainage density: the length of streams in a *drainage basin* divided by the basin area

driving variable (see *forcing variable*)

drowned landscape: a landscape, formerly above sea level, which has been submerged

duricrust: a hard crust formed at, or a little below, the land surface by processes of *weathering* or *soil formation*, commonly in arid and tropical regions

dust: any solid particle carried in suspension in the air. Sources include volcanic eruptions, wind erosion of *silt*-sized and *clay*-sized materials, fires, and pollution

dust veil index: a measure devised by Hubert H. Lamb to assess the possible influence of volcanic *dust* on climate. There are three forms of the index which use, in various combinations, the following variables: the greatest percentage depletion in monthly direct *solar radiation* in the hemisphere in which an eruption occurred; the greatest proportion of the Earth affected by the dust veil; the estimated lowering of average temperatures for the most affected year over the middle latitude zone of the hemisphere in which the eruption occurred; and the total time between the eruption and the last observation of the dust veil or its effect upon monthly temperatures (or radiation) in middle latitudes

eccentricity: the deviation of the Earth's orbit from a perfect circle towards an elliptical form (cf. *ellipticity*)

ecogeographical rule: an observed regularity between the geographical variation of size, colour, and so forth and an environmental variable such as temperature or humidity

ecological niche: the place of a species (or other *taxon*) in Nature, including its habitat, feeding habits, size, and so on; the constellation of environmental factors into which a species (or other *taxon*) fits

ecological race: a local race which owes its most conspicuous attributes to the selective effects of a specific environment

ecosystem: short for ecological system — a group of organisms together with the physical environment with which they interact

ecotone: a transition zone between two plant communities

ecotype: a local race that owes its most conspicuous character to the selective effects of local environmental factors

ectotherm: an organism whose body temperature is determined by the ambient temperature and who can control its body temperature only by taking advantage of sun and shade to heat up or to cool down; a poikilotherm or "cold-blooded" animal (cf. *homeotherm*)

effective rainfall: the proportion of precipitation that is not evaporated and therefore is "available"; defined as precipitation less evaporation, normally computed monthly or annually

El Niño: the appearance of warm water in the usually cold water regions off the coasts of Peru, Ecuador, and Chile

electromagnetic radiation: energy propagated through space or through a material as an interaction between electric and magnetic waves; occurs at frequencies ranging from short wavelength, high frequency cosmic rays, through medium wavelength, medium frequency visible light, to long wavelength, low frequency radio waves (see *insolation*)

ellipticity: the deviation of the Earth's orbit from a perfect circle towards an elliptical form (cf. *eccentricity*)

elytra (singular **elytron**): the leathery or chitinous forewings of a beetle or related insect

empirical relationship: a relationship between two variables established by experiment or observation

endogene: of, or pertaining to, the Earth's interior (cf. *exogene*)

energy balance model: a one-dimensional, mathematical model predicting land and sea-surface temperature at different latitudes from information on factors determining the Earth's surface energy balance in each latitudinal zone

Eocene epoch: a slice of geological time between 54.9 to 38 million years ago; the middle unit of the *Palaeogene period*

epeirogeny: the warping of large areas of the crust of the Earth without much crustal deformation

epharmonic convergence: the evolution of the same life form in unrelated groups of organisms living in similar environments

epicratonic: at the edge of a craton (stable shield area of a continent)

equinox: one of the two days in a year (presently 20 March and 22 September) when there are exactly 12 h of day and 12 h of night at all points on the Earth

erosion: the *weathering* (decomposition and disintegration), solution, corrosion, and transport of rocks and rock debris

erosion surface: a more or less flat plain created by *erosion*; a planation surface

erosional cycle: another term for *geographical cycle*

etchplain: a wide erosional surface, often interrupted by *inselbergs* ("island hills"), found in shield areas of the tropics

etchplanation: a double process by which *etchplains* are thought to form: a lower surface of *bedrock* is etched away by chemical attack while an upper land surface is trimmed by *surface wash*

Euphorbiaceae: the spurge family: a big family of flowering plants (herbs, shrubs, and trees) of predominantly tropical distribution

eustasy: a term referring to worldwide changes in sea level. Such changes may result from a change in the volume of the ocean basins (*tectono-eustasy*) or the depletion of the oceanic reservoir of water at the expense of *ice sheets* and *glaciers* (*glacio-eustasy*)

evaporation: the diffusion of water vapour into the *atmosphere* from sources of water exposed to the air (cf. *evapotranspiration*)

evaporite: a water-soluble mineral (such as *halite*, gypsum, and anhydrite), or a rock composed of such minerals, precipitated out of saline water bodies such as salt lakes

evapotranspiration: evaporation plus the water discharged into the *atmosphere* by plant transpiration

exogene: of, or pertaining to, the surface (or near the surface) of the Earth (cf. *endogene*)

external forcing (see *forcing variable*)

extinction: the demise of a *taxon*

extirpation: the local extinction of a *taxon*

fauna: all the animals living in an area or region

faunal turnover: the change in composition of animal species (or any other *taxon*) owing to origination (*speciation*) and *extinction*

ferrallitic soil: a *soil* of hot and humid climates showing signs of intense and prolonged *weathering*. A large part of the *silica* and virtually all the bases have been removed leaving iron and aluminium. The profile consists largely of quartz, *kaolinite, gibbsite,* and *hematite* or *goethite*

ferrallitization: the loss of *silica* and concentration of *sesquioxides* in the *soil*, with the formation of *gibbsite*, and with or without the formation of *laterite*; more or less synonymous with *soluviation, allitization, laterization,* and *latosolization*

ferricrete: an accumulation of iron oxides and iron hydroxides in the *soil*, usually as a zone of iron oxide cementation near the surface, as a result of *weathering* or soil-forming processes such as *laterization*

fersiallic soil: *soils* characterized by a red-coloured B *horizon* containing illuvial *clay*, a lack of acidification, the redeposition of calcium carbonate within the C *horizon*. Associated with a Mediterranean-type climate (cold, wet season followed by a hot, dry summer) (cf. *brown soils*)

fersiallitization: the dominant *weathering* process in subtropical climates with strong seasonal contrasts. It involves the *inheritance* and *neoformation* of montmorillonitic *clays* and the immobilization of iron oxides owing to the alkaline conditions

flood: the result of a river overflowing its banks and covering land which is not normally submerged

flood basalt: basalt (a dark-coloured, fine-grained basic volcanic rock) erupted from fissure vents to build up thick lava sequences which, as in the Deccan Traps of India, often form plateaux

floodplain: a surface of fairly flat land next to a river channel which is periodically covered with water when the river overflows its banks

flora: all the plants living in an area or region

fluvial: of, or pertaining to, flowing water (rivers)

fluvialism: the school of thought which contends that the action of flowing water (in rivers) is chiefly responsible for fashioning the Earth's topography

fluviatile: of, by, or pertaining to, rivers

footslope: a concave element of a *hillslope*, developed in colluvial material (debris washed down from higher slope positions) and linking a *backslope* to a *toeslope* or stream channel

foraminifer: a microscopic, unicellular marine animal

forcing variable: any constant or variable acting on a system from the outside; a driving variable or external forcing

formation, plant: the vegetational equivalent of a *biome*, that is, a community of plants of like physiognomy (life-form) occupying a climatically uniform area on a continental scale

formation type: a group of convergent plant formations occurring on different continents; the floral equivalent of a *zonobiome*

fossil soil (see *palaeosol*)

frost creep: the downslope movement of particles resulting from cycles of *frost heaving* (which lifts the particles in a direction normal to the ground slope) and thawing (which settles the particles in a nearly vertical direction); the overall effect of a cycle is a downslope transport of particles

frost heaving: the generally upwards movement of mineral *soil* during freezing resulting from the migration of water to the freezing plane and its expansion on freezing

fully developed slope: a slope containing a full complement of *hillslope elements*

Gaia hypothesis: the conjecture that the chemical and physical conditions of the surface of the Earth, the *atmosphere*, and the oceans are actively controlled by life, for life

gelifluxion: a form of mass wasting; a type of *solifluxion* occurring in *periglacial* environments underlain by *permafrost*

gene flow: the exchange of genetic factors between populations resulting from the dispersal of gametes (eggs and spermatazoa) or zygotes (fertilized eggs)

general circulation climate model: a three-dimensional, mathematical model of the *atmosphere* (atmospheric general circulation model; AGCM for short), or of the oceans (oceanic general circulation model; OGCM for short). Recently, attempts have been made to couple general circulation models of the *atmosphere* and the oceans

geobiosphere: all terrestrial *ecosystems*

geobotanical zone: used in a similar sense to *zonobiome*

geographical cycle: the cycle of erosion envisioned by William Morris Davis in which a block of uplifted land is gradually worn down to a peneplain, in doing so the landscape passing through three chief stages — youth, maturity, and old age. Davis thought the cycle in temperate climates — the fluvial cycle — was the norm. He also recognized an arid cycle and a glacial cycle. Other cycles, such as the savanna cycle, were added later by other geomorphologists

geographical pole shift (see *true polar wander*)

geoid: the shape of the Earth at mean sea level as extended under the continents

geomorphological process: any process involved in Earth sculpture — *weathering, erosion*, transport, and *deposition*

geomorphology: the study of the shape and development of landforms

geosphere: the solid Earth — core, mantle, and crust

geothermal energy: heat within the Earth

gibbsite: a form of *alumina* and a component of *bauxite*

glacial: a time when *ice sheets* expand and average global climates are cooler and drier than during an *interglacial*

glacial cycle (see *geographical cycle*)

glacial rocks (see *tillite*)

glaciation: the formation, movement, decay, and retreat of *ice sheets* and *glaciers*

glacier: a river of snow and ice, or, if there is not enough snow to maintain flow, a stagnant lake of snow and ice (glacier reservoir)

glacio-eustasy: worldwide changes of sea level caused by the depletion of the oceanic reservoir of water at the expense of *ice sheets* and *glaciers* (cf. *eustasy*)

Globigerina: a type of *foraminifer* whose shell consists of several globular chambers

gneiss: a coarse-grained, banded, crystalline rock with a similar mineralogical composition to *granite* (feldspars, micas, and quartz)

geothite: a brown-coloured, hydrated oxide of iron

Gondwana: a hypothetical, Late *Palaeozoic* supercontinent lying chiefly in the Southern Hemisphere and comprising large parts of South America, Africa, India, Antarctica, and Australia

gradation: the process of levelling a land surface by the wearing away (degradation) of hills and filling (*aggradation*) of valleys

granite: an *acid igneous rock*, consisting chiefly of quartz and alkali feldspars, and micas, in the form of coarse crystals

gravel: a rock fragment with a diameter in the range 2 to 60 mm (rock fragments >60 mm are called cobbles and boulders); the term also describes an unconsolidated mixture of largely gravel-sized rock fragments, though larger and smaller fragments may also be present

greenhouse climate: a general term for a time of global warmth in the geological past (cf. *icehouse climate*)

greenhouse gas: any gas, such as carbon dioxide or methane, which absorbs infrared (long-wave) radiation emitted by the Earth

greenhouse heating: the warming of the *atmosphere* owing to the absorption by *greenhouse gases* of infrared radiation emitted by the Earth

greenhouse state (see *greenhouse climate*)

grey-brown podzolic soil: any member of a group of *zonal soils* characterized by a thin mineral-or-ganic *horizon*, a greyish-brown leached *horizon*, and a brown illuvial B *horizon*. Formed under deciduous forest in humid and temperate climates

grey earth: a defunct term for *soils* of the northern cool temperate zone

grey wooded soil: also known as grey podzolic soils; any member a group of *soils* characterized by a thin layer of leaf litter, a thin mineral-organic *horizon*, a fairly thick albic (pale) *horizon* leached of *clay* and iron oxides which tongues into an argillic (illuvial *clay*-rich) or natric (sodium-rich argillic) *horizon*, a base saturation of more than 60% in the argillic *horizon*, and dry in some *horizons* for part of the year. Formed under cool to temperate, subhumid to arid climates

grus: an accumulation of ill-sorted, angular grains of quartz and clayey material derived from the local weathering of *granite*

gymnosperm: a vascular plant with ovules (the female reproductive organs which mature into seeds when fertilized) not enclosed in ovaries. A conifer is an example

gyre: the circular flow of water which occurs in the world's oceans owing to the joint action of prevailing winds and the Earth's rotation

Hadley cell: the large-scale, thermally driven circulation of air in tropical latitudes, one cell occurring in each hemisphere: air rises near the equator in the intertropical convergence zone, once aloft flows polewards, descends at about latitude 35°, thence splits, some air moving polewards, some returning towards the equator

halite: rock salt

halloysite: a *clay mineral*, similar to *kaolinite*, formed where aluminium and silicon are present in roughly equal amounts, providing hydronium concentration is high and the concentration of bases is low

heat balance: the budget of energy receipt, loss, transfer, transformation, and storage in the world climate system or part thereof; also called energy balance

heavy mineral: a detrital mineral in a sediment or sedimentary rock with a specific gravity consider-ably higher than some standard, usually 2.85 g/cm^3; examples are garnet, magnetite, ilmenite, and tourmaline

hematite: a blackish-red to brick-red oxide of iron

hemipelagic: of, or pertaining to, continental margins and the adjacent abyssal plains

hillslope: a slope produced by *weathering, erosion,* and *deposition*

hillslope elements: the components of a *hillslope: summit, shoulder, backslope, footslope,* alluvial *toeslope*

Holocene epoch: the most recent slice of geological time between 10 000 years ago to the present; the younger unit of the *Pleistogene period* (or *Quaternary subera*)

homeotherm: an animal which regulates its body temperature by mechanisms within its own body; an endotherm or "warm-blooded" animal (cf. *ectotherm*)

horizon (see *soil horizon*)

humic: of, pertaining to, or derived from *humus*

humus: a product of the transformation of plant litter by *soil* microorganisms

hyalosponge: any member of the order Hyalospongea, with a skeleton composed of six-rayed siliceous needle-like spines (spicules), without calcium carbonate or spongin

hydrobiosphere: all aquatic *ecosystems*

hydrolitic decomposition: the *weathering* of rock by water containing hydrogen ions

hydrological cycle (see *water cycle*)

hydrophyte: a plant which flourishes under very wet conditions

hydrosphere: all the waters of the Earth

hydrous: containing water, especially water of crystallization or hydration

hydrous mica: also called hydromica; virtually synonymous with *illite*

hypersaline: excessively salty

Ice Age: on old term for the *Pleistogene* glacial-interglacial sequence

ice age: a time when ice forms broad sheets at middle and high latitudes (often in conjunction with the widespread occurrence of sea ice and *permafrost*), and mountain *glaciers* form at all latitudes

ice core: a core of ice extracted from an *ice sheet* or *glacier*

ice-albedo feedback: the self-perpetuation and intensification of *glacial* conditions owing to the high reflectivity of ice and snow surfaces: if global temperatures drop, then bright white sheets of ice and snow will increase global albedo, cause more sunlight to be reflected back into space, and so lead to a further lowering of temperatures. When temperatures rise, the mechanism works in reverse

ice epoch (see *ice age*)

icehouse climate: a general term for a time of global coldness in the geological past (cf. *greenhouse climate*)

icehouse state (see *icehouse climate*)

ice-rafted deposit: debris carried in calved blocks of ice which is deposited when the ice melts

ice sheet: a large *glacier* shaped as a dome, up to 4 km thick

igneous rock: a rock formed by crystallization or by vitrification ("turning to glass") of *magma*

illite: a *2:1 clay* derived chiefly from the *weathering* of feldspars and micas under alkaline conditions with abundant ions of aluminium and potassium

infiltration: the movement or percolation of water into the *soil*

inheritance (of clays): the formation of *secondary clays* by the very slight *transformation* of primary minerals. Associated mainly with temperate and subtropical climates

inselberg: a large residual hill within an eroded plain

insolation: the *solar radiation* falling on a horizontal surface, or on a specified surface of known slope and aspect, per unit area per unit time

interglacial: a time of relative warmth between two *glacials* when *ice sheets* decay and retreat

internal forcing: a threshold within a system, the crossing of which causes a sudden change of state

internal heat (see *geothermal energy*)

interstadial: a warmer snap during the course of a major *glacial*, but not warm enough or prolonged enough to be dubbed an *interglacial*

intraspecific: among species

isophene: a line on a map connecting points of equal expression of a character in a population which varies along a *cline*

isothermal surface: a surface at all points on which the temperature is the same

isotope: a form of an element with a different number of neutrons, the number of protons being the same in all forms of the same element

isotope stage: a division of a deep-sea core on the basis of *oxygen-isotope ratios*. There have been 19 isotope stages since the reversal of the Earth's magnetic field 700 000 years ago

Jurassic period: a slice of geological time between 213 to 144 million years ago; the middle unit of the *Mesozoic era*

kaolinite: a *1:1 clay* mineral, essentially a hydrated aluminium silicate formed under conditions of high hydronium (hydrated hydrogen ion, H_3O^+) concentration and an absence of bases

karst: in general, a term signifying *limestone* areas with topographically distinct scenery including caves, dolines, and springs

lacustrine: of, or pertaining to, lakes

landmass: any large area of land, such as a continent

landform: any physical feature of the Earth's surface, such as a hill or a valley

landslide: a landslip; the downslope movement of a mass of rock or earth under the influence of gravity

Last Glacial Maximum: the acme of the last glacial stage some 18 000 years ago

laterite: a concrete-like layer found in some tropical soils consisting chiefly of ferric iron oxides with or without quartz and minor quantities of aluminium and manganese; variously styled ironstone, murram, and plinthite

laterite soil: any member of a group of *soils* characterized by a thin A *horizon*, and a reddish, leached B *horizon* resting on *laterite*; also, a general term for *soils* rich in iron and aluminium (or iron alone) with a hardened crust, or an indurated *horizon*, or a well-developed concretionary *horizon*; virtually synonymous with *latosol*

laterization: the loss of *silica* and concentration of *sesquioxides* in the *soil*, with the formation of *gibbsite*, and with or without the formation of *laterite*; more or less synonymous with *soluviation*, *ferrallitization, allitization,* and *latosolization*

latosol: any member of a group of *zonal soils* characterized by deep *weathering* and abundant *hydrous* oxides; virtually synonymous with *laterite soil*

latosolization: the loss of *silica* and concentration of *sesquioxides* in the *soil*, with the formation of *gibbsite*, and with or without the formation of *laterite*; more or less synonymous with *allitization*, *ferrallitization, laterization,* and *soluviation*

leached zone: a *soil horizon* from which soluble constituents have been washed out; an eluvial (washed out or impoverished) *soil horizon*

leaching: the washing out of water-soluble minerals from a *soil* body, usually the entire solum (the genetic *soil* created by soil-forming forces), by the downwards movement of water

leaf area index: the ratio of total leaf surface to ground surface. For example, a leaf area index of 2 would mean that if you were to clip all the leaves hanging over 1 m^2 of ground, you would have 2 m^2 of leaf surface

lessivage: the mechanical transport of small mineral particles from the A *horizons* to the B *horizons* in a *soil profile* producing an enrichment of illuvial ("washed in") clay in the B *horizons*

life zone: a climatically defined region where distinct biotic communities can be expected to live

limestone (see *calcareous rock*)

limnic: of, or pertaining to, lakes and ponds

lithology: the physical character of a rock

lithosphere: the solid crust of the Earth together with the solid portion of the upper mantle lying above the asthenosphere

lithostratigraphic unit: any body of sedimentary rock defined on the basis of lithological characteristics and stratigraphical position

Little Climatic Optimum: a time, lasting from about A.D. 750 to 1200, when the climate of Europe and North America was clement, even as far north as Greenland and Iceland

Little Ice Age: a cold snap lasting from about 1450 to 1850 (the dates vary from region to region) when temperatures were lower than at present by about 1 °C in the mid-latitudes of the Northern Hemisphere

littoral: of, or existing on, a shore

lixivation: *leaching*

loess: an unconsolidated sediment composed chiefly of *silt*-sized particles laid down by the wind

long profile (of a river): a line on a graph showing how the altitude of a river changes with distance downstream; the long profile is commonly concave and grades down to a local or regional *base level*

mafic: said of rocks composed chiefly of one or more dark-coloured, ferromagnesian minerals

magma: fused, molten rock material within the Earth, sometimes containing gases and solid mineral particles

magnetic-field reversal: a switch in the polarity of the Earth's magnetic field

magnetic susceptibility: the ratio of induced magnetization in rocks and sediments to the strength of the magnetic field used to effect the magnetization

marlstone: a rock formed by the consolidation of marl, an admixture of *clay* or *silt* and fine-grained crystals of calcium carbonate (aragonite or calcite)

mass balance (of glaciers): the difference between addition and loss (ablation) of ice: a zero mass balance describes a steady state in which gains and losses are matched; a positive mass balance describes an increase in glacier mass; and a negative mass balance describes a reduction of glacier mass

Maunder Minimum: a time, lasting from about 1645 to 1715, when *sunspots* were conspicuous by their scarcity

maximum valley-side slope: the gradient of the steepest section of a *hillslope*

meandering: a tendency of rivers to follow a sinuous, winding, and turning course

mechanical denudation: the laying bare of the land surface by physical means

mechanical weathering: the breakdown or disintegration of rocks by physical means

Medieval Minimum: a time, lasting from about 1120 to 1280, when *sunspot* activity was low

megafauna: in the context of *Pleistocene* extinctions, the large mammals, such as mammoths and sloths, contrasted with small mammals, such as rodents and insectivores

megatherm: a plant which flourishes in hot temperatures

megathermal: of, or pertaining to, hot temperatures

meridional: pertaining to a meridian or line of longitude (cf. *zonal*)

mesic: describes an environment with a moderate supply of water, not too dry and not too wet (cf. *arid* and *humid*)

mesophyte: a plant flourishing under *mesic* conditions

mesotherm: a plant which flourishes in moderate temperatures

mesothermal: of, or pertaining to, moderate temperatures

Mesozoic era: a slice of geological time between 248 to 65 million years ago comprising the *Triassic, Jurassic,* and *Cretaceous periods*

meteoritic dust: micron-sized extraterrestrial material which falls as continual rain into the *atmosphere*

microphyllous: small-leaved

microspherule: a tiny sphere or spherical body in a deep-sea sediment

microtektite: tiny particles of glassy material probably formed by the collision between a *planetesimal* and the Earth

microtherm: a plant which flourishes in cold temperatures

microthermal: of, or pertaining to, cold temperatures

mollisol: a *soil* with a thick, dark surface *horizon* rich in calcium and magnesium cations (a mollic *horizon*). One of the ten soil orders in the United States Department of Agriculture's Comprehensive Soil Classification System, Seventh Approximation

monosiallitization: the formation in the *soil* of secondary *1:1 clay* minerals such as *kaolinite* and *goethite* after the loss of alkalis and alkaline earths

monsoon: a strongly seasonal wind system

montmorillonite: a *2:1 clay* mineral requiring high ionic concentrations of *silica* and magnesium for its *neoformation*; also called smectite

morphoclimatic region: an area in which a particular type of climate is thought to exert decisive control over landscape development and to create a distinctive suite of *landforms*

morphoclimatic zone: synonymous with *morphoclimatic region*

morphometry: the measurement of form; in *geomorphology*, the measurement of the form of the Earth's surface — geomorphometry

mud: a wet, loose mixture of *clays* and *silts*

mudstone: consolidated *mud*; includes claystones and siltstones

multistratal forest: a forest consisting of several layers of vegetation, as in tropical rain forests which often have three distinct layers formed of trees of differing heights

nebula: any diffuse mass of interstellar dust, gas, or both

neoformation: the formation of *clays* by materials freed from *parent material* by complete *weathering.* Associated chiefly with tropical climates

Neogene period: a slice of geological time between 24.6 to 2 million years ago; the middle unit of the *Cenozoic era*

net primary production: primary production less respiration

net radiation (see *radiation balance*)

noncalcic brown soil: a member of a group of *zonal soils* characterized by a slightly acidic, light pink or reddish brown A *horizon* atop a light brown or dull red B *horizon*. Formed under subhumid climates in the transition zone between grassland and forest

nuclear winter: a severe climate, cold and dark, that would probably follow a major nuclear exchange owing to the blocking of *solar radiation* by the dense dust and smoke created (cf. *volcanic winter*)

obliquity of the ecliptic: the angle between the ecliptic (orbital plane) and the celestial equator; equal to the angle at which the Earth's rotatory axis is tilted from the vertical

oil shale: a fine-grained sedimentary rock with much original bituminous organic material

Oligocene epoch: a slice of geological time between 38 to 24.6 million years ago; the youngest unit of the *Palaeogene period*

Ordovician period: a slice of geological time between 505 to 438 million years ago

orogenesis: mountain building by folding and uplift as a purely structural concept; the effects of *erosion* in creating high peaks and deep valleys are excluded by the term

orogeny: the creation of mountains, especially by folding and uplift

outwash: stratified (layered) deposits of *sands* and *gravels* laid down by glacial meltwater beyond the ice margin

overkill: in discussions of *Pleistocene* extinctions, the killing by Man of so many members of an animal population that the number of individuals in the population cannot be maintained

oxisol: a member of a group of very deeply weathered, very *leached* acid *soils* with a diagnostic *horizon*, at least 30 cm thick, in which *weathering* has removed or altered a large portion of the *silica* and left *sesquioxides* of iron and aluminium and a concentration of *1:1 clays*. Associated exclusively with intertropical regions. One of the ten soil orders in the United States Department of Agriculture's Comprehensive Soil Classification System, Seventh Approximation

oxygen-isotope ratio: the ratio of oxygen-18 to oxygen-16 ($^{18}O/^{16}O$), used as an indicator of *palaeotemperatures* since it is related to ocean temperature

palaeobotanical record: pollen, spores, and the remains of plants found in *soils*, sedimentary deposits, and sedimentary rocks

Palaeocene epoch: a slice of geological time between 65 to 54.9 million years ago; the oldest unit of the *Palaeogene period*

palaeoclimatology: the study of climates of the past

palaeoecology: the study of past ecological systems

Palaeogene period: a slice of geological time between 65 to 24.6 million years ago

palaeogeography: the study of the geography of the past

palaeohydrology: the study of the hydrology of the past

palaeolatitude: the former latitude of a place

palaeomagnetism: the intensity, direction, and polarity of the Earth's magnetic field in the past as recorded in some iron-rich rocks

palaeosol: an old, fossil, or relict *soil* or *soil horizon*, usually buried

palaeotemperature: the temperature of the past as indicated by various environmental indicators

Palaeozoic era: a slice of geological time between 590 to 248 million years ago; comprises the *Cambrian, Ordovician, Silurian, Devonian, Carboniferous*, and *Permian periods*

palynology: the study of fossil spores and pollen grains

pandemic: an epidemic over a very wide area

parent material: the material in which a *soil* is formed; if the material is *bedrock*, rather than a superficial deposit, then the term parent rock is sometimes employed

partial pressure: the individual contribution of an atmospheric gas to the total pressure of the *atmosphere*

particulate load (see *suspended load*)

particulate radiation: low energy particles emitted by the Sun as the *solar wind*

pedalfer: any *soil* in which calcium carbonate does not accumulate anywhere in the *soil profile*, even though the *parent material* may originally have contained calcium carbonate

pediment: a smooth erosion surface, concavely sloping down from the foot of a highland area and grading in to a flat area corresponding to local *base level*

pediplanation: the processes which create *pediments*

pedocal: a *soil* in which *leaching* has been insufficient to wash out soluble salts so that free calcium carbonate occurs in its profile as concretions, layers, nodules, or veins

pedogenesis: the creation and development of *soil*

pedology: the study of *soil* as a natural body

pelagic: of, or pertaining to, the open sea

pelagic clay: *clay* deposited in the ocean deeps, under the open ocean

peneplain: a gently undulating surface close to *base level*, thought to be the ultimate stage of a cycle of erosion (cf. *geographical cycle*)

percolation: the essentially downwards movement of water in rock or *soil* lying above the *water table*

periglacial: the climatic and geomorphological zone peripheral to the *ice sheets* and *glaciers* at high latitudes, and occupying nonglacial environments at high altitudes

perihelion: the point on the orbit of the Earth (or any other body of the Solar System) which is nearest the Sun. For the Earth at present, perihelion occurs on 3 January

permafrost: technically speaking, *soil* or rock with temperatures below freezing point for two consecutive winters and the intervening summer; loosely, but inaccurately, defined as permanently frozen ground

Permian period: a slice of geological time between 286 to 248 million years ago

Phanerozoic aeon: the most recent geological aeon which started 590 million years ago

phyllite: one of the phyllosilicates (micas), common in igneous and metamorphic rocks, which occur as thin, flaky sheets

phytogeography: the study of the geography of plants

phytoplankton: the plant part of the *plankton*: the plant community in the marine and freshwater system which floats free in the water and contains many species of algae and *diatoms*

piping: subsurface channels (pipes), up to several metres in diameter, created by the dispersal of *clay* particles in fine-grained, highly permeable *soils*

planetary wave: wave motion in the *atmosphere* on a global scale taking the form of vast meanders of flowing air which are most clearly discernable at upper levels; a Rossby wave

planetesimal: an asteroid, comet, or meteorite

plankton: aquatic organisms which either drift in ocean currents or swim a little

plate motion: the lateral movement of a tectonic plate

plate tectonics: the theory that the Earth's crust consists of a number of large plates which are created along mid-ocean ridges and destroyed at subduction sites

playa: a closed depression in an arid or semiarid region which is occasionally flooded by surface runoff; also, the salt flat within such a closed basin

Pleistocene epoch: a slice of geological time between 2 million to 10 000 years ago; the older unit of the *Pleistogene period* (or *Quaternary subera*)

Pleistogene period: the most recent slice of geological time between 2 million years ago to the present (cf. *Quaternary subera*); subdivided into the *Pleistocene* and *Holocene epochs*

Pliocene epoch: a slice of geological time between 5.1 to 2 million years ago; the younger unit of the *Neogene period*

pluvial: a time of a wetter climate creating greater *effective precipitation* owing to increased precipitation, or reduced *evapotranspiration*, or both

podzol: *soils* with a spodic *horizon* (see *spodosols*)

podzolic soil: a *soil* with a well-developed A2 (leached) *horizon* and with signs of the downward movement of aluminium and iron (*podzolization*), though the *horizons* are not as clearly defined as they are in *podzols*

podzolization: a suite of processes involving the chemical migration of aluminium and iron (and sometimes organic matter) from an eluvial (leached) *horizon* in preference to *silica*

poikilotherm: an *ectotherm*

pollen profile: a diagram displaying change in the percentages of various types of pollen with time

pollen spectrum (see *pollen profile*)

polygenetic landscape: a landscape in which the elements have been created by more than one climatic regime

polymorphic: having more than one form

postdiluvian world: the world after Noah's flood

potential evaporation: the amount of water that would be evaporated from a land or water surface if the water supply were unlimited. The *actual evaporation* will fall below the potential evaporation rate when water at the evaporating surface is limited

prairie soil: any member of a group of *zonal soils* characterized by a decarbonated solum (A and B *horizons*) and no movement of iron and aluminium *sesquioxides*: dark coloured, grey to brown A *horizons* overlie a yellowish brown B *horizon*, and a yellowish C *horizon* with calcium carbonate at the base

Precambrian: a general term for all geological time before the start of the *Cambrian period*, 590 million years ago

precession: a complex motion of a rotating body, such as the Earth, in which the rotatory axis changes orientation when subject to an applied torque

precession of the equinoxes: a slow (about 50.27″ of arc per year) westwards shift of the equinoctial points along the orbital plane (ecliptic) owing to the *precession* of the Earth's rotatory axis. A result is that the poles of the rotatory axis describe circles in space over a cycle lasting about 22 000 years. Today, the winter *solstice* occurs near perihelion, whereas 11 000 years ago it occurred near aphelion. During the course of a precessional cycle, the distance between the Earth and the Sun, as measured on say 21 December, changes

primary production: the production of organic material by green plants

Priscoan aeon: the earliest part of geological time between 4.6 to 4 billion years ago

probiscidean: a member of the order of mammals known as the Probiscidea, which includes elephants, mastodons, and deinotheres

Proterozoic aeon: a slice of geological time between 2.5 billion to 590 million years ago

pyroxene: an anhydrous ferromagnesian silicate

Quaternary subera: the latest large slice of geological time which started 2 million years ago; the term *Pleistogene period* is preferable

race: an animal or plant population which differs from other populations of the same species in one or more hereditary characters

radiation balance: a budget of the components of incoming *solar radiation* and outgoing long-wave radiation. The difference between all incoming and all outgoing radiation is termed net radiation, a negative value indicating overall loss, a positive value indicating an overall gain

radiolarian: any marine protozoan of the order Radiolaria, with rigid skeletons of *silica* and radiating needle-like structures called spicules

rainsplash: *erosion* caused by falling rain drops: the energy released by a drop on impact lifts *soil* particles into the air to land a little downslope of their former resting place

red bed: a *soil*, sediment, or sedimentary rock coloured red by finely divided ferric iron oxides, mainly *hematite*

red earth: a virtually defunct term for a group of *soils* found largely in the tropical zone

red desert soil: any member of a group of *zonal soils* having a light, friable, reddish-brown surface *horizon* above a reddish-brown or red *horizon*, and a *horizon* of calcium carbonate accumulation. Formed under tropical to warm temperate, arid climates.

reddish brown lateritic soil: any member of a group of *zonal soils* formed in a red lateritic *parent material* and characterized by a reddish-brown surface *horizon* atop a red-clay B *horizon*

reddish brown soil: any member of a group of *zonal soils* having a reddish or light brown surface *horizon* surmounting a redder *horizon* and a light coloured *horizon* with lime (calcium carbonate) accumulation. Formed under warm, temperate to tropical, semiarid climates

reddish chestnut soil: any member of a group of *zonal soils* characterized by a thick surface *horizon* ranging from dark brown to reddish or pinkish, a reddish-brown subsoil *horizon*, and a *horizon* of calcium carbonate accumulation. Formed under warm to temperate, semiarid climates in the drier parts of temperate grasslands

reddish prairie soil: any member of a group of *zonal soils* characterized by a decarbonated (calcium carbonate free) solum (A and B *horizons*) and no movement of iron and aluminium *sesquioxides*: the upper *horizons* are *humus*-rich and grey brown in colour, the B *horizons* are reddish and clayey, and the C *horizons* are red or yellowish brown with small concretions of iron and manganese

red-yellow podzolic soil: any member of a group of *zonal soils* characterized by a well-drained, lime-free, acid profile with a black or chocolate coloured surface *horizon*, leached (slightly podzolized) lower A *horizons*, a bright coloured (red, yellowish red, or yellow) upper B *horizon* rich in *clay* and free iron and aluminium oxides, and a mottled lower B *horizon*. Formed on a range of materials under a moist, subtropical climate

refugia: favourable areas south of the ice front in which species and populations survived during a *glacial stage*

regolith: unconsolidated rock debris lying above *bedrock*

rejuvenation: the renewal of former activity, especially of a river whose erosional activity is reanimated by uplift or a fall of *base level*

relative relief: within a *drainage basin*, the ratio of basin relief (the difference between the highest and lowest points in the basin) to the length of the basin perimeter

relative time: the age of events or features expressed relative to other phenomena, and not in *absolute time* units

relief: variations in elevation

rhythmite: any individual unit of a rhythmic succession, or of beds created by rhythmic sedimentation; a *cyclothem* is an example

rock cycle (see *sedimentary cycle*)

salinization: the accumulation of water-soluble salts, such as chlorides and sulphates of calcium, magnesium, sodium, and potassium, in the *soil*, commonly as salty (salic) *soil horizons*

sand: fragments of minerals and rocks ranging in size from 2 to 0.02 mm

saprolite: weathered or partly weathered *bedrock* which has not been moved

sapropel: a *mud* or ooze formed chiefly from the decomposition of organic material in anaerobic (and usually aquatic) environments

savanna cycle (see *geographical cycle*)

schist: a metamorphic rock characterized by the development of flat, platy minerals (commonly micas) which give it a layered (foliate) texture

sea-floor spreading: the creation of new oceanic crust at, and its movement away from, mid-ocean ridges

secondary clay mineral: a *clay* mineral formed from the products of *weathering*

sediment yield: the total mass of sedimentary particles reaching the outlet of a *drainage basin*. Usually expressed as t a^{-1}, or as a specific sediment yield in t km^{-2} a^{-1}

sedimentary cycle: also called the rock cycle and geological cycle; the long-term cycle of lithification (rock formation), uplift, *weathering* and *erosion, deposition,* and renewed lithification

sesquioxide: an oxide which contains two metallic atoms to every three of oxygen; for instance, Al_2O_3

shoulder: a convex, upper element of a *hillslope* linking a *summit* and a *backslope*

sierozem: any member of a group of *zonal soils* characterized by brownish-grey surface *horizons* and light-coloured subsoil *horizons* resting on a layer of calcium carbonate accumulation and, sometimes, a hardpan. Formed under temperate to cool arid climates

silcrete: an indurated material rich in *silica* formed at, or near, the land surface by the silicification (processes leading to the enrichment of silica) of *bedrock, weathering* products, or by low-temperature physico-chemical processes

silica: silicon dioxide

silicate: any of the many compounds composed of silicon, oxygen, and a metallic or organic radical

siliceous: containing, composed of, or pertaining to *silica*

silt: mineral and rock fragments with a diameter in the range 0.02 to 0.002 mm; a deposit composed chiefly of such fragments

Silurian period: a slice of geological time between 438 to 408 million years ago

smectite: synonymous with *montmorillonite*

soil: rock at, or near, the land surface which has been transformed by the *biosphere* (see *soil profile*)

soil carbonates: carbonates, usually of calcium, found within a *soil profile*

soil creep: the slow, gradual, downhill movement of *soil* and loose rock fragments

soil-forming factor: any agent or force which may influence the development of a *soil* system; the four general factors are climate, organisms, *relief*, and *parent material* (to which some soil scientists add time)

soil horizon: a layer of *soil*, lying roughly parallel to the land surface, which has characteristics created by soil-forming processes

soil-landscape system: any landscape unit (such as a *hillslope* or a *drainage basin*) in which *landforms* and *soils*, and the geomorphological and pedological processes which create them, are seen as a unitary whole

soil organic matter: any organic material found in the *soil profile*

soil profile: all the *soil horizons* from the ground surface to the unaltered *bedrock* or *parent material*; usually divided into A ("topsoil"), B ("subsoil"), and C (weathered *parent material*) *horizons*, the A and B *horizons* together forming the solum

sol brun acide (see *brown forest soil*)

solar-cyclonic hypothesis: the postulate that the intensity of mid-latitude cyclones is influenced by the activity of the Sun

solar inertial motion: the movement of the Sun about the barycentre (centre of gravity) of the Solar System resulting from the changing configuration of the planets, and especially of Jupiter, Saturn, Uranus, and Neptune

solar luminosity: the brightness of the Sun

solar radiation: the total *electromagnetic radiation* emitted by the Sun

solar wind: the flux of low energy, charged particles emitted by the Sun (cf. *particulate radiation*)

solifluxion: a form of mass wasting; the slow, downslope movement of waste saturated with water under the influence of gravity

solstice: either of the two times in the year when the Sun appears to have no apparent northwards or southwards motion and to stand over the Tropic of Cancer or the Tropic of Capricorn

solute load (see *dissolved load*)

soluviation: the loss of *silica* and concentration of *sesquioxides* in the *soil*, with the formation of *gibbsite*, and with or without the formation of *laterite*; more or less synonymous with *allitization, ferrallitization, laterization,* and *latosolization*

speciation: the process of species multiplication

species diversity: the number of species of organisms in a specified area

species number (see *species diversity*)

species richness (see *species diversity*)

specific discharge: the discharge of a river per unit area

Spörer Minimum: a time, lasting from about 1400 to 1510, when *sunspot* activity was low

spodosol: any member of a group of *soils* all having a diagnostic *horizon* (a spodic B *horizon*) showing illuvial accumulation of *sesquioxides*, with or without organic matter, beneath a light-coloured (albic) A *horizon*. One of the ten soil orders in the United States Department of Agriculture's Comprehensive Soil Classification System, Seventh Approximation

state factor: any of the external variables (climate, organisms, *relief*, and *parent material*) which drive the *soil* system

state variable: one of any number of descriptors which define the condition of a system at a particular time

steady state: a condition in a system wherein inputs and outputs are balanced so that, despite continuous throughput, the system state remains constant (steady)

stenothermal: of, or denoting, organisms adapted to live within in a limited range of temperature

stratosphere: that part of the *atmosphere* lying between the troposphere and the mesosphere (between about 11 and 50 km)

stream frequency: the number of streams per unit area

stream length: the total length of rivers within a *drainage basin*

stream number: the total number of stream segments within a *drainage basin*

stream sinuosity: the length of a stream from start to finish as the crow flies compared with the actual length of the stream

subaerial: occurring at the land surface

succulent: a plant with thick, fleshy leaves or stems that conserve water. A cactus is an example

suid: a member of the pig family

summit: the relatively flat element at the crest of a *hillslope*

sunspot: a disturbance on the Sun's face seen as a dark region

supercontinent cycle: an hypothetical cycle in which continents coalesce to form a supercontinent and then break into smaller continents owing to the pattern of heat conduction and loss through the crust. A full cycle is thought to take about 440 million years

surface wash: material carried downhill by rainsplash (which detaches and transports) and surface flow

suspended load: the sediment carried in suspension by a river

synorogenic: said of a geological feature formed, or of a geological event or process occurring, during a period of *orogeny*

talus: rock fragments of any shape and size derived from, or lying at, the base of a cliff or steep rocky slope

taphonomy: a branch of *palaeoecology* dealing with the changes undergone by an organism from its death until its discovery as a fossil

taxon (plural **taxa**)**:** a population or group of populations (taxonomic group) that is distinct enough to be given a distinguishing name and to be ranked in a definite category

tectonic: of, or pertaining to, crustal processes (such as faulting, folding, and warping) producing the broad structural features of the Earth

tectono-eustasy: worldwide changes of sea level resulting from a change in the volume of the ocean basins (cf. *eustasy*)

Tertiary subera: a slice of geological time between 65 to 2 million years ago and the youngest unit of the *Cenozoic era*; formerly designated a geological period

tetrahedral layer: in *clay minerals*, a sheet comprising tetrahedra of four oxygen ions arranged around one silicon ion

texture ratio: the number of stream segments in a *drainage basin* divided by the length of the basin perimeter (watershed)

thermoluminescence: a method of dating a sample of quartz or zircon grains (or both) by measuring the intensity of the glow curve, that is, the light emitted by a sample which has been irradiated and subsequently heated, plotted as a function of temperature

till: a chiefly unsorted and unstratified (unlayered) deposit, usually unconsolidated, laid down directly by, and underneath, a *glacier* without reworking by meltwater

tillite: a consolidated sedimentary rock formed by the lithification ("turning to stone") of glacial *till*

tilt (see *obliquity of the ecliptic*)

timberline: the limit of altitude or latitude beyond which trees do not grow

toeslope: a concave or flat element formed on *alluvium* at the base of a *hillslope*

tolerance: the ability of an organism to survive a range of environmental factors

topography: the physical configuration and features of a place or region

tor: a high, isolated crag, pinnacle, or rock peak standing above the slopes which surround it, formed by differential *weathering* followed by mass wasting and stripping

tragulid: a member of the family which includes the chevrotain (mouse deer)

transformation (of clays): the formation of *clay* by the gradual transformation of primary minerals. Very slight transformation is called *inheritance*; considerable transformation is called *degradation*

tree ring: the annual growth ring of a tree

Triassic period: a slice of geological time between 248 to 213 million years ago; the oldest unit of the *Mesozoic era*

trilobite: a marine arthropod, any member of the class Trilobata

trophic: of, or pertaining to, nourishment or feeding

tropical ferruginous soil: a member of a group of *soils* with a thick, dark brown, *humus*-rich surface *horizon*, a brown-yellow leached *horizon* from which *clay* as well as bases have been partly removed, and a red, brown, or ochreous *clay*-enriched B horizon. Formed under humid tropical climates without a marked dry season and a tropical climate with a dry season

true polar wander: the movement of the entire globe about the Earth's rotatory axis causing a change in all latitudes; also termed geographical pole shift, Earth tumble, and evagation of the poles

tundra soil: any member of a group of *zonal soils* characterized by dark brown, highly organic surface *horizons* and greyish lower *horizons*. Formed over permafrost under a cold, damp climate with poor soil drainage

udoll: a suborder of *mollisols* associated with humid climates; essentially, *prairie soils* and *chernozems*

ultisol: a deeply weathered, leached acid *soil*. One of the ten soil orders in the United States Department of Agriculture's Comprehensive Soil Classification System, Seventh Approximation

underclay: a layer of fine-grained detritus, usually *clay*, lying below a coal bed or forming the floor of a coal seam

ustoll: a suborder of *mollisols* associated with subhumid to semiarid climates: essentially, *chestnut soils* and calcic *brown soils*

vapour pressure: the partial pressure of water vapour in the air

vapour pressure deficit: the different between the *vapour pressure* at saturation and the actual vapour pressure at the temperature in question; also known as the saturation deficit, it is a measure of the "drying power" of the air

varve: a bed (lamina) of sediment laid down in a body of still water within the course of a year

vascular plant: a plant containing vascular tissue, the conducting system which enables water and minerals to move through it

vermiculite: a *clay mineral* of the *hydrous* mica group

vertisol: a member of a group of dark clay *soils* associated with disturbance and inversion caused by the seasonal shrinking and swelling of *clays*. One of the ten soil orders in the United States Department of Agriculture's Comprehensive Soil Classification System, Seventh Approximation

volcanic winter: a severe climate, cold and dark, that would possibly follow a supereruption of a volcano or several volcanoes (cf. *nuclear winter*)

volcanism: processes involving intrusive and extrusive volcanic force or activity

Vostok ice core: an ice core extracted from the Antarctic ice sheet at the USSR's scientific research station at Vostok

water balance: a budget of the land phase of the *water cycle*. Usually expressed as precipitation equals evaporation plus runoff, but may include water storage terms

water budget (see *water balance*)

water cycle: the more or less continuous movement of water in all its forms (liquid, solid, and vapour) below, on, and above the surface of the Earth. Each round of the cycle involves *evaporation*, condensation, precipitation, runoff

water deficit: a situation obtained when *evaporation* exceeds precipitation over some prescribed time span

water table: the surface defined by the height of free-standing water in fissures and pores of saturated rock and *soil*

weathering: the biological, chemical, and physical decomposition and disintegration of rocks and sediments at and near the land surface (see *mechanical weathering, chemical weathering*)

weathering crust: a widespread zone of weathered material

wildfire: a raging fire that travels and spreads swiftly

wind stress: the force exerted at the ground surface by flowing air

Wisconsin glacial stage: the name for the last ice age in North America: the equivalent term in northern Europe is the Weichselian and in Great Britain the Devensian

xerophyte: a plant adapted for life in a dry climate

yellow earth: a virtually defunct term for a group of *soils* found chiefly in the subtropical zone

Younger *Dryas*: a cold spell after the last deglaciation lasting between about 11 000 to 10 000 years ago

zonal: pertaining to lines of latitude

zonal circulation: any flow along circles of latitude

zonal index: the average pressure gradient across circles of latitude

zonal soil: a *soil* which occurs over a large area roughly corresponding to a major climatic zone

zonobiome: one of nine, climatically defined, major community units of the Earth; the same as a biome type

Printing: COLOR-DRUCK DORFI GmbH, Berlin
Binding: Buchbinderei Lüderitz & Bauer, Berlin